UNDERSTANDING TRAFFIC SYSTEMS

01/06

UNIVERSITY OF
WOLVERHAMPTON

Harrison Learning Centre
City Campus
University of Wolverhampton
St Peter's Square
Wolverhampton
WV1 1RH
Telephone: 0845 408 1631
Online renewals: www.wlv.ac.uk/lib/myaccount

ONE WEEK LOAN

Telephone Renewals: 01902 321333 or 0845 408 1631
Please RETURN this item on or before the last date shown above.
Fines will be charged if items are returned late.
See tariff of fines displayed at the Counter. (L2)

Understanding Traffic Systems:
Data, Analysis and Presentation
Second Edition

MICHAEL A.P. TAYLOR
Professor of Transport Planning
University of South Australia

PETER W. BONSALL
Professor of Transport Planning
University of Leeds

WILLIAM YOUNG
Professor of Civil Engineering
Monash University

Ashgate
Aldershot • Burlington USA • Singapore • Sydney

© Michael A.P. Taylor, Peter W. Bonsall and William Young 2000

All rights reserved. No part of this publication may be reproduced, stored in a retrieval system, or transmitted in any form or by any means, electronic, mechanical, photocopying, recording or otherwise without the prior permission of the publisher.

Published by
Ashgate Publishing Limited
Wey Court East
Union Road
Farnham
Surrey, GU9 7PT
England

Ashgate Publishing Company
110 Cherry Street
Suite 3-1
Burlington
VT 05401-3818
USA

Ashgate website: http://www.ashgate.com

British Library Cataloguing in Publication Data
Understanding traffic systems : data, analysis and
    presentation. 2nd ed.
    1. Traffic engineering 2. Traffic engineering - Data
    processing 3. Traffic flow 4. Traffic flow - Data processing
    I. Taylor, M. A. P. (Michael Anthony Peter) II. Bonsall, P.
    W. (Peter W.), 1951- III. Young, W. (William), 1949-
    388.3'143

Library of Congress Control Number: 00-132607

ISBN 0 7546 1248 1

Transfered to Digital Printing in 2014

Printed in the United Kingdom by Henry Ling Limited,
at the Dorset Press, Dorchester, DT1 1HD

Contents

List of figures xv

List of tables xxi

Preface xxv

Acknowledgements xxvii

Part A: INTRODUCTION 1

1 Introduction and context 3

    1.1 Traffic impacts 5
    1.2 Traffic studies 5
        1.2.1 Principles of information gathering 6
        1.2.2 Important traffic parameters 7
    1.3 Levels of analysis 7
        1.3.1 Link level analysis 7
        1.3.2 Junction level analysis 11
        1.3.3 Area level analysis 12
        1.3.4 Relationships between link, junction and area 13
    1.4 Roles of the traffic analyst 14

2 The traffic analysis process 17

    2.1 Information, planning and decision making 17
    2.2 The traffic planning and decision making process 19
        2.2.1 Data collection 21
        2.2.2 Modelling and analysis 22
        2.2.3 Appraisal of policy options 24
        2.2.4 Monitoring 24

vi *Understanding Traffic Systems*

2.3 Statistics and data	25
2.3.1 Measurement	26
2.3.2 Continuous and discrete data	26
2.3.3 Quantitative and qualitative data	27
2.3.4 Stationary and time-dependent data	28
2.4 Analytical tools	30
2.4.1 Database managers	30
2.4.2 Spreadsheets	33
2.4.3 Statistical analysis packages	33
2.4.4 Geographic Information Systems (GIS)	34
2.4.5 Specialist models	36
2.5 A perspective on the traffic analysis process	37

Part B: BASIC TRAFFIC THEORY   39

3 Basic traffic flow theory   41

3.1 The traffic stream and its elements	41
3.1.1 The traffic lane	42
3.1.2 Traffic streams	42
3.1.3 Modifying factors	43
3.2 Basic relationships	43
3.2.1 Trajectory diagram	44
3.2.2 Speed distributions	46
3.2.3 Average speed and volume	49
3.2.4 Average speed and density	50
3.2.5 Volume and density	51
3.2.6 Ancillary characteristics	52
3.2.7 Combining traffic streams	53
3.3 Statistical modelling of traffic streams	54
3.3.1 Discrete distributions	55
3.3.2 Continuous distributions	58
3.3.3 Bunching of traffic	62
3.4 Vehicle interactions in traffic	63
3.4.1 Car following	64
3.4.2 Lane changing and overtaking	65
3.4.3 Bunches and platoons	67
3.4.4 Shock waves	67
3.5 Applications in traffic analysis	68

4 Theories of interrupted traffic flow 71

    4.1 Gap acceptance   73
        4.1.1 Gap acceptance mechanisms   74
        4.1.2 Basic results   75
        4.1.3 Multi-lane traffic flows   77
        4.1.4 Combined lanes   78
    4.2 Queuing and delay   79
        4.2.1 Arrival pattern   80
        4.2.2 Service mechanism   80
        4.2.3 Queue discipline   80
        4.2.4 Traffic intensity   81
        4.2.5 Derivation of simple queuing model   81
        4.2.6 Results of simple queuing theory   84
    4.3 Theory of traffic signal operation   86
        4.3.1 Basic parameters for signal operation   87
        4.3.2 Wardrop-Webster model   88
        4.3.3 Capacity of one movement   90
        4.3.4 Capacity of entire intersection   91
        4.3.5 Capacity analysis   94
        4.3.6 Degree of saturation   94
        4.3.7 Saturation flows   95
        4.3.8 Measures of performance   99
    4.4 Link congestion functions   104

5 Theories of area-wide traffic flow 109

    5.1 Principles of network analysis   110
        5.1.1 Trip generation   111
        5.1.2 Trip distribution   112
        5.1.3 Trip timing   112
        5.1.4 Modal choice   112
        5.1.5 Trip assignment   113
    5.2 The origin-destination matrix   114
        5.2.1 Synthesis of O-D matrices from link counts   116
        5.2.2 Modelling of O-D matrices from known trip end totals   119
    5.3 Network flow modelling   120
        5.3.1 Assignment strategies   121

5.3.2 Mathematical formulation of equilibrium assignment	123
5.4 Area-wide traffic control	124
5.4.1 Platoon dispersion	124
5.5 Congestion	126
5.5.1 Measuring the level of congestion	127
5.5.2 Generalised cost of travel	128
5.5.3 Congestion pricing	129
5.6 Fuel consumption and emissions	131
5.7 Exposure measures for accident analysis	132

Part C: DATA CAPTURE   135

6 Principles of survey planning and management   137

6.1 Elements of the survey planning and management process	137
6.1.1 Objectives	138
6.1.2 Availability of existing data	138
6.1.3 Specification of the requirement for new data	139
6.1.4 Available resources	139
6.1.5 Choice of survey instrument	139
6.1.6 Design of sample	140
6.1.7 Survey plan	140
6.1.8 Pilot survey	142
6.1.9 Conduct of main survey, data processing and archiving	142
6.2 What constitutes a good survey instrument?	143
6.3 Good practice in survey administration	144
6.4 Data capture without surveys	145

7 Experimental design and sample theory   147

7.1 An introduction to experimental design	147
7.2 Choice of variables	148
7.3 Alternative experimental designs	149
7.3.1 Comparison of one variable over two alternatives	149
7.3.2 Comparison of one variable over a number of	150

	alternatives	
	7.3.3 Analysis of situations with multiple variables	153
7.4 Sampling methods		155
	7.4.1 Target population	155
	7.4.2 Definition of sampling unit	155
	7.4.3 Selection of sampling frame	156
	7.4.4 Choice of sampling method	156
	7.4.5 The basis of statistical inference	159
	7.4.6 Sampling error and bias	159
	7.4.7 Sample size determination	159
	7.4.8 Requirements for sample design	162

8 Vehicle counting and classification surveys — 163

8.1 Measures of flow		163
	8.1.1 Classifications of flow	165
	8.1.2 Choice of survey technique	166
8.2 Manual counting methods		167
8.3 Counting methods involving video		172
8.4 Automatic detection of vehicles		175
	8.4.1 Pneumatic tube detectors	175
	8.4.2 Switch tapes	177
	8.4.3 Multicore cables	178
	8.4.4 Summary on axle detectors	178
	8.4.5 Inductive loop detectors	178
	8.4.6 Magnetic imaging sensors	180
	8.4.7 Electromagnetic beams	181
8.5 Automatic classification of vehicles		181
8.6 Data capture without surveys		182

9 Traffic condition data — 185

9.1 The need for data on traffic conditions		185
	9.1.1 Point velocities	185
	9.1.2 Vehicle headways	186
	9.1.3 Journey times and speeds	186
	9.1.4 Traffic incidents	189
9.2 Methods of collecting data on spot speeds		190

	9.2.1 Manual methods	190
	9.2.2 Methods involving automatic timing	191
	9.2.3 Methods employing the Doppler effect	192
	9.2.4 Methods involving video	194
9.3	Data on vehicle headways	195
	9.3.1 Instrumentation of a stretch of road	195
	9.3.2 Analysis of video	196
	9.3.3 Use of instrumented vehicles	196
9.4	Data on travel times	197
	9.4.1 Registration plate matching	197
	9.4.2 Remote or indirect tracking of individual vehicles	201
	9.4.3 Input-output methods	202
	9.4.4 Moving observer methods	203
	9.4.5 The use of volunteer drivers and fleets of probe vehicles	206
9.5	Data on delays	207
	9.5.1 Estimation of delay	207
	9.5.2 Direct methods of calculating delay	208
9.6	Off-line use of on-line data	209

10 Environmental impacts   211

10.1	Transport fuels	212
10.2	Pollutants from road transport sources	213
10.3	Estimating the environmental impacts of road traffic	216
10.4	Survey methods for fuel consumption and emissions	218
	10.4.1 Individual vehicle surveys	218
	10.4.2 Surveys of system-wide consumption and emissions	219
10.5	Traffic noise	222
10.6	Surveys of traffic noise levels	228
	10.6.1 Actual noise levels	228
	10.6.2 Community reactions	229
10.7	Environmental sensitivity	230
10.8	Role of the traffic analyst	230

Contents xi

Part D: TRAFFIC STUDIES — 233

11 Intersection studies — 235

    11.1 Turning movement flows — 235
    11.2 Delays and queuing — 241
        11.2.1 Queuing — 242
        11.2.2 Traffic delay — 244
    11.3 Intersection delay studies — 246
        11.3.1 Stopped delay — 246
        11.3.2 Overall delay — 247
    11.4 Saturation flow studies — 249
        11.4.1 Headway ratio method — 250
        11.4.2 Regression methods — 255
    11.5 Gap acceptance studies — 256

12 Origin-destination and route choice studies — 259

    12.1 Issues in O-D and route choice studies — 259
        12.1.1 Problems and difficulties — 259
        12.1.2 A categorisation of data sources — 260
    12.2 Methods of obtaining O-D data — 261
        12.2.1 Interviews and questionnaires — 261
        12.2.2 Registration plate matching — 263
        12.2.3 Use of vantage point observers or video — 265
        12.2.4 Tag surveys — 265
        12.2.5 Headlight surveys — 265
        12.2.6 Traffic counts — 266
        12.2.7 Vehicle tracking — 266
    12.3 Methods of obtaining route choice data — 267
        12.3.1 Observation of flows — 267
        12.3.2 Observation of individual vehicles — 268
        12.3.3 Interviews and questionnaires — 268
        12.3.4 Car following studies — 269
        12.3.5 The use of route choice simulators — 269
    12.4 Analysis of registration plate data for O-D surveys — 270
        12.4.1 Errors in data recording — 271
        12.4.2 Spurious matchings — 272

xii *Understanding Traffic Systems*

12.4.3 Analysis of results from registration plate surveys	273
12.4.4 Computer analysis of registration plate O-D data	275
12.5 Presentation of results	276

13 Traffic generation and parking studies — 281

13.1 An introduction to traffic generation and parking	281
13.1.1 Traffic generation	281
13.1.2 Parking	284
13.2 Supply of parking and entrance facilities	287
13.3 Traffic generation and parking demand surveys	289
13.3.1 Existing information	289
13.3.2 Interview surveys	291
13.3.3 Observational surveys	294
13.4 Summarising the requirements for parking surveys	302
13.5 Models of traffic generation and parking	303
13.5.1 Trip generation and parking rates	304
13.5.2 Regression models	305

14 Road safety studies — 309

14.1 Some definitions	311
14.2 Statutory crash records	311
14.3 Other sources of crash data	313
14.4 Site investigations	314
14.5 Conflict studies	314
14.6 Driver behaviour studies	316
14.7 Controlled crash tests	317
14.8 Analysis of crash data	318

Part E: DATA ANALYSIS AND MODELLING — 319

15 From data to information — 321

15.1 Tables	321

15.2 Diagrams	325
15.2.1 One-dimensional plots	325
15.2.2 The comparison of one-dimensional data using diagrams	334
15.2.3 Two-dimensional plots	335
15.3 Multi-dimensional plots	338
15.3.1 Interactive mapping in data analysis	338
15.3.2 Three-dimensional data	340
15.3.3 Three-dimensional graphics	341
15.3.4 More than three dimensions	342
15.4 Descriptive statistics	345
15.4.1 Statistics and parameters	345
15.4.2 Measures of central tendency	346
15.4.3 Spread of the distribution	348
15.4.4 Shape of the distribution	350
15.4.5 Proportions	351

16 Statistical analysis   353

16.1 Faulty data	353
16.1.1 What to do about data errors	354
16.1.2 Data editing	355
16.2 Hypothesis testing	356
16.2.1 Selecting the hypothesis	357
16.2.2 Principles in hypothesis testing	358
16.2.3 Testing the difference in means	361
16.2.4 Confidence limits	365
16.2.5 Testing the difference in dispersion	366
16.3 Analysis of variance	368
16.3.1 Testing for differences	368
16.3.2 Two-way analysis of variance	374
16.4 Non-parametric tests	377
16.4.1 The sign test for small samples	377
16.4.2 Two matched samples	378
16.4.3 Confidence intervals from ordinal data	379
16.4.4 Rank order correlation	381
16.4.5 Wilcoxon-Mann-Whitney test	382
16.4.6 Contingency table test	383

17 Statistical modelling — 387

    17.1 Model development process — 389
    17.2 Regression — 392
        17.2.1 Simple regression — 392
        17.2.2 Multiple linear regression — 398
        17.2.3 Investigation of residuals — 400
        17.2.4 Significance of regression coefficients — 402
        17.2.5 Stepwise inclusion of variables — 404
        17.2.6 Correlation — 408
        17.2.7 Generalisation of the approach — 409
    17.3 Maximum likelihood estimation — 410
        17.3.1 The mean of a normal distribution — 411
        17.3.2 The Poisson process — 413
        17.3.3 Normal regression — 414
        17.3.4 Other applications — 415
    17.4 Time series analysis — 416
        17.4.1 Time plots — 416
        17.4.2 Transformations — 416
        17.4.3 Analysing series that contain a trend — 417
        17.4.4 Autocorrelation — 417
        17.4.5 Cross-correlation — 419
        17.4.6 Probabilistic models of time series — 420
        17.4.7 An example of time series analysis — 421

Appendix A: Statistical tables — 423

Appendix B: Database of vehicle speeds on residential streets — 431

References — 435

Index — 445

List of figures

Figure no	Figure caption	Page
1.1	The traffic system seen in terms of issues, parameters and levels of analysis	10
1.2	Representation of strategic and dense road networks in the vicinity of a shopping centre	13
1.3	An isolated intersection and its place in the road traffic network	14
2.1	The Systems Planning Process (SPP) recommended for traffic planning and analysis purposes	20
2.2	Relationship between demand and supply modelling	23
2.3	Components of a time series	29
2.4	Intersection node components in TNRDB	32
2.5	Example spreadsheet, comparing traffic flows under different traffic management plans	34
2.6	Example of the superposition of data layers within a GIS	35
3.1	A typical trajectory diagram	45
3.2	Speed-flow envelope	50
3.3	Typical speed-density relationship	51
3.4	Flow-density relationship, showing regions A, B and C for traffic flow	52
3.5	Shock waves in an uninterrupted traffic stream due to a discontinuity at x0 (source: Lebacque and Lesort (1999))	69
4.1	Travel time-flow and speed-flow curves for uninterrupted and interrupted flows	72
4.2	Major and minor traffic streams at a T-junction	78
4.3	Sample junction plan and signal staging	87
4.4	Basic Wardrop-Webster model for a junction	89
4.5	Simple phasing arrangement at a cross intersection	91

4.6	Trajectory diagram of arrivals and departures for one movement	100
4.7	Queue formation by comparison of arrival and discharge rates	102
4.8	Typical form of a link congestion function	106
5.1	A traffic area defined by an external cordon line and showing types of through and local trips	115
5.2	Signal coordination principles and the 'green wave'	125
5.3	The economic price of congestion and travel demand	130
6.1	Stages in the design and conduct of a traffic survey	138
6.2	Typical survey schedule	141
8.1	Typical flow profiles on an urban radial road in the UK	164
8.2	Midlink classified traffic count form (reduced from size A4)	168
8.3	Midlink classified traffic count form using 'five bar gate'	169
8.4	Survey form for 'five bar gate' turning movement study	170
8.5	A typical bar code sheet (plastic coated for protection)	173
8.6	Selected vehicle detection technologies	179
9.1	Sample vehicle speed-time profile data	187
9.2	Time space diagram showing individual vehicle trajectories	188
9.3	An arrangement of mirrors to overcome the parallax problem	190
10.1	Conceptual model of effects of a pollutant (Brown, 1980)	215
10.2	Integrated modelling system for assessing environmental impacts of road traffic	217
10.3	Gaussian plume model of pollutant dispersion	221
10.4	Observed noise levels (a) time series (b) histogram (c) cumulative distribution	223
10.5	Flow chart for the CORTN procedure (UKDoT, 1986)	225
10.6	SIMESEPT output indicating links in a network with potential environmental problems	231
11.1	Turning movement flows at a junction (assuming left hand side driving)	237

List of figures xvii

11.2	Trajectory diagram at a signalised junction	242
11.3	Survey form for QDELAY	248
11.4	Observer location in saturation flow studies (Cuddon, 1993)	251
11.5	Saturation flow survey form (source: Adams and Hummer, 1993)	253
11.6	Estimation of critical gap and move-up time from observed accepted gaps for minor stream traffic leaving a standing queue	257
12.1	Typical roadside interview form	262
12.2	Setting a cordon line for a study area	264
12.3	The VLADIMIR simulator (Bonsall et al, 1997)	270
12.4	General procedure for correction of errors in registration plate surveys	274
12.5	Example of a desire line diagram	277
13.1	The trafficgraph for a traffic generating event	283
13.2	Combination of traffic events	285
13.3	Simple inventory map showing available parking facilities	288
13.4	Example of a parking interview form	292
13.5	Typical reply paid questionnaire form	294
13.6	Typical form for patrol survey	299
13.7	Vehicles missed by patrol survey (Young and Thompson, 1990)	300
13.8	Comparison of parking duration for input-output and video patrol methods	301
13.9	Possible effects of extrapolation outside the range of the observed data	307
15.1	Quantile plot of vehicle speeds	326
15.2	Dot diagram of the vehicle speed data	326
15.3	Stacked dot diagram of the vehicle speed data	327
15.4	Jitter plot of the vehicle speed data	327
15.5	Histogram of vehicle speed data using 2 km/h class intervals	328
15.6	Histogram of vehicle speed data using 1km/h class intervals	329
15.7	Continuous distribution	331

15.8	Comparison of histogram and continuous function	331
15.9	Cumulative frequency distributions	332
15.10	Box plot of the vehicle speed data	332
15.11	Stem and leaf diagram for the vehicle speed data	333
15.12	Pie chart of fatally injured pedestrians by blood alcohol content (BAC)	334
15.13	Box plot of monthly variations of arterial road flows	335
15.14	Comparison of median speed values using box plots	336
15.15	Relationship between vehicle speed and per cent local traffic	337
15.16	Change in average vehicle speed over time with (a) unlabelled and (b) labelled axes	338
15.17	GIS maps showing vehicle ownership and dwelling density by zone in an urban area	339
15.18	Comparison of three variables using pair-wise comparisons	340
15.19	Partitioning of one axis of a graph (Chambers et al, 1983)	341
15.20	Three-dimensional display of a dependent variable and two independent variables (in this case a car engine map for hydrocarbon emissions)	342
15.21	Multiple pair-wise comparisons	343
15.22	Example star plot axes	344
15.23	Comparison of speed limit performance using star plot	345
15.24	Unimodal histogram of accident frequencies	346
15.25	Bimodal histogram of accident frequencies	346
15.26	Symmetry of a distribution: (a) bell-shaped curve (b) right skew (c) left skew	351
15.27	Peakedness or flatness of a distribution: (a) bell-shaped curve (b) lepto-kurtosis (c) platy-kurtosis	352
16.1	Possibilities of Type I and Type II errors	360
16.2	Seven-point semantic scale	378
16.3	Confidence intervals for non-parametric data	380
17.1	Specification error and measurement error in models	388
17.2	The model development process	390
17.3	Model errors comparing observed and estimated data	393
17.4	Scatterplot of vehicle mass/fuel consumption data	394
17.5	Regression output for the vehicle mass/fuel consumption data	397

List of figures xix

17.6	Scatterplot of speed and gap data	399
17.7	Plot of residuals	401
17.8	Forward stepwise regression applied to the speed data in Appendix B	405
17.9	Backward stepwise regression applied to the speed data in Appendix B	407
17.10	Correlation matrix for the speed data	409
17.11	Time series plot of the traffic flows of Table 17.9	422
17.12	Averaged traffic flows in consecutive four-week periods over four years (see Table 17.9)	422

List of tables

Table no	Table caption	Page
1.1	Inventory of principal traffic parameters	8
1.2	Survey methods for given traffic parameters	9
2.1	Properties associated with the levels of data classification	27
4.1	Base lane saturation flows (s_b tcu/h) for Australian conditions	96
4.2	Burrow's (1989) comparison of traffic signal overflow delay expressions	104
4.3	Representative parameters for Akcelik's congestion function	108
5.1	Typical structure of an O-D matrix for a traffic area	115
7.1	Possible blocking situations when number of tests equals the number of alternatives	152
7.2	Possible blocking situations when number of tests is less than the number of alternatives	152
7.3	Latin squares design for car emissions using 4 drivers and 4 cars	153
7.4	3x3 factorial experiment investigating vehicle speeds on residential streets	154
7.5	2x2x2 factorial experiment of vehicle speeds in residential streets	154
7.6	Values of $\alpha^2(2+u^2)$ for use in sample size determination, when determining percentile estimates with level of confidence α per cent	161
8.1	Automatic detectors	176

xxii *Understanding Traffic Systems*

9.1	Various measures of travel time	189
11.1	Schematic turning movement flows matrix	237
11.2	Turning proportions for Toronto (from Hauer, Pagitsas and Shin (1981))	240
11.3	Vehicle numbers over time at a sample intersection	247
12.1	Data sources for origin destination and route choice studies	260
12.2	Worked example of Furness Technique	278
13.1	Information needs for parking analysis (Bonsall, 1991)	286
13.2	Data items obtained from different observational parking survey methods	295
14.1	Components of typical crash report form	312
15.1	Vehicle speeds (km/h) in residential streets	322
15.2	Table of speeds (km/h) sorted in increasing order	322
15.3	One-dimensional table	323
15.4	Two-dimensional table: survey results for blood alcohol levels of drivers involved in accidents at five locations	324
15.5	Three-dimensional table: motor vehicles – summary of selected items	324
15.6	Frequency table for vehicle speed data (km/h)	328
16.1	Possible consequences in hypothesis testing	359
16.2	Speeds along test roads entering a town centre	369
16.3	Average speed of vehicles on the same road	369
16.4	The analysis of variance table	372
16.5	Comparison of mean speeds	373
16.6	One-way analysis of variance table for the speed data	374
16.7	Speeds on roads by day of week	375
16.8	Components of two-way ANOVA table	376
16.9	Two-way ANOVA table	376
16.10	Traffic flow on arterial roads	377
16.11	Perception of traffic noise	379
16.12	Ordered traffic flows on arterial roads	380
16.13	Sample speeds of vehicles on two roads	381
16.14	Sample traffic flows from two local government areas	382

List of tables xxiii

16.15	Combined ranking for the W-statistic	383
16.16	Alcohol level of accident victims at five locations (percentages)	384
16.17	Alcohol level of accident victims at five locations (number)	384
16.18	Expected accident number for accident locations	385
17.1	Vehicle mass and fuel consumption over a standard driving cycle	394
17.2	The analysis of variance table	396
17.3	Data on vehicle speeds and gaps	398
17.4	Analysis of variance table for initial regression	400
17.5	Parameter estimates and model fit for the quadratic model	402
17.6	Analysis of variance table to determine model improvement	403
17.7	Investigation of exclusion of constant term (a)	404
17.8	Distribution of accidents	413
17.9	Traffic flows over consecutive four-week periods over four years	421
A.1	Area under normal curve, to the right of the mean ($z = 0$)	424
A.2	Percentage points for t-distribution	425
A.3	Percentage points for the F-distribution (5 per cent upper line, 1 per cent lower line)	426
A.4.1	Rank order correlations of d^2 significant at 5 per cent level	428
A.4.2	Wilcoxon-Mann-Whitney two sample test	429
A.5	Percentage points for χ^2 distribution	430

Preface

This is the second edition of *Understanding Traffic Systems*. In the short while since the publication of the first edition in 1996, there have been continuing developments in the technology available to the traffic analyst – and in the social, ecological and economic environment in which analysts work. This new edition of *Understanding Traffic Systems* reflects those changes. And yet, the basic message of *Understanding Traffic Systems* remains the same – the need for proper thought, care and attention in the collection, analysis and presentation of traffic data.

Traffic and its problems affect all aspects of modern life, leisure and industry, particularly in urban areas. This book aims to introduce and demonstrate techniques that the analyst, engineer or planner can apply to examine and solve traffic problems. The underlying theme is that an *understanding* of traffic systems and their performance can only come from *information* which will in turn come from the intelligent processing, refinement, appraisal and evaluation of *observed data*. This requires the collection and statistical analysis of traffic data, soundly-based inference and deduction, and the application of modelling techniques.

The book is arranged in five parts which, between them, offer an integrated approach to tackling and resolving road traffic problems.

Part A introduces the theme of the work, which is how to gain information and understanding about traffic systems through the derivation of traffic information from traffic data, and then how to apply this information in the assessment of the performance of traffic systems and their interaction with land use activities. The importance of traffic safety, of traffic congestion, and of the environmental impacts of traffic operations is emphasised and the role of analyses at various levels of detail is outlined. The technical endeavour of traffic impact analysis is then placed in the context of public decision making and current policy concerns. The relationship between traffic analysts and decision makers is discussed and the importance of the means by which traffic information and forecasts are presented to decision makers and the community is emphasised. The concepts of scenario generation and the use of 'what-if' analysis through computer modelling are introduced, leading to consideration of 'pro-active' analysis that anticipates future problems and proposes solutions before the problems can emerge.

Part B introduces the key theories relating to the operation of traffic systems. These range from the simplest theories concerning the behaviour of traffic on isolated links (uninterrupted flow models), through theories relevant to the performance of junctions (interrupted flow models), and on to models dealing with the performance of complete networks.

The use of traffic flow and transport network theory requires the acquisition and application of traffic data. Part C describes the principles of good survey planning and management, stressing that data collection should only be undertaken after careful analysis of the problem to be investigated, definition of specific data needs, and the determination of resources available for data collection and analysis. The principles of experimental design and survey sample design are then introduced. A variety of methods for observing and estimating traffic parameters such as flows, speeds and queue lengths are described in detail.

Part D describes specific types of traffic studies, including intersection studies – for turning flows, delays, saturation flow rates and gap acceptance, origin destination and route choice, traffic generation, parking needs, patterns of demand and road safety.

Part E is concerned with the development and use of analytical techniques for transforming raw data into useful information. Separate chapters deal with data presentation techniques and descriptive statistics, statistical analysis techniques and the development of basic models of the traffic system.

Acknowledgements

The authors would like to thank the many people who helped in the preparation of this book. Rocco Zito, Troy Young, Peter Gipps, Frank Montgomery, Kylie Cook and especially Rob Alexander deserve particular mention for their inputs into the completion of the work.

Part A

INTRODUCTION

1 Introduction and context

Traffic systems ought to provide high levels of accessibility without undue cost in terms of finance, accidents or environmental detriment. The issues of greatest current concern are: traffic safety, which is increasingly recognised as a significant problem in contemporary public health; traffic congestion, which can disrupt the living processes within an area and may place severe constraints on social well-being and economic development; and the environment, since road traffic is known to be a significant source of air and noise pollution which can have a deleterious effect on the physical environment and amenity of an area and on the health of its inhabitants.

Contemporary transport planning is based on a change of emphasis away from large scale construction of new facilities and towards a more incremental approach with an increasing concern with travel demand management (TDM) and traffic calming which aim to reduce the impact of vehicles and traffic in an area and reflect a growing concern with environmental quality.

The major thrust of the development of transport and traffic planning in recent years has been the widespread development and deployment of advanced information technology, telecommunications technology and computer technology in the operations and management of transport systems. Collectively these initiatives are known as Intelligent Transport Systems (ITS), or sometimes as Advanced Transport Telematics (ATT). ITS enables real time monitoring and control of transport systems, the provision of up-to-date information for systems managers and the travelling public, and enhanced mobility, safety and pollution control. ITS development such as electronic tolling and electronic road pricing (ERP), advanced traffic management systems (ATMS) and advanced traveller information systems (ATIS) provide transport planners and engineers with significant new capabilities to explore new policies and practical implementations in TDM and traffic calming. ITS is being used, for instance, to provide information on the best travel route, to increase road system capacity through traffic signal coordination, and to reduce capacity constraints at toll booths.

The worldwide interest in TDM is a response to a widespread belief that transport systems, especially the private car, have an insatiable appetite for new infrastructure. Many communities now question the wisdom of continued new investment in that infrastructure, especially if its provision is only required to cater for traffic movement in a few hours of each day. Alternatives to providing new infrastructure include encouraging alternative patterns of usage of the existing infrastructure, such as the provision of better information so that travellers can make more informed choices, spreading of peak demand periods, or limitations on private vehicle access to some zones (e.g. the city centre), through physical restraints or through the use of charging systems such as ERP.

The role of traffic analysis in TDM is to examine the degree to which existing infrastructure can support alternative patterns of demand and to provide information for the successful implementation of TDM initiatives. Quality in traffic analysis is made more necessary by the advent of ITS, which yields large volumes of dynamic data about the state of a traffic system. Traffic engineers and planners are thus increasingly concerned with marginal changes and fine tuning of the system. Traffic data must therefore be more reliable, detailed and comprehensive so that small differences in traffic performance may be detected. Analysis must be performed more frequently, more rapidly and at a lower cost.

The increasing reliance on private sector funding for transport infrastructure developments also places more stringent demands on traffic analysis, especially in terms of demand and revenue predictions on which investment decisions will be based.

Another trend is for a wider range of bodies, in both the public and private sectors, to have interests in, or responsibilities for, traffic planning and management. This is in part due to the belated acceptance of the close interaction between traffic and land use and the increasing recognition that formal Traffic Impact Analyses (TIAs) should precede any major land use development. It is also due to new relationships between private developers and the public authorities and the increasing use of the private sector to provide technical services, such as traffic planning, for public sector bodies. Finally, there is also an increasing interest by the public at large in traffic matters. The resulting situation is a large number of bodies having interests in traffic information processing, and analysis and interpretation. The difficulty is that many of the agencies with statutory responsibilities for traffic data can only afford small departments to address these responsibilities. Fortunately, however, data collection technology has advanced dramatically in the last few years and modern computer-based systems with networking and database communications permit continuous monitoring of key sites, efficient conduct of ad hoc surveys and offer the prospect of comprehensive network-wide data. Personal computer software provides a universal capability for sophisticated analysis of traffic data at low cost, using the latest and best practices of mathematical statistics, information theory and database management.

The theme of this book is to clearly define and espouse these best practices for data collection, data analysis and information management in traffic systems analysis, as the only way to proper understanding of traffic systems.

1.1 Traffic impacts

Traffic can have impacts in a number of ways, for example we can consider performance in terms of: traffic efficiency, accessibility and movement; environment and amenity; safety; public transport and pedestrian movement, and road pavement and bridge life. Traffic efficiency mainly involves the performance of networks in terms of their ability to handle volumes of moving traffic. Environmental impacts largely relate to pollution, noise, vibration and intrusion. Amenity is mainly a concern on minor roads, affecting accessibility to properties and the ability of inhabitants to undertake a range of activities without undue interference from traffic. Safety is a concern on all roads, and should include qualitative, as well as quantitative aspects. Public transport operations and the efficient and safe movement of vulnerable road users (e.g. pedestrians and cyclists) are integral parts of traffic systems and have their own special needs. Damage to road pavements, structures and bridges can occur on all classes of roads, but is usually only of importance where there are large numbers of heavy vehicle movements.

1.2 Traffic studies

Most traffic control and design problems demand a fairly detailed knowledge of the operating characteristics of the traffic concerned and hence traffic studies have long been an important aspect of the traffic engineer's work. The results of traffic studies are used in traffic planning, traffic management, economic studies, traffic and environmental control, road safety studies, land use-transport interaction studies, and for establishing and revising design standards and systems models. Traffic analysis makes use of a variety of theories, statistical tools and mathematical models. These need to be tested

6 *Understanding Traffic Systems*

against observations of behaviour in real world traffic systems. Thus the purposes for which traffic data are required may be summarised as:
- *monitoring*, the collection of information about the traffic conditions prevailing at any time, and as they change over time;
- *forecasting*, the use of data on existing traffic systems as one of the inputs to a procedure for estimating what the traffic would be like under different conditions, either now or in the future;
- *calibration*, the use of traffic data to estimate the values for one or more parameters in a theoretical or simulation model, and
- *validation*, the verification of a theoretical or simulation model against information independent of that used to calibrate the model.

The first two items are primarily the concern of practitioners, whereas the latter two items have tended to lie more in the interests of researchers and model developers. The new emphasis on traffic volume and revenue prediction in infrastructure investment decision making is changing this situation rapidly, as the need for relevant and dependable modelling becomes more and more part of normal professional transport practice.

Traffic surveys are the means of collecting the base data from which will come traffic information. The main types of information that are commonly required are: flows of vehicles and pedestrians; numbers of waiting vehicles or pedestrians; numbers of parked vehicles; numbers of occupants in vehicles; speeds of vehicles; travel times; accidents; fuel consumption and emissions; vehicle weights; and origins and destinations of journeys. This information may be sought at a number of levels of detail; from broad indications of traffic conditions over a region to detailed measurement of individual vehicle movements at a given location.

1.2.1 Principles of information gathering

This book seeks to demonstrate the need for high level of integration of the collection and analysis of traffic data, and the other elements of modern traffic science, (such as the theory of traffic flow and the use of traffic models) for proper understanding of the operation of traffic systems. This has led us to adopt a *systems approach* to traffic studies and to stress the importance of new technologies in traffic data collection and analysis. The systems approach permits individual surveys to be placed in a proper context so that their results may be fully appreciated. New technologies can bring improvements in both quantity and quality of data. Data may be acquired directly using automatic recording devices, observers, interviews and video recording. Alternatively, they may be acquired indirectly; for example by using traffic flows to predict traffic noise, accidents and emissions, or by using models to estimate patterns of demand.

1.2.2 Important traffic parameters

Information on traffic flows, speeds, congestion, delays, parking conditions, safety, fuel consumption, loads carried and environmental impacts is necessary for diagnosing problems, finding appropriate solutions and studying the effects of implemented schemes. Other data, such as origin-destination demands, are particularly useful in forecasting the effects of planned projects.

Table 1.1 provides a summary of the most commonly required traffic parameters and some of their associated measurement quantities. Further subdivision of the information in each dimension may be made. For example, vehicle volumes may be disaggregated by the type of vehicle and the purpose for which it is being used. Table 1.2 lists of some of the commonly used methods for collecting data on these traffic parameters. Chapter 6 indicates how a systematic approach to survey planning can provide an integrating methodology to draw the techniques together.

1.3 Levels of analysis

Given the wide range of traffic parameters and the commensurate variety of levels of detail that can occur, the traffic analyst is presented with a particular problem in the choice of analytical method to apply to a specific problem. We can define three levels at which investigations may be directed. These levels relate to physical elements of the traffic system; they are the *link* (or road) section, the *junction* or intersection, and the *area* (or network). The interactions between each of these elements, and with the major issues in traffic analyses, is indicated schematically in Figure 1.1.

1.3.1 Link level analysis

The link represents a length of road between two junctions. Traffic flow on the link may be comprised of one or two broad traffic streams, depending on whether the link is available for one-way flow only, or for flow in each direction. A traffic stream along a link may have one or more lanes on the carriageway available to it.

Table 1.1 Inventory of principal traffic parameters

Traffic parameter	Measurement quantities
Volume	Number of vehicles per unit time by vehicle type Number of pedestrians or cyclists per unit time Toll revenue, congestion pricing revenue
Speed, delay and queuing	Time, distance, speed (time/unit distance), delay, queue length
Density (concentration)	Vehicles/unit length
System inventory	Numbers of control devices, signals, signs, street length by road class, lane length by road user class
Generation	Trips/person or household, activity unit by vehicle type
Parking supply	Number of spaces, costs per unit time and per unit space, maximum allowable duration
Parking demand	Occupancy, mean duration, turnover rate, parking volume, compliance with parking restrictions
Safety	Numbers of accidents by severity class, numbers of traffic conflicts, amount of travel (veh-km or volume)
Environmental factors	Noise: dB (A,B,C) scales, L_{10}, L_{50}, L_{eq}. Air pollution: concentration per unit volume (ppm), air temperature, wind speed. Vibration: frequency, acceleration and amplitude
Energy	Fuel consumed per unit of travel, by fuel type and road user class/vehicle type

Table 1.2   Survey methods for given traffic parameters

Traffic parameter	Survey methods
Volume	Link-based counts by time of day, junction turning movements, vehicle classification surveys, pedestrian and bicycle counts
Speed, delay and queues	Radar surveys, vehicle detection surveys, video surveys, floating car, chase car and registration plate surveys, input- output surveys, path trace, internal-based queue length, event-based queue length
Heavy vehicles	Vehicle classification, axle weight, vehicle dimensions
Density	Volume-density relationships, lane occupancy
System inventory	Street characteristics, junction control, parking restrictions inventory
Trip generation	Household interview, postcard, residential and employment densities by area
Parking supply	Capacity, costs by time or location
Parking demand	Parking duration, spatial distribution of demand, temporal duration of demand
Safety	Traffic conflict, driver behaviour, regulation observance, accident database records
Environmental factors	Spatial and temporal distributions of noise, air quality, and vibrations
Energy	On-road fuel consumption, dynamometer, driver diaries
Origin-destination	Registration plate, cordon identification, roadside interview, postcard
Public transport	Travel time and delay on-board, passenger loading and unloading, passenger origin-destination

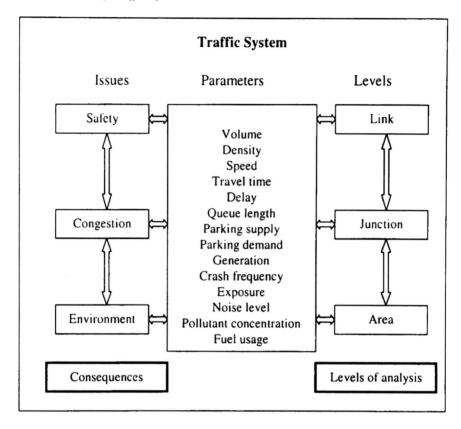

Figure 1.1 The traffic system seen in terms of issues, parameters and levels of analysis

A common first order approximation is that flow in a traffic stream along a link can be regarded as 'uninterrupted' flow, in that the operational characteristics of the stream depend on its own parameters (such as traffic volume, composition, and speed) and the physical characteristics of the link (e.g. roadway width, pavement surface type and condition), and are independent of the influences of external factors (such as other traffic streams or abutting land uses). This provides the basis of the traffic flow theory introduced in Chapter 3.

For instance, delays incurred in travelling along a link may then be seen in terms of the increases in travel times occurring as traffic volume on the link increases. Strictly, however, this assumption can only be applied to those links with limited access and physical separation between opposing flows, such as a motorway or high capacity divided carriageway arterial road.

In the real world, traffic flow on most road sections is subject to external influences: vehicles overtaking in the face of opposing traffic, vehicles parking and unparking at the kerbside, traffic turning into or emerging from car parks and land use developments at sites along the road.

Nevertheless, the essence of traffic movement on links is the directional flow of traffic over a length of road. The progression of vehicles along links gives our basic measures of the amount of travel taking place, such as parameters of vehicle-kilometres of travel (VKT, the product of the traffic volume traversing a link and the length of the link, summed over the links in a network), and vehicle-hours of travel (VHT, the product of the traffic volume and the travel time along the link). The set of link flows on a network provides us with basic information about the actual distribution of traffic in the network: which are the busiest parts? which roads attract the largest amounts of traffic? which are the important routes for traffic movement? which roads and streets provide access to land use activities rather than cater for traffic movement? etc.

1.3.2 Junction level analysis

Junctions occur at the intersection of road sections (*links*). At these points, and only at these points, traffic can change direction of travel. In the traffic network terminology, junctions form *nodes*, the connection points between links. Traffic streams come into conflict at junctions, for the streams on the intersecting roads have to cross or merge with each other. Junctions represent the points in a traffic system where capacity problems may first emerge, because multiple traffic streams have to use the same physical road space. It is therefore the junctions which usually dictate urban network capacity; mid-block capacity (i.e. that applying to a link) usually exceeds the capacity of the junction at the downstream end of the link, unless there are some peculiar and severe features on the link (e.g. a narrow bridge or other constriction). Safety problems may also be more pronounced at junctions, because the intersecting traffic streams may be competing for the same road space, and because drivers may have to face more challenging and complex tasks in negotiating an intersection. Environmental problems may result from noise generated in acceleration and braking at the junction, and from the presence of long queues and their resultant exhaust emissions.

Traffic flow at junctions provides good examples of 'interrupted' flow. The concepts, theories and models required to analyse and explain the behaviour of traffic at a junction and the performance of that junction need to take account of the interactions between the competing traffic streams, the junction's geometric and physical characteristics, and its traffic control regime (e.g. signals, roundabout or priority road). The traffic control regime used at a junction will have a strong bearing on its operational performance, and selection of the appropriate control regime for a given junction is an important task in traffic engineering. Capacity concerns focus on the throughput of vehicles (or other flow units, e.g. people) at a junction and on its approaches. Junction performance is determined by the delays associated with passage through the junction, and with the extent of queuing that can be attributed to it. Secondary measures of performance include excess fuel consumption and pollutant emissions. The 'excess' is gauged by considering the extra fuel or emissions resulting from delays incurred at the junction compared to travel along equivalent lengths of road with no junction present.

The most efficient traffic performance at a junction may come from considering the relationship between that junction and its neighbours, and this leads us to considerations of area-wide traffic analysis.

1.3.3 Area level analysis

Traffic impacts often spread beyond the immediate vicinity of the cause of those impacts: they are felt not just along a road or at a critical intersection, but also in an area around the site of generation. Traffic management schemes must also be considered at the area level (where the area is an agglomeration of links and intersections). A solution to a traffic problem that involves shifting traffic from one street to an adjacent street may not have solved the problem, merely changed its location. Travel patterns across an area, including the choice of routes by drivers, may be affected by changes to one or more elements of a network. Thus there are often strong needs to consider the spatial distribution of traffic movement when assessing traffic impacts and implementing traffic controls, especially when environmental and safety issues are of primary concern.

Modern traffic control principles and theory rely on the coordination of traffic signals at neighbouring junctions, or at a sequence of junctions along a road. Area-wide considerations of the relationships between junctions thus figure strongly in traffic systems planning and control. The theories utilised to consider area-wide impacts rely on descriptions of the patterns of travel demand over an area, and the relationships between the traffic operations of links and junctions.

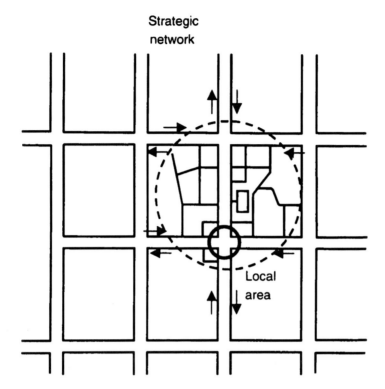

Figure 1.2 Representation of strategic and dense road networks in the vicinity of a shopping centre

1.3.4 Relationships between link, junction and area

Figures 1.2 and 1.3 offer a simple view of some of the possible relationships between the three spatial levels of analysis. Figure 1.2 shows a local area road network (or *dense network*), comprising major and minor roads and the land uses abutting them, surrounded by the regional (or *strategic*) network of major roads representing the surrounding area.

Figure 1.3 shows some of the elements of the local traffic system in more detail, highlighting the access and parking arrangements associated with a particular site (perhaps a shopping centre), the roads to which the site is attached, and a nearby intersection whose traffic control system and physical layout may be relevant to the functioning of the site.

14 *Understanding Traffic Systems*

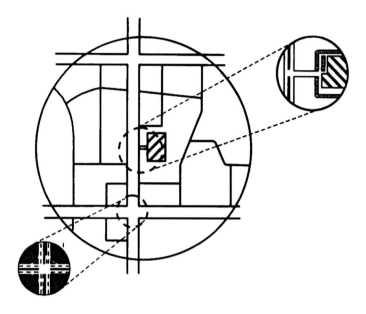

Figure 1.3 An isolated intersection and its place in the road traffic network

1.4 Roles of the traffic analyst

The traffic analyst provides information for the planning, design and implementation of improvements to the traffic system and, more recently, with various forms of demand management. To fulfil this role the analyst requires a sound grasp of the needs and aspirations of the community, of the characteristics of the existing traffic and transport system and of available policy options. Increasingly, however, the traffic analyst may be required to contribute to more specialist areas such as the conduct of formal Traffic Impact Analyses (TIAs), Environmental Impact Assessments (EIAs) and safety audits. It is perhaps useful to consider each of these areas in a little more detail.

TIAs have an important role in local area planning and it has become the norm to require such a study prior to any significant new land use development. The study will consider the likely pattern of traffic associated with the new development and the ability of the local network to support it. The findings may result in a decision to forbid the development, require some modification to the proposal or some contribution by the developer to any necessary upgrading of the network. A TIA should involve a two-way exchange of ideas and cooperative consideration of solutions to any problems identified. TIAs are normally commissioned by the developer rather than by the public authority.

EIAs are specialist assessments of the environmental impacts of proposed developments or schemes and are concerned with many factors in addition to traffic. The traffic analyst might expect to contribute to an EIA through an assessment of the effects of a proposed change in the transport system or of the traffic effects of a new development. Relevant factors will include social effects, such as disruption or dislocation of an established community or the curtailment of social activities; physical and biological impacts through emissions of pollutants and the degradation of physical environments, and economic effects through the consumption of scarce resources such as fossil fuels, land and (in congested conditions) time. Environmental impacts have traditionally been seen as a local issue raising questions such as; how much additional traffic on the surrounding network?, how much noise pollution?, what level of photochemical smog?, etc. Increasingly, however, there is a concern with the global dimension; giving rise to questions about the possible contributions of a local facility or traffic network to global warming, greenhouse gas emissions and resource consumption. This is clearly something of a departure for the traditional concerns of the traffic analyst but cannot be ignored. Further details can be found in Chapter 10.

Safety audits are increasingly popular as a means of concentrating attention on safety issues. They may be conducted on existing systems or as part of the design process. Safety audits of existing systems will involve local evaluation of accident risk via analysis of accidents record and measures of exposure and will involve detailed site surveys. Chapter 14 gives further details of the techniques involved.

In all these areas the trend is for traffic analysts, whatever their particular role, to be making increasing use of advanced facilities for the collection and analysis of data and for modelling the traffic system.

2 The traffic analysis process

Traffic systems are complex entities, and proper analysis of data on the behaviour of traffic systems is fundamental to the investigation of a system's performance. The processes of traffic data collection, analysis and dissemination need to be understood and applied in a systematic fashion. Good organisation is the key to success. The major phases in traffic surveying and data analysis which involve traffic engineers and planners are:
- specification of data requirements;
- design and conduct of surveys;
- verification and analysis of data, and
- application of data to design and decision making.

This chapter describes an overall systems approach that outlines the processes in planning and decision making and identifies the role of traffic data collection and analysis in that process. Major components of the planning process are identified, and the interactions and protocols between these components are highlighted. This requires a consideration of the role of information in decision making, the nature and properties of data observations, a description of state-of-the-art methods for the assembly, management and use of databases, and a discussion of the application of data to traffic planning problems involving scenario generation, sensitivity analysis and 'what-if' studies.

2.1 Information, planning and decision making

Planning involves the generation of alternative proposals and the assembly of supporting information to be presented to the decision makers who have to consider the consequences of the alternatives and select an alternative for implementation. The role of information in planning may be reflected in the following propositions:
- *history is made as a result of decisions*, not as a result of plans;
- *planning is only effective if it is based on information of direct relevance to those who make decisions.* Good decisions require proper understanding of the short and long term consequences of the

alternatives under consideration but this may not always be appreciated by the decision makers themselves;
- *all decisions involve the 'evaluation' of alternative scenarios of the future and the selection of the most highly valued feasible alternative.* This evaluation may be quite cursory or may involve sophisticated analysis but, even where the evaluation is cursory, decision making requires an *agenda* (outlining the alternative scenarios and some analysis of their likely implications) and a *valuation scheme* (outlining the preferences for the characteristics of the likely outcomes of any one decision);
- *the degree of uncertainty associated with items on the agenda affects the evaluation and decision strategies and the quality of the decisions themselves.* Decisions concerning future actions are based on assumptions about the likely consequences of choosing alternatives and the future state of the system in which the decision will be implemented. Thus, the greater the degree of uncertainty associated with these assumptions, the higher the value that should be placed on decisions that leave future options open;
- *the products of the planning process should be designed to increase the probability of making better decisions.* Thus the planning process needs to examine a wide range of agendas, and to consider alternative goals and objectives. Much can be learned by examining past decisions that were not successful, and from watching out for the early warning signs of emerging problems, each of which will benefit from monitoring system performance before and after implementation of supposed solutions, and
- *the outcomes of the planning process are but a small part of the information required by decision makers.* The usefulness of planning information will be increased if attempts are made to adapt the products of the planning and analysis process to fit the substantive and interpretative requirements of the decision makers. To want to use the planning information, the decision makers will need to be convinced of its accuracy and relevance. This can only be accomplished by presenting the information in forms acceptable to and easily understood by them.

The primary implication of these propositions is that (traffic) planning and investigation work should focus on the information needs of decision makers, and recognise the limited capabilities of individuals unfamiliar with technical analysis to accept and interpret the information produced. At the same time, the process must do more than merely provide the information apparently desired by the decision makers. The full range of information needed to provide a more complete understanding of the problem and the implications of the various alternatives must also be included. To do this requires the establishment of a strong degree of acceptance of the respective roles of the planner-analyst and the decision makers, and the development of mutual trust between the two groups.

2.2 The traffic planning and decision making process

Figure 2.1 represents the overall traffic planning and decision making process as a 'Systems Planning Process' (SPP). This process diagram is adapted from that proposed by Richardson, Ampt and Meyburg (1995).

The first observation from Figure 2.1 is that there are no obvious starting or finishing points to the SPP! The process is continuous. The likely place to enter the process is 'problem definition', that is, the realisation that there is some deficiency in an existing system or state. This means some perceived difference between the actual state and the desired state. Problem definition depends on the values and aspirations of the affected community, which are manifest in a set of goals (sometimes conflicting) that the community will establish, wittingly or otherwise. The careful definition of the problem will greatly assist in suggesting possible solutions. Indeed, the proper enunciation of the problem may well be a crucial step in its solution.

Following the realisation of the existence of a problem, there is the investigation of the extent of that problem, i.e. the 'system boundaries'. Who, where and what does the problem affect? Given the community's goals and the system boundaries for the problem, specific goals may be refined into particular objectives, and criteria may be established to indicate the success in solving a problem. For instance, a community may believe it endures too much traffic noise, that its problem is one of excessive noise pollution from road traffic. The objective might then be to achieve 'acceptable traffic noise levels on sub-arterial roads in inner urban areas'. The criterion for determining success in this objective might be specified very precisely, for example 'an L_{10} (18h) noise value of 68 dB(A) one metre from the building facade of residences along Main Road in Hushville'.

Solution of a problem requires the allocation of (often scarce) resources such as time, money and people. The levels of these resources that can be made available will determine:
- the type, scale and number of alternative solutions to be investigated;
- the types and quantities of data that can be collected, and
- the level of modelling and data analysis that can be undertaken.

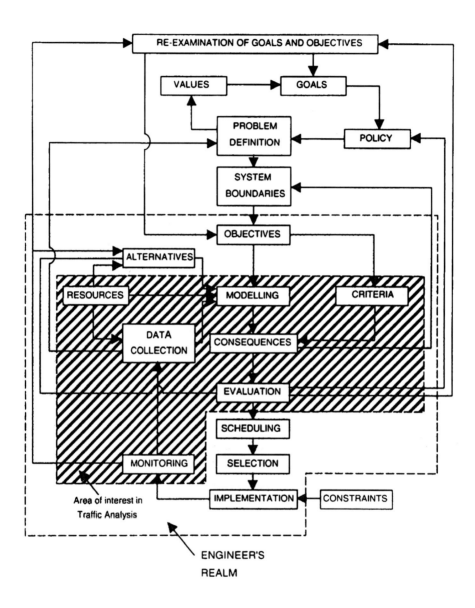

Figure 2.1 The Systems Planning Process (SPP) recommended for traffic planning and analysis purposes

The analyst-planner will then specify a number of alternative approaches to solving the perceived problem or reaching the desired standard, and will use appropriate techniques to predict the likely consequences of each of the proposed solutions. In drawing up the list of candidate solutions the analyst will have regard to externally imposed constraints such as financial limits, political acceptability and institutional practicality.

The planning process is an ongoing one with many linkages between its components. Evaluation involves comparing predicted consequences of possible 'solutions' against previously established criteria and constraints. If none of the proposed solutions meet the criteria and constraints, it may be necessary to define and assess alternative options. Failing this it may become necessary to amend the criteria to attempt to overcome the constraints, or even reconsider the underlying goals and objectives.

The planning process does not properly finish with the implementation phase. Monitoring the performance of the selected alternative will indicate whether the chosen alternative really provides the predicted level of performance. The results of such monitoring will also provide information with which to improve the forecasting techniques and more accurately specify the system boundaries and so contribute to more accurate predictions in the future. An ongoing monitoring process will also alert the planner-analyst to the emergence of new problems and will reveal any shift in the community's goals and values system, perhaps as a result of their experiences with the selected policy option.

2.2.1 Data collection

The data collection phase of the SPP represents one of the principal areas of interest in this book, and as such receives significant coverage elsewhere in it. Chapter 6, in particular, describes data collection and the extraction of information from the data. Two important points are however needed at this stage. Firstly, the purpose of traffic analysis within the traffic planning process is to provide valid information for use in decision making. This information is derived from data. The process of analysis is to convert data into information, by adding explanation and inference to the data. Secondly, the collection of new data sets is resource intensive; it can be expensive in terms of money, time and labour, and thus the collection of new data should only be undertaken when existing data and information resources have been examined and found to be inadequate. State-of-the-art techniques and tools for database interrogation, especially the use of Geographic Information

Systems (GIS), provide substantial means to examine existing data sets and to glean information from them. An understanding of such techniques and their use is an important skill to be acquired by the traffic analyst.

2.2.2 Modelling and analysis

Modelling and analysis provide the basic means by which assembled raw data may be transformed into useful information for decision making. The choice of the models to be used in the planning process is governed by the objectives of the analysis as well as the available resources. Models may be divided into three basic types: supply models, demand models, and impact models. Supply modelling might generally be seen as a process of determining the effects of changes in the usage of a system on the operating characteristics of that system. For instance, an increase in the traffic volume using a road link may increase the travel time to cover that link. A supply model would indicate the likely increase in travel time resulting from the increased volume. Conversely, demand modelling may be seen as the process of estimating the effects of changes in operating characteristics on the usage of the system. In the case of the congested road link, the travel time along the link might influence the decisions of drivers to take a route containing it. A demand model would predict the usage of the link in the light of its travel time. Since both of these modelling efforts are concerned with the same pair of variables, with the output of one model being the input to the other, it is obvious that demand and supply modelling form a recursive system, as shown in Figure 2.2.

The supply model takes account of the physical characteristics of the system and its current usage in generating a set of system operating characteristics. The demand model then assesses the reaction of the users (and potential users) of the system to these operating characteristics, and generates a new forecast of system usage. The modelling process continues until some form of equilibrium is reached between system operating characteristics and system usage.

Although there is a close relationship between these two modelling efforts, the modelling techniques employed are distinctly different. For the traffic analyst, *supply modelling* is generally concerned with the modelling of physical relationships between elements of the system. Thus, for example, the determination of operating characteristics for a bus route would involve consideration of such factors as bus size and performance, stop spacing, loading and unloading rates, distribution of passenger demand along the route and general traffic system characteristics. *Demand modelling*, on the other hand, involves some representation of social behaviour in that it attempts to assess the response of individuals to changes in the physical system. The scope of such system changes and behavioural responses is quite considerable. Thus, for example, in traffic systems planning, demand modelling is concerned not only with the changes in the usage of individual links in the network, but also with changes in the demand for particular land uses, in terms of trip destinations or residential locations, as a result of changes in the transport or land use systems.

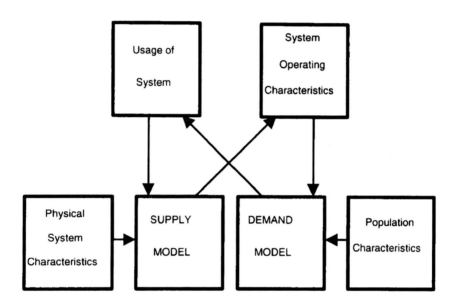

Figure 2.2   **Relationship between demand and supply modelling**

If it can be assumed that a given impact will not materially affect demand, it is possible to circumvent the equilibration process inherent in using supply and demand models and instead to use *impact models*. For example, traffic noise prediction models estimate the noise resulting from a given traffic volume and traffic composition along a road. Since it is unlikely that this noise will significantly deter drivers from using that road, the traffic noise models can be said to be an impact model.

The form of model used may vary widely. In many cases the 'model' might be no more than the experience or guess of the particular analyst. In other situations, the model may be a complex set of mathematical equations or a computer program that can take account of the many system interactions. In all cases, the model simply makes predictions of the likely consequences of the alternatives to be analysed. All models are simplifications of real world processes, and no models are perfect.

2.2.3 Appraisal of policy options

The resources devoted to the appraisal of policy options should be commensurate with the scale of the project, or, more particularly, with the anticipated costs (however defined) of making the wrong decision. The resource cost of the appraisal process can easily get out of hand so it is wise to include a preliminary phase during which checks are made to ensure that the candidate policies are feasible and meet external constraints and that the 'boundaries' of the appraisal have been appropriately specified.

By sieving out policy options which do not meet the basic requirements, or which are obvious 'no-hopers' in terms of their performance against the agreed criteria, the resources available for appraisal can be targeted on the options which stand some chance of being acceptable. Sensitivity analysis can be a powerful technique in this 'sieve' approach to appraisal (Bonsall et al, 1991).

An initial appraisal may reveal that the boundaries originally defined for the appraisal process are inappropriate; the candidate solutions may affect a wider/narrower geographical area, a wider/narrower range of people or firms, a wider/narrower range of factors or they may have impacts over a longer/shorter time scale. It is wise to begin the appraisal by setting wide boundaries with a view to drawing them in later rather than setting them too narrow from the outset because, if they are initially too narrow, the occurrence of effects outside the boundaries may go unnoticed.

2.2.4 Monitoring

The planning process does not end with the implementation of an alternative. The final phase of the SPP is *monitoring* of the performance of the implemented alternative. Monitoring involves continued data collection and analysis, and is important as it:
- provides data on the actual operation of the alternative including any unexpected impacts;
- may provide the basis for recalibration, or reformulation, of models and procedures for better prediction of future consequences;

- may suggest changes that should be made to the selected alternative to improve its operation;
- can alert the analyst-planner to the emergence of new problems and issues at an earlier stage then might otherwise be possible and, by providing a more comprehensive picture, can help avoid decisions being taken solely in response to issues which happen to have had media attention, and
- can improve analysts' understanding of the operating of the system and so enhance their credibility as technical experts and help them to propose more imaginative and effective policies in future.

The inclusion of the monitoring step is essential and highlights the fact that planning and professional practice are continual processes that do not finish with the generation of a set of plans for implementation. These plans must be continually revised in accordance with changes in community goals, changing economic conditions and developing technology, and are an essential part of the successful operation of the new system.

2.3 Statistics and data

A knowledge of statistical theory and method is important in many areas of the traffic systems planning process. Statistical procedures help to determine the data requirements, monitoring procedures and survey resources required in traffic studies. Statistics are also relevant in determining the methods and criteria used to evaluate schemes. The most obvious role of statistics is in defining an adequate sample on which to base our understanding of a part of the traffic system. Samples may be drawn for a number of reasons, such as:
- the impracticality of recording data for the entire population; for example, home interview surveys for urban-wide transport plans cannot collect data on the entire urban population, due to the high cost and long time period involved;
- the impossibility of obtaining a total population because of the rarity of the events; for example, road accidents involving casualties are rare events in terms of the total flow of vehicles and the total number of accidents, and
- the futility of testing a commodity with a destructive test which would eliminate the entire population. The expected life of light bulbs for traffic lights would be one such commodity.

Statistical theory and inference provide mechanisms for determining the size of the sample and analysing the data obtained. Chapters 7, 9 and 16 explore these mechanisms in some detail.

2.3.1 Measurement

The fundamental building blocks for traffic data collection and analysis are the data items to be recorded and analysed. A number of different types of data can be observed, each with their own properties which place some bounds on the manipulations and interpretations that can be applied. Before entering into a discussion of the statistical techniques that can be used to analyse data, an introduction to the various levels of classification of data is therefore required.

Measurement involves the assignment of a score or value to the observation. The rules defining the assignment of the value determine the level of measurement. Four levels are commonly used to describe data (Carterette and Friedman, 1974). These are the nominal, ordinal, interval and ratio measures. The *nominal* is the lowest level since no value is associated with the data. Rather, each value is placed in a distinct category. Examples of this group are the numbers placed on buses to distinguish their route, and numerical codes used to identify (say) the sex or profession of respondents. The *ordinal* level ranks data in some order. Examples of this group include the ordering of buses on their arrival at a terminal, the ranking of alternative transport plans and the ranking of individuals' preferences. *Interval* scales have the property that the distance between each category is defined in terms of actual units of measurement. An example of this level of measurement is the actual time of arrival of buses at a terminal. The final level, *ratio* scales, have all the properties of the above scales, as well as having a definite zero. Examples of this measure are the engine capacity of vehicles, the speed of vehicles and the money cost of particular transport alternatives. Specific properties of each of these levels of measurement are given in Table 2.1.

2.3.2 Continuous and discrete data

Data may also be divided on the basis of whether the observations represent continuous or discrete variables.

Continuous variables are those for which there are no breaks in the sequence of possible values. Examples of these are vehicle speeds where, for example, an infinite number of possible values lie between 20 km/h and 50 km/h; no matter how small the difference between two speeds (e.g. 30.00

The Traffic Analysis Process 27

km/h and 30.01 km/h) there will always be the possibility of a value between them. This is what is meant by 'no breaks', the observation scale is continuous. It is not necessary to measure the full range of possible values; for example, a radar gun may only show speeds to the nearest kilometre per hour.

Table 2.1 Properties associated with the levels of data classification

Scale Type	Central tendency	Statistics variability	Individual position	Permissible uses	Permissible transforms
RATIO	Geometric mean	Coefficient of variation	Absolute score	Find ratios between	Multiply and divide
INTERVAL	Arithmetic mean	Variance standard deviation	Relative score	Find difference between	Addition and subtraction
ORDINAL	Median	Range	Rank percentile	Establish rank order	Any that preserve order
NOMINAL	Mode	Number of categories	Belonging to category	Identify and classify	Substitution within category

Note: The higher level scales assume all the features of the lower level scales.

Discrete variables are variables whose values have breaks or jumps. Examples of discrete variables are counts of the number of accidents or the number of cars passing a given point in a given time period. There are distinct jumps between the number of cars observed passing a point in a given time interval (e.g. per minute), and the number of accidents observed at a site per year. Such counts will always be integer values.

2.3.3 Quantitative and qualitative data

Yet another distinction between data items is whether they represent quantitative or qualitative observations. Quantitative data can be expressed numerically. For example, speed, income, number of cars and kilometres driven by a vehicle are quantitative. Qualitative data are expressed by some

28 *Understanding Traffic Systems*

non-numerical property. For example, vehicle type, defective or non-defective light bulbs, or the satisfaction or dissatisfaction of a resident with vehicle noise, are qualitative.

2.3.4 Stationary and time-dependent data

Conventional methods for data analysis often assume constancy of behaviour over time. For example, when observing the speeds of vehicles at a point on the road system, we might expect that average speeds will be much the same from one day to the next. We recognise that different drivers may wish to travel at different speeds and that weather, environmental and traffic factors may influence observed speeds to different degrees on different occasions, but, in the long run we can assume that the overall parameters of speeds remain fixed. This assumption is not tenable however, for all traffic data particularly if it is influenced by the size of the population (which is likely to be changing over time) or the level of human activity (which is likely to vary according to the hour of the day, the day of the week, and the season of the year). The loads on a transport network reflect the time-dependent variations in activity, and so traffic data such as hourly or daily traffic volumes which is indicative of these loads must be recorded as time dependent data.

The distinguishing feature of time-dependent data is that they come from processes that are undergoing continual change. If we are to understand the data we must extract the components of change that are involved, and separate the random effects from the trends and cycles that influence the data. Stationary processes (those whose parameters are stable over time) offer the possibility of repeating observations in order to uncover the degree of variability existing in the data. Time series data do not permit repetition of observations, for data collected at one point in time will, of necessity, differ from those collected at other times.

Special methods of time series analysis are needed to make the proper comparisons of time-dependent phenomena. These methods are beyond the scope of this book but it is useful to consider the elements of time series analysis that may impinge on general traffic data analysis. Figure 2.3 shows four major components of time-dependent processes.

The time series components in Figure 2.3 may be described as follows:
- *trend* (Tr). This is the long-term change in the average quantities, such as the growth of traffic over time;

- *seasonal variations (Sc)*. These result from different levels of flow at different times of the year, such as the level of recreational traffic on a rural highway;
- *cyclic variations (Cy)*. These result from cycles in activity (such as time of day or day of week influences on working or shopping behaviour). Cyclic variations implies that the pattern is repeated over relatively short time spans. Several cycles may exist concurrently, such as the mix of hourly, daily and monthly variations in traffic volume at a site, and
- *random component (Ra)*. This may stem from short-term variations in behaviour (e.g. influenced by weather), or special events (such as construction works, or community events such as market days).

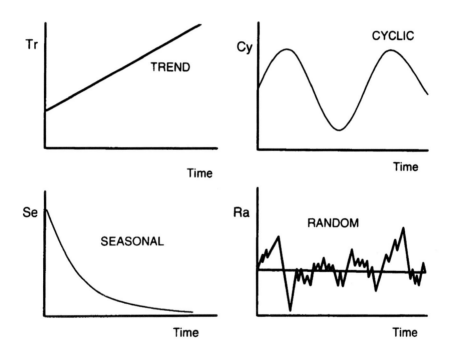

Figure 2.3   Components of a time series

The components may be summed to give an overall effect, e.g. variable $X(t)$ at time t is given by

$$X(t) = Tr(t) + Sc(t) + Cy(t) + Ra \qquad (2.1)$$

30 *Understanding Traffic Systems*

The ordered sequence $X(t_i)$ of observations of variable X at times t_i, where $i = 1, 2, ..., n$ is called a time series. The best examples of time series data in traffic analysis are traffic counts, as described in Chapter 8. Chapter 17 provides an introduction to the analysis of time series data. The concepts of time series involve the ideas of variations of data observations for a given variable, including the notion of random effects. A process involving some random element is sometimes termed a stochastic process. An introduction to the study of stochastic processes in traffic analysis is provided by Chapter 3, where the use of probability distributions is discussed, and by Chapter 4, where an introduction to queuing theory is provided. For the moment, we should turn our attention to the general-purpose tools available for managing and using traffic data.

2.4 Analytical tools

Traffic analysis involves the manipulation of large amounts of data. Computers and associated technologies have provided the analyst with powerful and readily accessible generic tools for data manipulation. These include database managers, spreadsheets, statistical analysis packages, geographic information systems (GIS) and specialist modelling software.

2.4.1 Database managers

New data are entered into a database, edited, sometimes merged with existing data, transformed (e.g. in the computation of secondary variables such as the calculation of speeds using measured time to cover a known distance), analysed, reported and archived. These processes form the major information management tasks to be addressed by traffic engineers and analysts. Database management software provides the means for entry, editing, storage, retrieval and reporting of data in the most efficient and user-friendly manner. This software, which relieves the analyst of writing special programs for most tasks in data handling, provides a means for efficient file storage based on standard file formats and enables data to be shared between different applications.

The essential components of a database manager are a user interface and a data storage and retrieval framework. The user interface provides the means for the entry of data and the instigation of database queries. Data entry is used to establish new data files and to add data to existing files. Database queries are used to tap the stored information, by extracting desired pieces of information from a database (e.g. how many heavy commercial vehicles used a particular road link over a specified time period, or how many vehicles exceeded a particular speed at an observation site). Most user interfaces now use windows-based interactive screen displays, in which the user can choose the items to be entered, amended or chosen at will. Most of the generic database management software packages provide opportunities for the developer of a database to establish a customised interface (an 'on-screen' data entry form) which further enhances ease of use.

Data storage in a database manager is defined by a data structure, perhaps consisting of linked tables of information, made up of fields and records. A record is a set of data items from one individual entity in the database, such as the traffic counts from one location over time, or the answers provided on a questionnaire by one respondent. Each record may contain a number of fields, each corresponding to an individual data item in the record. Each field normally has an associated 'type' and 'length'; the field type will indicate what category of data (e.g. integer or real number, text characters, etc) is held in the field, while the field length indicates the maximum number of digits or characters to be expected. Knowledge of these two items allow the database manager to store the data in the most efficient way, and to help detect any errors in the input data.

Data needs are of obvious importance in transport systems modelling. The needs to share data between different applications also suggest the needs for common database structures, if not formats. There is a general requirement to use a data storage system that is compatible with many computer-based applications, from control programs to design and analysis tools. At present there are many such applications in use; each one typically using its own proprietary format for data input, manipulation and storage. Thompson-Clement, Woolley and Taylor (1996) developed a Transport Network Relational Database (TNRDB) structure, to enable commonality of format making the exchange of data between models a relatively simple exercise, even when the various models use different levels of detail to describe network elements and traffic flows. The TNRDB is particularly useful as the foundation for information transfer in the model hierarchy.

The representation of an intersection under TNRDB is shown in Figure 2.4. In strategic level studies the intersection would be represented as a single node. For more detailed studies, say for intersection capacity analysis, additional information required to describe the intersection. This includes the specific turning movements and their characteristics, the traffic lanes on each approach to the intersection, and the turning

32 *Understanding Traffic Systems*

movements that can use each lane. Turn bans, turn penalties or delays, channelisation and physical characteristics (e.g. lane width) are required. Figure 2.4 indicates the relational data structure adopted in TNRDB. The node is connected to its neighbouring nodes by 'links'. The flow directions for the links are indicated by one-way 'arcs', where a two-way link has a pair of directional arcs. At the intersection node, the arcs are connected through 'vertices', defining the turning movements available. Each approach and exit arc may have a number of traffic lanes, with each lane turning movement connected through a set of 'nibs'. The node is thus a data object, with a large number of internal data elements. Sufficient data elements to describe the intersection at the required level of detail for the particular model level may then be extracted from the TNRDB. The TNRDB also allows for a hierarchy of nodes. At the strategic level the network connections will be between major intersections, whilst at the more detailed levels minor intersections (along the road sections between the major intersections) will be required.

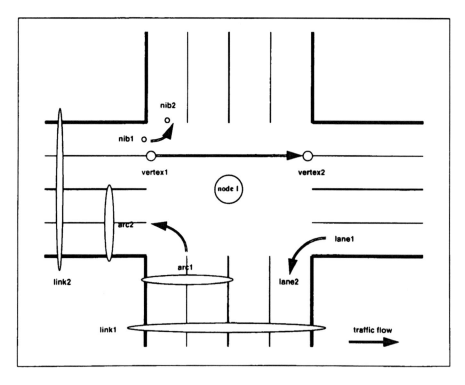

Figure 2.4   Intersection node components in TNRDB

2.4.2 Spreadsheets

Spreadsheets are powerful, generic analysis and presentation tools widely used in personal computing. The spreadsheet is a large calculation sheet, such as those often used in accounting and financial analysis, but capable of much more besides. The worksheet is set up on a grid basis. The overall display is a neatly tabulated array of data, with text headers and identifiers to make it a legible document. Charts and graphs may also be included.

Mathematicians and engineers will recognise the spreadsheet as a large matrix. People from less quantitative disciplines can understand and use spreadsheets without realising that they are engaged in matrix manipulations. Many of the tasks which previously required special software are now easily accomplished in a spreadsheet environment, which also offers significant inbuilt analytical and display capabilities. Figure 2.5 provides an example of a simple spreadsheet, comparing modelled traffic flows between two alternative traffic management plans for an area.

2.4.3 Statistical analysis packages

Although spreadsheets can do much data reduction and analysis, including descriptive statistics (e.g. computation of means, standard deviations and ranges, forming histograms, and performing simple curve fitting), they may not be able to perform some of the more complicated statistical procedures involved in traffic analysis. Statistical analysis packages are more powerful than spreadsheets in this respect; they enable inferences to be made about items in one or more data set and enable hypotheses about the underlying phenomena that the data set represents to be tested.

Personal computing revolutionised the capability to perform statistical analysis. In particular, the combination of interactive data analysis and interactive colour graphics gave great improvements to data handling and for understanding the contents of a data set. Many of the techniques now widely available are not new, but until recently there were no realistic means to apply them. Exploratory Data Analysis (EDA), to be described in Chapter 15, is a prime example. Widespread access to sophisticated analytical procedures is, however, not always matched by expertise and knowledge about the application of those procedures and the circumstances in which they can be applied. Some guidance on techniques known to be valid and useful in traffic analysis is provided in Chapters 15, 16 and 17, but the engineer or analyst should be prepared to seek expert advice, especially when tackling a novel or unfamiliar problem.

34 *Understanding Traffic Systems*

TrafikPlan modelling for North Terrace options - June-July PM peak hour					East-bound traffic			
			Existing			(D) = (B) + North Tc transit mall		
Section	dist (m)	q (veh/h)	travel time (min)	v (km/h)	q (veh/h)	travel time (min)	v (km/h)	Δq
West Tc								
	180	1502	0.13	83.1	831	0.18	60.0	-671
Gray St								
	340	1402	0.38	53.7	724	0.45	45.3	-678
Morphett St								
	100	1500	0.09	66.7	829	0.11	54.5	-671
Greater Union carpark								
	90	1605	0.09	60.0	934	0.09	60.0	-671
Convention Centre carpark								
	170	1870	0.16	63.8	1197	0.25	40.8	-673
Bank St								
	30	1869	0.05	36.0	1194	0.06	30.0	-675
Festival Dr								
	130	1869	0.36	21.7	1234	0.36	21.7	-635
King William St								
	140	1487	0.12	70.0	565	0.22	38.2	-922
Stephen Pl								
	50	1402	0.15	20.0	391	0.1	30.0	-1011
Gawler Pl								
	150	1430	0.27	33.3	302	0.2	45.0	-1128
JM carpark								
	170	1615	0.3	34.0	741	0.27	37.8	-874
Pulteney St								
	210	1917	0.5	25.2	1548	0.45	28.0	-369
Frome St								
	260	1986	0.39	40.0	1535	0.49	31.8	-451
East Tc								
	Totals	2020		2.99	40.5		3.14	38.6	

Figure 2.5 Example spreadsheet, comparing traffic flows under different traffic management plans

2.4.4 Geographic Information Systems (GIS)

Another important tool in traffic data analysis is Geographic Information Systems (GIS) software. Traffic data, like many other data sets in civil engineering and the social sciences, often have spatial attributes; traffic counts come from a site or set of sites, travel time data refer to a route, origin-destination data refer to an area. A conventional database package cannot make much use of the spatial or locational attributes, other than hold reference details for it. A GIS, on the other hand, takes the database and relates the spatial attributes to maps of the region to which the data relate. It has interactive map display capabilities to display the data and can indicate spatial patterns of the distribution of variables.

In addition, GIS packages have the capability to link different databases, where these refer to the same region (i.e. there are common attributes between the databases from the locational information they contain). This is normally done through introducing a series of data layers in the GIS analysis, with each layer referring to a different database. Thus traffic data may be combined with land use, socio-economic, environmental and topographical data for a given region, as in Figure 2.6.

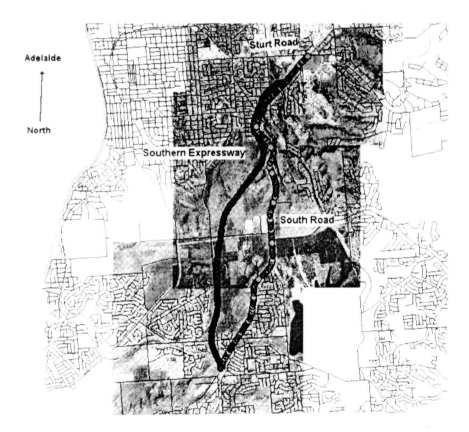

Figure 2.6 **Example of the superposition of data layers within a GIS**

2.4.5 Specialist models

The various levels of investigation involved in traffic impact analyses call for a range of traffic models. The hierarchy of traffic network models described below provides a means by which to select analytical techniques and models appropriate to a wide variety of circumstances ranging from the appraisal of multi-modal metropolitan transport systems to detailed analysis of traffic at an individual intersection. Different levels in the hierarchy represent modelling at different scales, and no one model can be expected to provide useful answers to the full range of problems (see Young et al, 1989). The hierarchy of traffic models provides a means to reconcile the differences between alternative modelling concepts and theories, by defining the specific realm of application of each model and the possible connections between different models (Taylor, 1991). Starting at the most detailed (micro-) level, the hierarchy is:

- *microscopic simulation*, of individual units in a traffic stream. For example, for the assessment of individual vehicle or driver performance at an intersection or along a link;
- *macroscopic flow models*, in which the flow units are assumed to behave in some collective fashion
- *simulation models of flows in intersection clusters*, for the optimisation of network performance (e.g. delays at traffic signals when the flows on each road section or link are fixed);
- *dense network models*, which simulate flows in small-scale networks where the level of flow on each link can vary in response to changes in the traffic control system and traffic congestion levels. These models focus on short time periods (e.g. a peak hour);
- *strategic network models*, which simulate or optimise network flows in the large-scale networks which represent a regional or metropolitan transport system. These models focus on long time periods (e.g. 24 hour flows);
- *land use-transport-environment impact models*, that focus on the extent of changes to new land use facilities (e.g. a retail centre) and use a rudimentary description of the transport system serving that facility in predicting its impacts on the surrounding region, and
- *sketch planning models*, of land use-transport interactions.

Given that the analyst has a range of models with which to describe (simulate) an existing system to an acceptable degree, then those models may be used to predict the consequences of each of a series of policy options (including, perhaps a 'do nothing' policy). This process can be described as 'what-if' modelling; testing the sensitivity of the system's performance to changes in its structure or parameters. Such modelling is normally associated with short term, immediate or near-future effects. Longer term forecasts have greater inherent uncertainty and it has become common practice to attempt to deal with this by using a 'scenario approach' whereby different forecasts are produced for each of a set of possible future situations (for example, one might forecast for a 'low growth' and a 'high growth' future).

Long term analysis is increasingly associated with 'pro-active' planning where an attempt is made to anticipate, and hopefully solve, problems before they are manifest. This style of planning is aided by scenario-based forecasts but can also benefit from the use of prescriptive models that seek to determine optimum policy rather than 'simply' predicting the effect of each of a set of candidate options. One approach to the optimisation problem is to use predictions from an appropriate model (e.g. a network simulation model) to calibrate a regression model whose dependent variable is the quantity to be optimised (e.g. the total VHT in the network) and whose independent variables include the available policy levers (e.g. expenditure on each of a range of alternative traffic management measures). Inspection of the calibrated coefficients will indicate how the objective funding can be optimised within given budget constraints – see Fowkes et al (1998).

2.5 A perspective on the traffic analysis process

The focus of this chapter has been on a general description of traffic analysis as a component of an overall planning and decision making framework. It has also indicated why it is necessary to place the technical work of traffic impact analysis within such a framework. The analysis is a means to an end, and that end is quality decision making based on the availability of relevant and understandable information about traffic systems performance, and the likely impacts of proposed changes to a traffic system. The relationship and interaction between the analyst and the decision maker are of paramount importance in this regard; the credibility of the analyst in the eyes of the decision maker will depend on the analyst's track record in making sound analysis and accurate predictions, in recommending successful solutions and in having a sound grasp of political and financial realities.

The process of traffic analysis is to transform available data ('lists of facts') about a traffic situation into information, which provides understanding of the situation and insights into the underlying phenomena.

The available data may be gathered from existing sources, or may be obtained by surveys specifically commissioned for the particular study. The transformation of the data into information requires examination of the data and the possible application of models to assist the understanding.

All this points to our central theme; good traffic planning is based on sound traffic analysis. The hallmark of good traffic analysis is the timely provision of comprehensive information about the performance of a traffic system and its abilities to cope with changes. This information comes from an understanding of the ways in which traffic systems work and the availability of relevant and reliable data. The remaining parts of this book provide the concepts and theories that lead to such understanding and the means to acquire and validate the necessary information resources.

Part B

BASIC TRAFFIC THEORY

3 Basic traffic flow theory

This chapter provides an introduction to the characteristics of traffic flow and the theories, models and statistical distributions used to describe many of the phenomena and processes that traffic engineers and analysts must consider. This field of knowledge is known as traffic science. The theory presented in the chapter is applied to individual links in a traffic network, and is concerned primarily with uninterrupted traffic flows, which are flows that are unaffected by external influences, such as other traffic streams.

3.1 The traffic stream and its elements

Traffic flow is concerned with the movement of discrete units (such as vehicles or people) around a transport system. In general, these units move independently of each other, although they may interact. Each unit is usually under the control of a human operator, and the processes by which a traffic stream works can often be described in terms of 'random' behaviour. The randomness originates from the multitude of individual decisions that occur in a traffic stream, where each human operator has some personal freedom of choice and action. The three main components of road traffic systems are: the driver, the vehicle, and the environment. These components all interact with each other. Consequently the moving traffic stream has characteristics that are quite different to those of the individual elements. For this reason traffic engineering assumes a most important role in highway planning, design and management.

In traffic management and control, knowledge of the characteristics of the traffic stream provides the means for understanding traffic behaviour in situations involving manoeuvres such as queuing, car following, turning, crossing, merging and weaving. These characteristics, in conjunction with travel speeds, traffic volumes and traffic density, are the major determinants of road capacity in a given traffic environment. A knowledge of traffic behaviour and parameters provides the means for the quantitative justification of traffic control systems and devices.

3.1.1 The traffic lane

A traffic lane is defined as that portion of a carriageway used for the passage of a single stream of vehicles. The traffic lane is thus the basic element in the provision of road space. The carriageway consists of a number of traffic lanes. Traffic lanes are usually delineated by lane lines or markings on multi-lane or one-way roads, or by separation and barrier lines on two-way roads.

The minimum width of a traffic lane is about 2.7 m on low-speed urban roads. The maximum allowable physical width of vehicles is about 2.5 m so that this minimum width is just adequate to allow two wide vehicles to pass each other. A lane width of between 3.5 m and 3.7 m is generally accepted as a standard for satisfactory high-speed operation. Increasing lane widths above 3.7 m may lead to poor lane discipline by drivers, which can adversely affect traffic flow and capacity.

3.1.2 Traffic streams

Traffic flow may be classified according to the number of lanes available:
- single lane flows only permit one-way flow. They do not allow for overtaking, so that the speed of an individual unit is restricted to that of the slowest vehicle in the traffic stream ahead;
- two-lane roads may permit either one-way or two-way flows. Under one-way operation, overtaking is usually possible up to a high traffic density at which point large platoons of vehicles will form. Under two-way operations, overtaking opportunities exist, but are limited by prevailing sight distance and by the gaps encountered in the opposing traffic stream. Platoons will form at a much lower density than that for one-way operation. Sections 3.3.3 and 3.4.3 extend the discussion of two-lane two-way traffic streams, and
- three-lane and four-lane roads allow easier lane changing and thus traffic operations are smoother. The phenomenon of peak period tidal flows often seen on urban roads (i.e. where the traffic flow is predominantly in one direction at a given time of day) means that traffic strategies can be used in which the centre lanes are exclusively occupied by vehicles travelling in the major direction. This can permit more efficient flows.

Different classes and types of vehicles have different physical, geometric and performance characteristics. Speeds and accelerations can vary greatly from on vehicle type to another, with heavy commercial vehicles possessing inferior acceleration performance to other vehicle types. This becomes significant in considerations of traffic queuing. Likewise, the ability of heavy vehicles to maintain highway speed when climbing hills can be present problems for the maintenance of smooth traffic flows on trunk roads. Knowledge of the distribution of vehicles in traffic lanes is important in understanding traffic behaviour and in estimating the resultant capacity of a given section of road.

3.1.3 Modifying factors

There is considerable variability in the physical and performance characteristics of road user groups. Each vehicle may be considered as an independent unit having only limited coherence with other units on the roadway. Traffic stream performance and thus roadway capacity is affected by physical conditions such as:

- carriageway width
- vertical and horizontal alignment
- geometric design features
- type and condition of the road surface
- width, number and separation of lanes.
- gradient
- sight distance
- frequency and form of junctions
- terrain and attractiveness of a route

Parking, bus stops, light rail tracks, abutting land uses and other factors also cause considerable variation in lane formation, lane discipline and vehicle distribution, particularly on urban roads. In addition the performance of a traffic stream may be affected by signs, traffic signals, pavement markings, traffic regulations, pedestrians and other road users. Street lighting, weather and other environmental conditions may also affect traffic performance. The conventional analysis will be based on a determination of traffic stream performance under normal operating conditions, and then examine the changes to the expected performance that would result when the additional physical and environmental factors are introduced. Design or remedial measures can then be examined, to ensure that performance at some minimum standard is maintained.

3.2 Basic relationships

The three basic parameters used to describe a traffic stream are:
- *volume* (or flow rate) (q), the number of vehicles passing a fixed point per unit time. Typical units for q are veh/day, veh/h or veh/s;

- *speed* (or velocity) (*v*), the distance travelled by a vehicle per unit time. Typical units of speed in traffic engineering are km/h or m/s, and
- *density* (or concentration) (*k*), the number of vehicles per unit length of lane or roadway at a given instant of time. The typical unit for density is veh/km.

Average speed and volume are the more commonly-used descriptors of a traffic stream, as they can be measured quite easily. Density is more difficult to measure directly, but there are simple methods of estimating it indirectly. The three basic parameters are related to each other through the continuity of flow equation, which is

$$q = k\bar{v}_s \qquad (3.1)$$

and includes $\bar{v}_s$, the mean *space* speed (see Section 3.2.2). As will be seen in that section, there is particular cause to consider the specific definition of the mean speed used in this equation. Equation (3.1) only applies for uninterrupted traffic flows, i.e. traffic streams moving independently of external influences. Uninterrupted conditions are best represented by traffic flows on main highways (away from junctions) and freeways.

Supplementary traffic flow parameters are also used. These include *headway*, *spacing* and *occupancy*, and they can be related to flow, speed and density (see Section 3.4). Headway is the time gap between successive vehicles in a traffic stream (strictly it is the time gap between the same point on two vehicles, e.g. the front bumper bar or the front axle, so that headway is independent of vehicle length). The mean headway for a traffic stream is the reciprocal of q. Similarly spacing is the distance between the same physical point on two successive vehicles. Mean spacing is the reciprocal of density. Occupancy is the proportion of time that a designated point in a traffic lane is covered by vehicles. Occupancy is directly related to density.

3.2.1 Trajectory diagram

A trajectory diagram is a graph showing the position of a vehicle along a length of road (X_r) over a period of time (T). The vehicle trace on the trajectory diagram provides the complete history of each vehicle's journey along the road, and may be used to determine and illustrate the traffic flow parameters for a traffic stream. Figure 3.1 is a trajectory diagram for a one-directional uninterrupted flow. Each solid line on the diagram is the trace of the position of an individual vehicle over time. Thus the trace for vehicle 'A' in Figure 3.1 indicates the changing position of that vehicle from time $t = 0$ to time $t = T$. The slope of the vehicle trace (tan∅) is the speed of the vehicle at a given instant of time. The diagram shows vehicles changing speed, catching up to slower vehicles, sometimes overtaking them, sometimes tracking behind them.

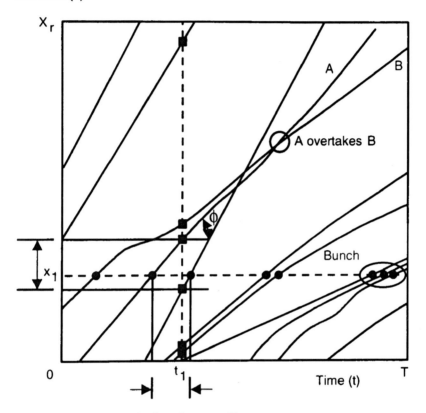

Figure 3.1 A typical trajectory diagram

Headways and spacings for individual vehicles can be read from the trajectory diagram, as indicated in Figure 3.1. The trajectory diagram may also be used to determine volume and density. The volume at the point x_l can be determined by counting the vehicles that pass that point over the time period T, say $N(x_l)$,

$$q = \frac{N(x_1)}{T} \qquad (3.2)$$

Similarly the density can be determined by observing the number of vehicles on the road section X_r at time t_1, for instance by taking an aerial photograph of the road at that time and counting the number of vehicles, say $N(t_1)$. Then

$$k = \frac{N(t_1)}{X_r} \qquad (3.3)$$

Mean speeds can also be taken from the trajectory diagram. If we measure the speeds of the $N(x_1)$ vehicles passing point x_1 over the time interval T, then the mean speed of those speeds ($\bar{v}_t$) is the mean spot speed, sometimes known as the mean time speed. Similarly, if we took the mean of the speeds of the $N(t_1)$ vehicles on the road section X_r at time t_1, the mean space speed would result. One of the complications of traffic flow analysis is that $\bar{v}_t$ and $\bar{v}_s$, and the speed distributions they represent, are not identical even though they both consider the same traffic stream. This is discussed in Section 3.2.2, following the definitions of the different types of speed measurements.

3.2.2 Speed distributions

The speeds of the individual vehicles in a free flowing traffic stream are rarely identical. Individual speeds vary in response to the characteristics of drivers and vehicles and environmental conditions. The most suitable way to describe speeds is in terms of statistical distributions. Speeds are often represented by the normal distribution (Section 3.3.2), while the log-normal and gamma distributions are favoured for travel times and delays. There are empirical and theoretical reasons why these distributions should be considered, as described below, and in Chapter 9.

A common hypothesis in traffic flow theory and traffic analysis is that speeds, particularly for free-flowing traffic, are normally distributed. A further hypothesis is that each driver has a *desired speed*, which is the speed they would choose to travel at if unimpeded by other traffic. Desired speeds depend on road conditions, weather and environment, and perhaps some characteristics of the driver and vehicle. Desired speed is useful as a concept, but in practice we can only observe *free speeds*, i.e. the speeds adopted by those drivers who are travelling freely. Although free speeds should be strongly related to desired speeds, free speed and desired speed distributions are not necessarily the same, as we can only observe that proportion of all drivers able to travel freely. Only the drivers of isolated vehicles or those at the head of a platoon will be able to adopt their free speeds.

Free speeds are usually assumed to be distributed according to the normal distribution. Further, the coefficient of variation (the ratio of standard deviation to mean) of the speed distribution seems to be constant over a wide range of speeds. The assumption of a constant value of the coefficient of variation of free speed distributions is useful, for it means that knowledge of a mean speed leads to knowledge of the complete speed distribution as well.

A complication arises because of the two different ways that speeds may be observed (see Section 3.2.1). We can measure the speeds of either a set of vehicles passing a point along a road (the 'spot' or *time speeds*), or a set of vehicles on a road section at a given instant of time (the *space speeds*). The distribution of spot speeds and the distribution of space speeds are not the same, even though they represent the same traffic stream. There are relativistic differences between them. Spot speeds will be biased in favour of the faster vehicles on the road, as is shown by the following analysis.

The spot speed distribution is the distribution representing the speeds of vehicles passing a point by the road (probability density function $f_t(v)$), while the space speed distribution represents the distribution of speeds of all the vehicles on the road at an instant of time. The probability density function for the space speeds is given by $f_s(v)$. The definition of a probability density function requires that the integral of that function, over the range of all possible values (i.e. speeds from zero to infinity), must equal one, so that:

$$\int_0^\infty f_t(v)dv = \int_0^\infty f_s(v)dv = 1 \qquad (3.4)$$

In addition, the mean value of a probability distribution $f_t(v)$ of a positive variable $v \geq 0$ is defined as

$$\bar{v}_t = \int_0^\infty v f_t(v)\,dv \qquad (3.5)$$

and the variance of that probability distribution is

$$\sigma^2 = \int_0^\infty v^2 f_t(v)\,dv \qquad (3.6)$$

Consider the proportion of vehicles flowing past an observation point whose speeds lie in the range $(v, v + \delta v)$. This proportion may be approximated by $f_t(v)\delta v$ for small δv, and consequently the flow rate for those vehicles with speeds in the range will be $qf_t(v)\delta v$. Similarly, by considering the vehicles on the road section at a given time, with speeds in the range $(v, v + \delta v)$, it is possible to approximate the density of vehicles with speeds in the interval as $kf_s(v)\delta v$. The relationship between the two distributions can then be derived by considering the flow of vehicles past a fixed point during a time interval T. All the passing vehicles whose speeds lie in the interval $(v, v + \delta v)$ are within a distance vT upstream of the point at the start of the time interval. From the spatial density function this number is $kvTf_s(v)\delta v$. On the other hand, the temporal traffic count function indicates that the number of vehicles in this speed ranging passing the observation point is $qTf_t(v)\delta v$. Equating the two results yields

$$q\,f_t(v)\,\delta v = k\,vf_s(v)\,\delta v$$

As q and k are average parameters for a particular flow regime we can assume that under equilibrium conditions (i.e. stable flows) q/k is a constant, C. Therefore,

$$f_t(v)\,\delta v = C\,vf_s(v)\delta v \qquad (3.7)$$

and, as δv approaches 0, equation (3.7) may be integrated over the range of possible speeds to yield

$$\int_0^\infty f_t(v)\,dv = C\int_0^\infty v\,f_s(v)\,dv$$

so that, using the results of equations (3.4) and (3.5),

$$C = \frac{1}{\bar{v}_s}$$

By substituting in equation (3.7), we find that

$$f_t(v)\delta v = \frac{v}{\bar{v}_s} f_s(v)\delta v \qquad (3.8)$$

This result indicates that the spot speeds are biased in favour of the faster vehicles in the traffic. In fact, by integrating equation (3.8) over the range of speeds, we obtain

$$\bar{v}_t = \bar{v}_s \left[1 + \left(\frac{\sigma_s}{\bar{v}_s} \right)^2 \right] \qquad (3.9)$$

where σ_s is the standard deviation of the space speed distribution. The ratio $(\sigma_s / \bar{v}_s)$ is the coefficient of variation of the space speed distribution. A typical value of this ratio for free speeds is about 0.14. Using this value of the coefficient of variation means that, from equation (3.4), $\bar{v}_t \approx 1.02 \bar{v}_s$, i.e. the mean time speed will be about two per cent higher than the mean space speed. Space speed is the appropriate speed definition to use in the continuity of flow equation (3.1). However, most speed observations are time speeds because spot speed data is much easier and cheaper to collect. See Chapter 9 for an account of the methods for collecting speed data.

Sub-classifications of speeds may also be made, for both spot and space speeds. The speeds of all vehicles in a traffic stream may be termed the actual speed distribution. These speeds will include interaction effects between vehicles, and may thus be biased towards the slower vehicles on the road. They may also be influenced by transient environmental conditions such as the weather. For design purposes traffic engineers need to know about the speeds that unimpeded vehicles will want to adopt on the road. These unimpeded speeds form the desired speed distribution. Desired speeds can be inferred by observing free speeds (i.e. the speeds of vehicles that can be observed to be travelling independently of other traffic), although some care is required in interpreting the results because free speeds are likely to be biased in favour of slower vehicles.

3.2.3 Average speed and volume

Figure 3.2 shows the general relationship between mean speed and volume. The relationship is actually an envelope, with the horseshoe curve defining a feasible region for traffic flow under different combinations of volume and speed. The curve defines the limits to this region. The two arms of the horseshoe represent quite different traffic flow conditions. The upper arm is for freely flowing traffic, with a mean free speed of v_f and a significant proportion of the vehicles travelling at their desired speeds. This proportion decreases as the nose of the envelope is approached. The lower arm represents congested flow conditions, when most of the traffic is strongly influenced by the vehicles in front of them. This implies stop-start flow conditions.

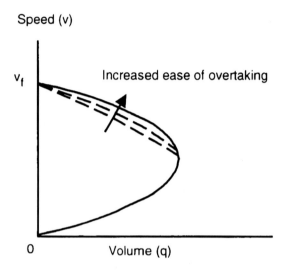

Figure 3.2   Speed-flow envelope

The speed-flow envelope is widely used in determining the design capacity of a road, through the use of 'levels of service'. Note that, from the envelope, there are two limiting speeds for any volume, except for the maximum volume (q_m). Thus volume is not a particularly useful indication of the quality of traffic flow, unless an indication of speed or density is also available. A flow of 100 veh/h may represent light traffic, or a severe traffic jam! The mean speed of moderate volume free flowing traffic will increase slightly as overtaking is made easier (e.g. through improved sight distance, or special overtaking zones).

3.2.4 Average speed and density

The relationship between mean speed and density, for uninterrupted flow, is given by a monotonic (single-valued) decreasing curve, which meets the v-axis at the mean free speed (v_f) and the k-axis at the jam density (k_j). Figure 3.3 shows a typical v-k curve. A common simplifying assumption is that this curve can be represented by a straight line between the points (0, v_f) and (k_j, 0). This assumption is reasonable for most densities, apart from the very small (close to zero) and very large (close to k_j) densities.

The speed-density curve may be considered as the fundamental diagram displaying the performance of a traffic stream, given the continuity of flow relationship (equation (3.1)).

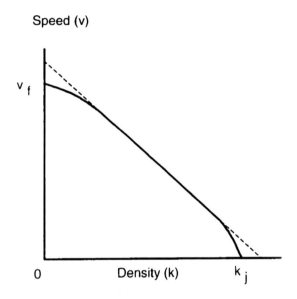

Figure 3.3 Typical speed-density relationship

3.2.5 Volume and density

Figure 3.4 shows a typical q-k curve. This is another horseshoe curve, but this time it is a single value function, not an envelope. The curve rises to a maximum volume (q_m) at a density of k_m, then decreases to zero volume at the jam density (k_j). A given volume rate may exist at two separate densities, but a given density has only one volume associated with it. The left arm of the curve (i.e. that for density $k < k_m$) represents free flow conditions, whilst the right arm is for congested flows. Thus density is the traffic parameter that can provide a full description of the state of a traffic stream.

The q-k curve of Figure 3.4 is split into three regions. These regions are for light, freely flowing traffic (region A), moderate but stable, freely flowing traffic (region B), and unstable, forced flows under congested conditions (region C). Region B is the one sought in traffic design, for flows in this region achieve an optimum efficiency for a given construction standard and operating environment.

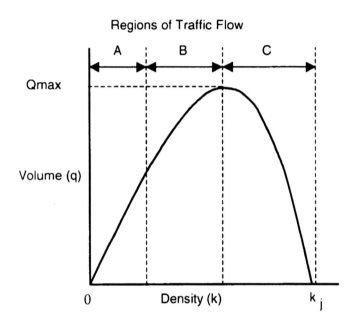

Figure 3.4   **Flow-density relationship, showing regions A, B and C for traffic flow**

3.2.6 Ancillary characteristics

The traffic parameters headway, spacing and occupancy are related to volume, density and speed. These ancillary measures are useful for some theoretical and practical investigations of traffic flow.

Headway is defined as the time gap between the same physical point (e.g. front bumper, or front axle) on two successive vehicles in a traffic lane or stream. The mean headway is given by the reciprocal of the traffic volume.

Spacing is defined as the distance between the same physical point on two successive vehicles in a traffic lane or stream. The mean spacing is given by the reciprocal of the traffic density.

Occupancy (Ω) is the proportion of time that vehicles are physically present at a point in a traffic lane. Occupancy can be measured by placing a vehicle presence detector (e.g. an inductive loop, see Section 8.4.5) in a lane, and recording the amount of time that the detector records vehicles crossing it. This time divided by the total elapsed time, is the occupancy of the lane. Occupancy is directly proportional to density:

$$\Omega = Lk \qquad (3.10)$$

where L is the mean length of vehicles in the traffic stream. Occupancy can be monitored continuously, so thus so can density.

3.2.7 Combining traffic streams

Given that traffic flow on a road section may consist of a number of component traffic streams, there is often need to combine the components into a single stream, and then to determine the parameters for the combined stream. For example, the components might be the different types of vehicles travelling in one direction on the road, the flows in two or more lanes, or the combined two-way flow on a single carriageway road. The basic traffic flow parameters volume and density are found simply by adding the parameters for the components. A stream of n vehicle types, with volume q_i, $i = 1, 2, ..., n$, and density k_i, $i = 1, 2, ..., n$, has a total volume (q) given by

$$q = \sum_{i=1}^{n} q_i \qquad (3.11)$$

and density (k) given by

$$k = \sum_{i=1}^{n} k_i \qquad (3.12)$$

The determination of the average speed measures for the combined flow is a little more complicated. Of course, the continuity of flow equation (3.1) can be applied immediately to yield the overall mean space speed $\bar{v}_s = q/k$, but the mean spot speed is not revealed by this result. A return to the definitions of these mean speed measures is needed, as follows. Assume that all type i vehicles are travelling at the same speed v_i. Then, over an interval of time T, q_iT such vehicles will pass a stationary observer by the roadside. The mean spot speed over the time interval is given by the average speed of all the vehicles passing the stationary observer, and this is given by:

$$\bar{v}_t = \frac{\sum_{i=1}^{n} v_i q_i T}{\sum_{i=1}^{n} q_i T} = \frac{\sum_{i=1}^{n} v_i q_i}{\sum_{i=1}^{n} q_i} \qquad (3.13)$$

as the total number of vehicles passing the observer in time T is qT. Similarly, at an instant of time, there will be k_iX_r type i vehicles on a road section of length X_r. The mean space speed for the combined traffic stream is thus given by

$$\bar{v}_s = \frac{\sum_{i=1}^{n} v_i k_i X_r}{\sum_{i=1}^{n} k_i X_r} = \frac{\sum_{i=1}^{n} v_i k_i}{\sum_{i=1}^{n} k_i} = \frac{q}{k} \qquad (3.14)$$

for the total number of vehicles on the road section is kX_r.

3.3 Statistical modelling of traffic streams

The previous section defined some basic parameters that can be used to define the average or representative characteristics of a traffic stream. Along the way to those definitions we had to introduce the idea of describing a traffic stream in terms of some statistical distribution. This was done, for instance, to consider the speeds of vehicles in the stream and how these speeds could be represented or used in an analysis. Statistical distributions have proven to be useful for describing many of the processes that take place in a traffic stream. Statistical theory provides an excellent way to account for the differences in individual behaviour amongst drivers and other road users. Statistical distributions may be used to provide models of the occurrence of particular events in traffic. They may also be used to describe the detailed characteristics of the traffic stream so that models of traffic performance (e.g. delays and capacities) may be developed. Two broad categories exist for statistical distributions: discrete or counting distributions, and continuous distributions. Both of these categories find wide application in traffic engineering and traffic analysis.

3.3.1 Discrete distributions

In the discrete distributions the number of events is the variable. Some examples are the number of vehicles in a given time interval or on a given length of road, the number of accidents in a given time period, or the number of violations of a particular traffic law. In each of these cases the variable is a whole number. A discrete distribution indicates the probability that a certain number of the events will occur. Four discrete distributions of relevance in traffic engineering are the binomial distribution, the Poisson distribution, the geometric distribution, and the Borel-Tanner distribution.

Binomial distribution. A binomial process is one that has the following characteristics. It consists of a series of n independent trials, each of which has only two possible outcomes. On each trial, the probability of obtaining a particular outcome is constant. An example is a coin tossing experiment, in which a coin is tossed (say) ten times. There would thus be ten trials in this experiment. Each trial has only two possible outcomes (heads or tails). Further, the probability of obtaining a head remains constant from trial to trial.

In general terms, one of the possible outcomes is defined as a 'success'. Let the probability of this outcome be p. Then the probability of the other event is $q = 1 - p$. The binomial distribution gives the probability (P_r) of the number of successes, r, in n trials, and this is given by

$$P_r = C_r^n p^r q^{n-r} \qquad (3.15)$$

where

$$C_r^n = \frac{n!}{r!(n-r)!}$$

and $r = 0, 1, 2, 3, ..., n$. C_r^n is called the binomial coefficient. It is the number of ways that n things can be placed into groups of r things.

The expected or mean number of successes in n trials of a binomial process, when p is the probability of success on any trial, is np and the variance of r is npq. The binomial distribution may be used to predict such events as the number of cars making a specified turn at a junction or the number of vehicles exceeding the speed limit on a section of highway.

A simple example of the use of the binomial distribution is a consideration of vehicle types in a traffic stream. Assume that a traffic stream contains 15 per cent trucks and 85 per cent cars. Then the probability of observing three trucks in a sequence of any ten vehicles is given by the binomial probability P_i with $n = 0$, $r = 3$ and $p = 0.15$. From equation (3.15), this probability is

$$P_3 = \frac{10!}{3!(10-3)!} (0.15)^3 (0.85)^7 = 0.130$$

Poisson distribution. The Poisson distribution is the most frequently used discrete distribution in traffic flow. It is a limiting case of the binomial distribution, and applies in cases where n (the total number of events) is very large, and p or q is small. Under these conditions the binomial formula is cumbersome to apply, if not impossible. For example, although we can observe the number of accidents that occur at a site over a period of time, we cannot tell how many 'non-accidents' occurred in that period. The Poisson distribution provides a convenient method for determining the probability of r 'successes' occurring if we know the expected (or mean) number of 'successes' in the process under investigation. The Poisson distribution is defined by the equation

$$P_r = \frac{1}{r!} m^r e^{-m} \qquad (3.16)$$

where r is the actual number of successes in a given time period, m is the expected number of successes, and e is the base of natural logarithms. Besides applications in the analysis of accident frequency data, the Poisson distribution may be applied to problems such as the prediction of the number of arrivals at an intersection, the number of vehicles delayed at an intersection during the red phase, and delays to right turning traffic.

The following example indicates the use of the Poisson distribution. Consider an intersection that has averaged two accidents per year for the past ten years. What is the probability that there will be five accidents at the intersection in any one year, if there is no change in the true mean? For this example, $m = 2$ (the expected number of accidents in a year) and $r = 5$. Applying equation (3.16) yields $P_5 = 0.036$. This means that five accidents could occur at the intersection in about one year in 28, if the true mean number of accidents per year is two.

Geometric distribution. The geometric distribution is used to describe situations in which the probability of observing $(r + 1)$ events is a constant multiple of the probability of observing r events, independent of the value of r. This distribution arises from simple queuing theory (see Chapter 4), where it defines the probability of having r customers waiting in a queue. The distribution has the form

$$P_r = \rho^r (1 - \rho) \qquad (3.17)$$

where $r = 0, 1, 2, 3, ...$ The mean of the geometric distribution is $\rho/(1 - \rho)$ and the variance is $\rho/(1 - \rho)^2$.

A typical application of the geometric distribution is in assessing the probability of observing a sequence of r similar events, such as r successive gaps less than a certain size, or tossing r successive heads with a coin. If each event is independent of the others, then the probability of r successive events is the probability of observing a sequence of the r like events ('heads' or 'short gap') followed by the alternative event ('tails' or 'long gap'). In equation (3.17) ρ is the probability of the 'success' and $(1 - \rho)$ is the probability of the alternative event, as for the binomial distribution.

In the first example for the binomial distribution we considered the probability of observing three trucks in a sequence of any ten vehicles, when the proportion of trucks in the traffic was 0.15. The probability of observing three trucks *in succession* (i.e. three trucks followed by a car) is given by the geometric distribution. This probability is $P_3 = (0.15)^3 \times (1 - 0.15) = 0.003$, applying equation (3.17). This means that a sequence of exactly three trucks would occur about 3 times in 1000. Compare this to the probability of observing three trucks in a group of any ten vehicles, which was 0.13.

Borel-Tanner distribution. This distribution provides a useful model for describing the distribution of bunch sizes in traffic on two-way, two-lane rural roads (Taylor, Miller and Ogden, 1974). The probability of a bunch of size r is given by

$$P_r = \frac{1}{r!} \left(rze^{-z} \right)^{r-1} e^{-z} \qquad (3.18)$$

for $r = 1, 2, 3, \ldots$ and mean bunch size $m = 1/(1 - z)$, so that the parameter z is given by $z = (m - 1)/m$. The application of this distribution is described in Section 3.3.3.

3.3.2 Continuous distributions

Continuous distributions are used to describe variables that can assume any value within a specified range of values. Examples of continuous variables include gap sizes in a traffic stream, the speeds of individual vehicles in a traffic stream, and the time between accidents at a location. Since the variable can assume any value within the specified range, it can assume any of an infinite number of possible values. This means that the probability of a continuous variable assuming any one particular value is zero. Rather, the continuous distribution will provide the probability that they value of the variable will lie in a designated finite range. For example, the probability of occurrence of a gap of size exactly four seconds is zero, but there is a defined, finite probability that a gap may fall in the range 3.95 to 4.05 seconds. Some useful continuous distributions for traffic flow analysis are the normal distribution; the negative exponential distribution, the shifted exponential distribution, and the mixed exponential distribution.

A continuous distribution for a random variable x is defined by its probability density function (pdf), denoted as $f(x)$. The pdf is analogous to the discrete probability function P_r considered in Section 3.3.2. The probability that the value of the variable will fall in a small range $(x, x + \delta x)$ is $f(x)\delta x$. The usual method of citing probabilities from a continuous distribution is in terms of the cumulative density function (cdf) $F(x)$, which indicates the probability of observing a value of the variable that is less then the specified value for x. The cdf is given by the integral

$$F(x) = \int_{-\infty}^{x} f(t)\,dt \qquad (3.19)$$

Normal distribution. The normal distribution is the continuous distribution most commonly found in statistical theory. It has been used to model many physical and human phenomena. In traffic theory the normal distribution is used to describe distributions of vehicle speeds, as discussed in Section 3.2.2. Despite its ubiquitous nature, the normal distribution has a particular difficulty associated with it, for its cdf cannot be explicitly defined. The pdf for the normal distribution is given by equation (3.20), the integral of which cannot be determined analytically. Note that the '*exp(X)*' in equation (3.20) is an alternative representation of the exponential function e^X. Tables of the area under the normal curve are available instead, as in Table A.1 of Appendix A.

$$f(x) = \frac{1}{\sqrt{2\pi}\,\sigma} \exp\left[-\frac{(x-\mu)^2}{2\sigma^2}\right] \qquad (3.20)$$

Negative exponential distribution. An important continuous distribution in classical traffic theory is the negative exponential distribution. The pdf for this distribution is

$$f(x) = \frac{1}{q} e^{-qx} \qquad (3.21)$$

where q is the mean value. The variance of this distribution is q^2, and the corresponding cdf is

$$F(x) = 1 - e^{-qx} \qquad (3.22)$$

which is the probability of occurrence of an 'event' (e.g. a headway) $\leq x$. The negative exponential distribution represents the distribution of headways in the Random Traffic Model, and the Poisson distribution is its discrete analogue. This model has provided the basis for the most commonly applied results of traffic flow theory, including expressions for gap sizes and delays. The Random Traffic Model depends on two assumptions:
- the time of arrival of a vehicle at a fixed point on a road is independent of the time that the previous vehicle arrived, and
- the probability of a vehicle arriving in a small interval of time (δt) is $q\delta t$, where q is the traffic volume.

The negative exponential distribution may also be used to represent the distribution of vehicle spacing under the same conditions. The Random Traffic Model has been widely used in traffic engineering since the 1930s, and continues to provide useful results, despite the restrictive nature of its two basic assumptions. In principle and in practice, these can break down under many common traffic situations. The model is generally considered to apply for uninterrupted flow rates of: 500 veh/h total flow on two-way two-lane roads, and 400 veh/h/lane one-way on a multi-lane carriageway.

The random Traffic Model becomes less applicable when overtaking and lane changing become more difficult due to the traffic and environmental factors on a given road. The assumption of independence of vehicle arrival times (hence of vehicle headways) is a particular problem. Firstly, the model assumes that zero headways are possible. Indeed the maximum (or modal) frequency of headways is zero in the negative exponential distribution. Zero headways can occur on a multi-lane carriageway where vehicles can travel side by side, but cannot occur in single lane flows, where the finite length of each vehicle will set a minimum headway greater than zero - remember that the definition of headway is that of the time duration between the arrival of the same physical point (e.g. the front bumper bar) on two successive vehicles. Zero headway between two vehicles in a single lane thus implies that the vehicles have zero length. In urban networks with many signalised intersections, traffic flows are often cyclic in nature, as platoons are released from upstream signals. Cyclic flows will not be random. Thus some care is needed in applying the Random Traffic Model in practice. Nevertheless, the model is easy to use and gives useful if approximate results.

Consider the following example. At a priority road junction with a stop sign for the minor road, minor stream vehicles need a six second headway in the major stream to cross that stream. If the major stream volume is 400 veh/h, what is the probability that any given headway will be six seconds or greater?

In this case q = 400 (veh/h), i.e. q = 1/9 (veh/s), and the critical headway size t_a is six seconds. Equation (3.22) gives the probability of a headway being less than or equal to a specified size, so the probability of a headway (t_h) greater than t_a will be $Pr\{t_h > t_a\} = 1 - [1 - exp(-qt_a)] = exp(-qt_a)$. Thus $Pr\{t_h > 6\} = exp(-6/9) = 0.513$, so that according to the negative exponential distribution (and the Random Traffic Model), approximately 51.3 per cent of the headways will be longer than six seconds.

Shifted exponential distribution. A number of problems associated with the application of the negative exponential model to real traffic streams are apparent. One of these problems concerns the minimum following headway for vehicles travelling in a single traffic lane. The shifted exponential distribution provides a more realistic model for single lane flow.

This model includes a finite minimum time interval (τ) between events. The basic effect is to shift the exponential distribution to the right by an amount τ. As the mean headway must still be equal to $1/q$, the pdf of this distribution is

$$f(t_h) = 0 \qquad t_h < \tau$$
$$f(t_h) = \frac{q}{1-q\tau}\exp\left[-\frac{q(t_h - \tau)}{1-q\tau}\right] \qquad t_h \geq \tau \qquad (3.23)$$

The cdf for this function is given by equation (3.24):

$$F(t_h) = 0 \qquad t_h < \tau$$
$$F(t_h) = 1 - \exp\left[-\frac{q(t_h - \tau)}{1-q\tau}\right] \qquad t_h \geq \tau \qquad (3.24)$$

If, in the example given previously for the negative exponential distribution, we assume a single lane one-way stream with a minimum tracking headway of two seconds, what then is the probability that a given headway will exceed six seconds? In this case $q = 1/9$ (veh/s) and $\tau = 2$ s. Thus $Pr\{t>6\} = 1 - F(t)$ where $F(t)$ is defined by equation (3.24), and so

$$Pr\{t > 6\} = \exp[-(1/9)(6 - 4)/(1 - 2/9)] = \exp[-2/7] = 0.751$$

so that in this case about 75 per cent of the headways will exceed six seconds. Compare this to the result found the negative exponential distribution (51 per cent).

Mixed exponential distribution. Troutbeck (1986) suggested that a more general form of the shifted exponential distribution can provide a useful description of traffic flow on urban roads. This distribution is based on the idea that a traffic stream can be broken into two distinct populations of vehicles free-flowing vehicles and tracking vehicles. Free-flowing vehicles move independently of other traffic in the stream, while tracking vehicles are constrained to follow behind the vehicle in front of them. A tracking vehicle follows its leader at a constant headway of τ is, and the tracking vehicles form a proportion (θ) of the total traffic volume in the stream. The free-flowing vehicles are the remaining (1 θ) proportion of the traffic, and follow a shifted exponential distribution with headways $t > \tau$. The cdf of this distribution is

$$F(t_h) = 0 \qquad\qquad t_h < \tau$$
$$F(t_h) = \theta \qquad\qquad t_h = \tau \qquad (3.25)$$
$$F(t_h) = 1 - (1-\theta)\exp(\alpha(t_h - \tau)) \qquad t_h > \tau$$

where the coefficient α is given by $\alpha = q(1 - \theta)/(1 - q\tau)$.

In the example used in the previous two sections, assume that the proportion of tracking vehicles is 0.3. What is the probability that a headway will then exceed six seconds? The previous values for q (= 1/9 veh/s) and τ (= 2 s) apply. Then α = (1/9)(1 - 0.3)/(1 - 2/9) = 0.1, and using equation (3.25), $Pr\{t > 6\}$ = $1 - (1 - 0.3)\exp[-0.1(6 - 2)]$ = 0.531. This suggests that about 53 per cent of headways will exceed six seconds if this is the appropriate model for headways in the traffic stream.

3.3.3 Bunching of traffic

An alternative model of traffic flow on two-lane, two-way roads (especially in rural areas) is a vehicle bunching model. This assumes that vehicles travel in clusters (bunches) of size r (where r = 1, 2, 3, ...) at a tracking headway within the bunch, and with inter-bunch headways separating the bunches. The bunching model requires two sub-models as its components: (1) a discrete distribution for bunch size, and (2) a continuous model of inter-bunch headways.

The Borel-Tanner distribution (see Section 3.3.1) has been found to provide a reasonable representation of the distribution of bunch sizes on two-way, two-lane rural highways, while the negative exponential distribution has been found to fit observed distributions of inter-bunch headways (Taylor, Miller and Ogden, 1974). To apply this combined model requires knowledge of the mean bunch size (m) and an assumed mean tracking headway (τ). The mean flow rate of bunches (q_b) is then given by

$$q_b = \frac{q}{1 - q\tau} \qquad (3.26)$$

Bunching is a useful indicator of the quality of flow in a traffic stream, particularly in rural areas. It reflects the degree of constraint experienced by drivers, and the extent to which the demand for overtaking opportunities exceeds their supply. In bunched traffic the vehicle leading a bunch is assumed to be travelling at its desired speed, whilst the vehicles behind it in the bunch are tracking behind it, presumable at less than their desired speeds.

Basic Traffic Flow Theory 63

If we observe vehicles passing a point by the road, then the proportion of following vehicles (θ) is given by

$$\theta = \frac{m-1}{m} \qquad (3.27)$$

If $m = 2$ then half of the vehicles are following in bunches. The extent of bunching is usually measured by the 'percentage of vehicles following' (*PVF*), which is equal to 100θ (per cent), or the 'percentage of time spent following' (*TSF*). *PVF* is easy to observe but is measured at a point on the road only, and may fluctuate along the road (as overtaking opportunities vary). On the other hand *TSF* is a lineal parameter, relating to flow along a finite length of road. This makes it a better descriptor of the quality of flow, but it is also harder to measure. There are methods to estimate *TSF* from *PVF*, as described in HCM (1985).

Surveys of bunching require techniques that measure the separation between consecutive vehicles and assess the likelihood that a vehicle is tracking behind the one in front, or is travelling freely. There are several different criteria that may be used for this process of bunch discrimination. Subjective assessment of bunch membership by trained observers is possible but may lead to inconsistencies when comparing data from different sources. Minimum free headways are often used, with a recommended minimum headway of four to five seconds used to distinguish bunches. Vehicles travelling with smaller headways are assumed to belong to a bunch. A combined headway-speed differential criterion may be used as an alternative.

3.4 Vehicle interactions in traffic

The statistical models provide a useful framework for describing the traffic and movement characteristics of the individual units in a moving traffic stream. These models work best under those conditions where free flowing movement is possible, for only then can the basic assumptions of statistical independence between successive events be properly valid. As traffic volume increases into region C of the q-k curve of Figure 3.4, the opportunities for individual vehicles to travel freely diminish. Interactions between vehicles, including minimum spacings and headways, speed differentials and the impacts of accelerations or decelerations, play an increasingly important role in determining the nature of the traffic flow. Alternative descriptions are needed, including models for car following, overtaking, lane changing and platoon dispersion.

3.4.1 Car following

Car following models of traffic flow describe the traffic performance of individual vehicles in a single lane, when vehicles are travelling in close proximity to each other and cannot overtake. Under these conditions a following vehicle needs to respond quickly and adequately to the acceleration-speed performance of its leader. One means of modelling car following behaviour is through application of the well-known stimulus-response law in behavioural psychology, i.e.

Response at time $t + \Delta t$ = sensitivity coefficient × Stimulus at time t

This law may be interpreted in general terms in the following way. An organism is subject to a stimulus (some action on it) at time t. The organism produces a response or reaction to that stimulus after a small time lag (a reaction time) of Δt. The magnitude of the response is influenced by some sensitivity coefficient, that depends on the individual characteristics of the organism. In a traffic situation the stimulus may be an acceleration or deceleration by the leading vehicle, the response may be the corrective action of the second driver, and the sensitivity will relate to the characteristics of that driver and vehicle.

Two very useful car following models resulted from research at the General Motors Detroit Research Laboratories in the 1950s and 1960s. The *first General Motors model* assumes that the acceleration of a following vehicle is directly proportional to the speed differential between it and the vhicle in front. Thus, if the vehicles travelling together in a lane are numbered sequentially, starting with the leading vehicle,

$$a_n(t + \Delta t) = \Lambda_n \left[v_{n-1}(t) - v_n(t) \right] \tag{3.28}$$

where $a_n(t + \Delta t)$ is the acceleration of the nth vehicle at time $(t + \Delta t)$, Δt is a reaction time, Λ_n is a sensitivity coefficient and $v_i(t)$ is the speed of vehicle i at time t.

A problem with the model is that it does not include any reference to the spacing between the vehicles. A driver can be expected to be more responsive to changes in speed if the leading vehicle is only ten metres in front, rather than 100 metres. The *second General Motors model* took the separation between vehicles into account, as follows:

$$a_n(t + \Delta t) = \Lambda_n \frac{v_{n-1}(t) - v_n(t)}{x_{n-1}(t) - x_n(t) - l_{n-1}} \qquad (3.29)$$

where l_{n-1} is the effective length of vehicle *(n-1)* and $x_i(t)$ is the position of vehicle *i* on the road at time *t*. This model indicates that as the separation between a pair of vehicles increases, so the level of interaction between them will decrease.

There have been many subsequent developments in the study of car following behaviour since the advent of the General Motors models. The more recent models have allowed greater accuracy in predicting vehicle movement at the microscopic level, though this has been accomplished by making the models more complex and hence more difficult to apply. The basic concept of a stimulus-response relationship dictating vehicle behaviour in close following traffic is the main idea that the traffic engineer needs. Gipps (1981) and Hidas (1998) provide descriptions of the development and use of behavioural car following models that closely mimic observed driver behaviour in congested traffic.

3.4.2 Lane changing and overtaking

The individual driver can react in one of two ways having caught up to a slower vehicle travelling in the same lane. The driver can track behind that vehicle, or can change lanes. The transfer of a vehicle from one lane to another is known as 'lane changing'. This process is more complicated than that of car following, for the driver has more decisions to make. For instance, on a two-lane, two-way road, the driver must allow for the presence of on-coming traffic when deciding to change lanes and overtake. On a multi-lane carriageway the driver must assess if there is a sufficient gap in the target lane to permit the lane change to occur. Two basic types of lane change are recognised:
- a forced lane change, which occurs when the headway between two vehicles in the same lane is approaching zero more rapidly than the driver of the following vehicle can decelerate, and
- optional lane changes, which are made under other, less urgent conditions.

When the average speed is 50 to 65 km/h observations suggest that the average length of roadway required to accomplish a lane change varies from about 40 to 80 m, depending on the type of lane change to be made. The average time required for a lane change varies from 2.5 to 4.5 seconds. The average optional lane change requires a distance of perhaps 70 m, and takes about 4 seconds. A forced lane change requires about 45 m at these speeds.

The demand for lane changing is at a minimum when all drivers desire to travel at about the same speed. As the relative differences in vehicle speeds increase, so does the frequency of demand for lane changing. The desire to change lanes will also increase with increasing traffic volume. This desire is, however, often thwarted by a corresponding decrease in the number of acceptable gaps available in the traffic stream.

Overtaking is a special form of lane changing that assumes particular importance on two-way, two-lane roads. This manoeuvre is of special importance in the design and operation of two-lane highways. The desired overtaking rate per unit distance (OT) has been shown to be proportional to the square of the density. Indeed, if speeds are normally distributed then the relationship is approximated by

$$OT = 0.56\sigma k^2 \qquad (3.30)$$

where σ is the standard deviation of the speed distribution. Under the assumption that the standard deviation is about 14 per cent of the mean speed (see Section 3.2.2) and applying the flow continuity equation (3.1), equation (3.30) becomes

$$OT = 0.0784qk$$

or, if we replace k entirely by q (and $\bar{v}_s$),

$$OT = 0.0784 \frac{q^2}{\bar{v}_s}$$

If q is in veh/h, k is in veh/km and $\bar{v}_s$ is in km/h, the units for overtaking rate OT are veh/h/km. On two-lane, two-way roads the level of service provided by the road is affected by the number of times that a slow vehicle is overhauled by a faster vehicle, and the time spent following related to the availability of overtaking opportunities. At higher traffic volumes, the relationship between the percentage of the road providing overtaking opportunity and level of service can be derived, using the methods outlined in HCM (1985). Under low volumes, a slow vehicle is caught up by a faster vehicle infrequently enough that the level of service is not seriously affected. The time of enforced tracking (time spent following) is, however, more significant to the faster drivers than their overall average journey speed, so that the availability of overtaking opportunities along the road applies as a 'quality of service' concept.

3.4.3 Bunches and platoons

By observation, bunches and platoons would appear to be the same traffic phenomenon. Both represent clusters of vehicles travelling together in close proximity at about the same speed. There is, however, an important difference between them that has strong implications for traffic control strategies. Thus there is good reason for adopting a different terminology. Bunching and platooning represent traffic flow in different types of road environments.

Bunching may be taken to represent the clustering of traffic under uninterrupted flow conditions, where the clusters arise through limited opportunities for overtaking (as discussed in Section 3.4.2). These conditions are best represented by traffic flow on two-way, two-lane rural highways. Traffic design for this type of flow usually aims at reducing the amount of bunching, e.g. by improved sight distance, provision of overtaking sections, or even duplication of the road.

Platooning represents the clustering of traffic in urban arterial road networks with many signalised intersections. These traffic conditions represent interrupted traffic flow. The signals break up the through traffic flow along a road. Pulses of traffic (the platoons) are released periodically during the green time at each signal. These platoons then travel along the link to the next set of signals. Traffic design under these conditions aims at preserving the platoons, for efficient signal coordination and network traffic control.

3.4.4 Shock waves

The dynamics of traffic flow along a road is an important consideration in traffic performance monitoring and incident detection. A longstanding model for the dynamics of an uninterrupted traffic stream is the macroscopic flow model based on 'kinematic waves' introduced by Lighthill and Whitham (1955). This model is based on the differential equation (3.31) for continuity of flow where volume and density may both vary with respect to position x along the road and over time t:

$$\frac{\partial q(x,t)}{\partial x} + \frac{\partial k(x,t)}{\partial t} = 0 \tag{3.31}$$

This equation may be used with equation (3.1) which defines the mean speed ($\bar{v}_s$) of the stream as the ratio q/k. Lebacque and Lesort (1999) demonstrate how the assumption of some specific relationship between $\bar{v}_s$ and $k(x,t)$ can then lead to a complete dynamic model of traffic flow along a road, such as Payne (1979).

The Lighthill-Whitham model may be used to study the effects of discontinuities in flow, such as an incident or lane blockage or at the transition from one road configuration to another (e.g. a change in the number of lanes provided on the road). For example, the model can explain the observed phenomenon of shock waves in a traffic stream where such a discontinuity exists. If x_d and x_u are points downstream and upstream of a discontinuity (at x_0, i.e. $x_d < x_0$, x_u) then Lebacque and Lesort (1999) indicate that the speed of the shock wave $U_s(t)$ is

$$U_s(t) = \frac{q(x_d,t) - q(x_u,t)}{k(x_d,t) - k(x_u,t)} \tag{3.32}$$

Figure 3.5 shows examples of the shockwave phenomenon for three different cases:
- a capacity restriction, where demand is below capacity. The downstream speeds are less than the upstream speeds;
- a capacity restriction at which the demand exceeds the downstream capacity but is less than the upstream capacity. A standing shock wave forms at x_0, with traffic at the wave front coming to rest, then feeding away from the constriction, whilst the wave front moves back along the traffic stream, and
- a capacity expansion, where the demand exceeds the upstream capacity and is less than the downstream capacity. This case is symmetric to (b).

3.5 Applications in traffic analysis

This chapter has provided a starting point for consideration of traffic movement and performance along the links in a network. Its concepts and theories relate primarily to the link as the level of analysis. Further, this level of analysis has only been considered in the form of uninterrupted traffic flow. No account has yet been taken of external influences on traffic flow (e.g. the effects of other, intersecting traffic streams). Significant complications result when the performance of one traffic stream is affected by one or more other streams. Chapter 4 provides consideration of interrupted traffic flows, with emphasis on the performance of traffic signals at junctions. Chapter 5 then extends this analysis to consider traffic signal linking and area-wide traffic control.

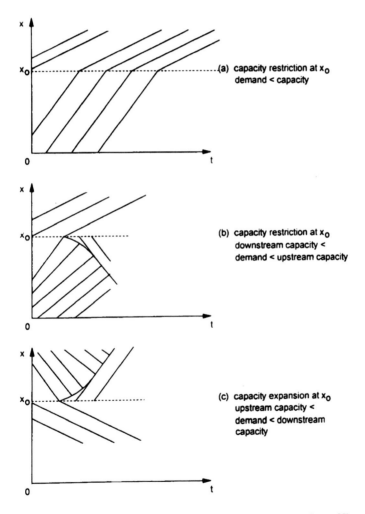

Figure 3.5   Shock waves in an uninterrupted traffic stream due to a discontinuity at x0 (source: Lebacque and Lesort (1999))

4 Theories of interrupted traffic flow

The traffic flow theory presented in the previous chapter applies to traffic streams moving free from external influences: the only factors affecting the performance of those traffic streams are those that relate directly to the streams themselves. This type of traffic flow is known as uninterrupted flow. In many situations, particularly in urban areas, factors external to the traffic stream will have an effect on traffic flow. Such factors include other, intersecting traffic streams (e.g. at intersections), other modes (e.g. pedestrians, bicycles, buses and trams, railway level crossings, pedestrian crossings), parking and parked vehicles. Traffic flow under these conditions is known as interrupted flow.

The distinction between uninterrupted flow and interrupted flow is important because it provides a dichotomy between the model types needed to describe flows in which delay and congestion is generated by the internal interactions in the traffic stream (e.g. freeway traffic), and delays and congestion generated by the interaction between intersecting traffic streams (e.g. a signalised intersection). The freeway traffic represents uninterrupted flow, the signalised intersection is an example of interrupted flows. Figure 4.1 shows stereotypical speed-volume and travel time-volume curves for each of the flow regimes. The figure clearly indicates one significant difference between uninterrupted and interrupted flows: for uninterrupted flows there can exist more than one travel time or speed value for a given traffic volume, as was described in Section 3.2.3. This is not the case for interrupted flow, where the lower arm of the v-q curve will dominate (as vehicles are usually moving away from a stationary queue for example when a traffic signal changes from red to green) - and thus the starting speed is zero, not the free speed v_f. Note that Figure 4.1 includes graphs of unit travel time (the time taken to travel a given distance, e.g. minutes per kilometre, which is the inverse of speed, the distance travelled per unit time, e.g. kilometres per hour) versus volume. There are distinct advantages in using unit travel times in preference to speed when undertaking network analysis. Unit travel time can be applied directly to calculate delays and travel costs, and is compatible with many applications in travel demand modelling and network analysis, such as the modelling of modal choice, destination choice and route choice.

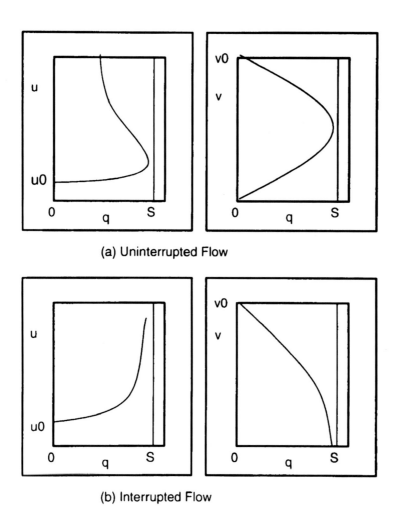

Figure 4.1 Travel time-flow and speed-flow curves for uninterrupted and interrupted flows

The most common circumstances for interrupted flows occur when two or more traffic streams intersect. This is clearly seen at a road junction, where traffic streams using each of the intersecting roads must cross the paths of other streams, and much of our attention will need to be focused on how to describe and analyse this general situation and its specific instances, and how traffic systems can be organised to minimise the inevitable delays and maximise the crossing opportunities for traffic at the junction. There are other manifestations of interrupted flows, that may not be so obvious. For example, interrupted flow situations can occur for traffic ostensibly travelling in the 'same' stream when those vehicles reach a junction and some of them wish to turn out of the road they presently occupy. For example, if the turning vehicles must make their manoeuvres in the face of oncoming traffic and are then obliged to slow down, stop and wait to make the turn, then through vehicles sharing the same lanes as the intending turners will also be delayed. In this circumstance the single traffic stream has split into two separate streams. Similar circumstances emerge on a multilane road when one lane is dropped, either permanently, due to road narrowing, or temporarily when a lane is blocked by roadworks, a breakdown or an accident. The compression of the flow into a smaller number of lanes is regarded as interrupted flow because of the external influence placed on the traffic stream by the removal of a lane. Dealing with intersecting or interacting traffic streams is a common problem for traffic engineers. The basic operation is that one stream must filter through or into the other, and the solution usually involves application of one of the following treatments:

- *setting a priority rule* so that drivers can assess which vehicles have precedence in traversing the intersection;
- *separating the traffic streams in time*, which is the basis of traffic signals control, and
- *separating the traffic streams in space*.

The most common of these treatments, reckoned in terms of the number of situations where it will be applied, is that of setting a priority rule, and it is this form of traffic filtering control that we will tackle first. This is done through the mechanism of *gap acceptance*, and makes use of the statistical models of vehicle headways examined in Chapter 3.

4.1 Gap acceptance

Many traffic manoeuvres involve an individual road user (e.g. driver, rider or pedestrian) selecting a break or *gap* in a traffic stream of suitable size for the safe accomplishment of the manoeuvre. Typical examples are:
- a pedestrian crossing a road;
- a vehicle entering a road from a car park;
- a vehicle making a filtered turn through an oncoming traffic stream, and

- one vehicle overtaking another.

Analytical methods based on the assumed distribution of gaps in a traffic stream and assumed rules of human behaviour have been developed, and are known as gap acceptance methods. They permit the prediction of the likelihood of delay and the probable duration of delays. We start by considering gap acceptance when the main stream headways follow a negative exponential distribution.

4.1.1 Gap acceptance mechanisms

Consider the case of pedestrians waiting to cross a road. The mechanism for gap acceptance by an individual pedestrian involves the following stages:
- the pedestrian arrives at the kerbside, and begins to scan the gaps (breaks) in the traffic stream. The first 'gap' in the traffic is the time from the arrival of the pedestrian to the arrival of the next vehicle to pass by. This gap is known as the *lag*;
- if the lag is of an acceptable size, the pedestrian crosses the road immediately and continues, unimpeded;
- if the lag is too small, the pedestrian is delayed and must wait for a subsequent gap that is large enough to allow the crossing manoeuvre to take place, and
- a delayed pedestrian may have to wait for r (= 1, 2, 3, ...) gaps before a suitable gap arrives. This means being delayed while a sequence of r successive gaps less than the acceptable gap occur, followed by a gap which is of acceptable size. The time incurred while waiting for the r gaps to pass (which equals the sum of those gap sizes) is the waiting time delay experienced by the pedestrian.

Assuming that individuals behave consistently, then for a given traffic situation they will require a gap greater than or equal to their minimum acceptable gap before proceeding. Gaps smaller than this gap will be rejected. Gap acceptance is based on a process of human observation and estimation, in which a person observes an approaching vehicle, makes an assessment of the distance to that vehicle and its speed of approach, and then performs a mental calculation to see if there is sufficient time to make the required manoeuvre in safety, or at least with an acceptable minimum perceived risk. This time is the *critical gap*. Different individuals have different critical gaps, and as humans are imperfect measuring instruments, they may occasionally misjudge the size of a gap or end up accepting a gap smaller than one previously rejected. In most cases it is acceptable to use a mean value of the critical gap (t_a) to describe the behaviour of all individuals.

Theories of Interrupted Traffic Flow 75

Note that a group of pedestrians may all use the same gap simultaneously, as they can stand side-by-side (i.e. queue 'in parallel'), so that all of the group can take advantage of the entire gap when it appears.

On the other hand, vehicles in a minor stream must queue one behind the other in a lane, so that only one vehicle at a time can use an acceptable gap. The next vehicle may be able to use the residual part of that gap once it reaches the head of the queue. An additional parameter is needed in this case. This is the *move-up time* (or *follow-up headway*) t_f, which is the minimum headway between minor stream units. When the vehicle at the head of the queue departs, the next vehicle will reach the 'stop line' t_f seconds later. This translates into the following set of possible events for the vehicles in the minor stream, assuming that t_a and t_f are constants:
- gaps less than t_a will not be accepted;
- gaps between t_a and $t_a + t_f$ will be used by one minor stream vehicle;
- gaps between $t_a + t_f$ and $t_a + 2t_f$ will be used by two minor stream vehicles, and
- in more general terms, gaps between $t_a + (i-1)t_f$ and $t_a + it_f$ will be used by $i = 1, 2, 3, ...$ minor stream vehicles,

on the assumption that there are always vehicles queued in the minor stream to take full advantage of every possible gap. This means that, in theory, there is then an infinite queue on the minor stream.

In practice the critical gap is difficult to measure (see Section 11.5). If a driver rejects a number of gaps before accepting one then all that can be said is that, assuming consistent behaviour, the driver's critical gap is larger than the largest rejected gap and smaller than the accepted gap. The theoretical results that follow are based on the assumption of known mean acceptable gap t_a and follow-up time t_f.

4.1.2 Basic results

For random traffic, the cumulative density function (cdf) defined by equation (3.22) defines the probability that a gap will be less than a certain size. The probability of being delayed is thus the probability that the first gap (the lag) is less than t_a. This probability is, from equation (3.22), $Pr\{delay\} = Pr\{lag < t_a\} = 1 - exp(-q_p t_a)$ where q_p is the flow rate for the priority stream. The probability of no delay is the probability that the lag is greater than or equal to t_a. This probability is $Pr\{no\ delay\} = Pr\{lag \geq t_a\} = exp(-q_p t_a)$. These results indicate the proportion of minor stream road users that will be undelayed, or will suffer some delay.

76 Understanding Traffic Systems

The theoretical maximum rate at which minor stream units can enter or cross the major stream is an important traffic parameter. This rate is known as the *absorption capacity* (*C*). It is found by considering an infinite queue on the minor stream, and assuming that each and every suitable gap will be used to its maximum potential (i.e. the maximum possible number of minor stream units will use every gap which exceeds t_a and in large gaps vehicles follow-up at headways of t_f). Absorption capacity in the Random Traffic Model is then given by the equation

$$C = \frac{q_p \exp(-q_p t_a)}{1 - \exp(-q_p t_f)} \qquad (4.1)$$

Field values of absorption capacity may be higher or lower than this theoretical value at a given site, usually due to site conditions, the inaccuracy of the Random Traffic Model, or 'pressure' on drivers due to the degree of saturation at which the intersection is operating.

The amount of delay can also be predicted. The mean delay (w_h) to pedestrians or isolated minor streams vehicles in waiting for a suitable gap under the gap acceptance mechanism described in Section 4.1.1 is given by Adams's formula

$$w_h = \frac{1}{q_p} \exp(q_p t_a) - \frac{1}{q_p} - t_a \qquad (4.2)$$

This result gives the mean delay to a minor stream unit in looking for gaps in a major stream. Note that it does not include any queuing delays (i.e. time spent in a queue before the unit reached the head of the queue and can start scanning for gaps). Not all minor stream vehicles are delayed, and the mean delay to those who are delayed (w_{hd}) is equal to $w_h/Pr\{delay\}$, which may be expressed as

$$w_{hd} = \frac{1}{q_p} \exp(q_p t_a) - \frac{t_a}{1 - \exp(-q_p t_a)}$$

The mean delay to minor stream vehicles when queuing is included (w_m) is given by the equation

$$w_m = \frac{w_h + \eta \rho}{1 - \rho} \tag{4.3}$$

where ρ is the degree of saturation (or 'utilisation') and is equal to the minor stream flow rate q_m divided by the maximum absorption capacity C (i.e. $\rho = q_m/C$). The factor η is given by the expression

$$\eta = \frac{\exp(q_p t_f) - q_p t_f - 1}{q_p (\exp(q_p t_f) - 1)} \tag{4.4}$$

Equation (4.3) provides a useful form of the delay equation because it relates the actual mean delay to minor stream vehicles to the Adams' delay and to the absorption capacity of the minor stream.

4.1.3 Multi-lane traffic flows

The previous discussions have implied that the major stream flow is in a single lane (or direction) in which minor stream units seek acceptable gaps.

Consider the case of two major road streams (either two-way traffic or two-lane one-way traffic) with flow rates q_1 and q_2, each with negative exponential headway distributions. The probabilities of occurrence of suitable headways in each of the streams are, from equation (3.22)
- stream 1, $Pr\{t_h \geq t_a\} = \exp(-q_1 t_a)$, and
- stream 2, $Pr\{t_h \geq t_a\} = \exp(-q_2 t_a)$.

Assuming that the two streams move independently of each other, the probability of the simultaneous occurrence of a suitable headway in both streams is given by the product of the above probabilities, i.e. for the combined stream, $Pr\{t_h \geq t_a\} = \exp(-q_1 t_a) \exp(-q_2 t_a) = \exp[-(q_1 + q_2)t_a]$. This result is in the same form as that for a single lane stream, implying that when the Random Traffic Model is employed for the purposes of gap acceptance analysis, a number of separate major road traffic streams may be combined into a single stream with overall flow rate equal to the sum of the flow rates of the individual streams. Note that this simple result does not necessarily follow for those situations where the Random Traffic Model does not apply.

The other situation where multi-lane flows are considered is when there are two or more lanes available for the minor stream. In this case the total absorption capacity for the stream is equal to the sum of the absorption capacities of the individual lanes.

4.1.4 Combined lanes

Often more than one minor stream may share the same lane on an approach. For example, left and right turners may share a single lane, as in Figure 4.2.

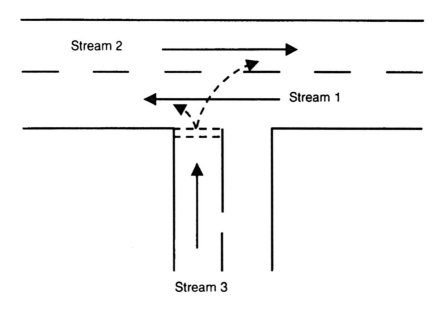

Figure 4.2   **Major and minor traffic streams at a T-junction**

Now, each of these movements yields to different combinations of traffic streams. Figure 4.2 is drawn for driving on the left hand side of the road, as in Japan, the UK and Australasia, where the left turners in stream 3 yield to stream 1 only, whereas the right turners give way to streams 1 and 2. Thus the absorption capacities for the two movements are different. In addition, the presence of right turning vehicles in the queue ahead of a left turner reduces the opportunities for the left turner to find and utilise acceptable gaps.

For the general case where there are multiple movements sharing a lane, each looking for suitable gaps in a major stream, the overall absorption capacity C of the lane is given by equation (4.5), in which C_i is the absorption capacity of the ith movement in the stream and p_i is the proportion of the total flow in the lane that is making movement i:

$$\frac{1}{C} = \sum_i \frac{p_i}{C_i} \qquad (4.5)$$

4.2 Queuing and delay

An alternative approach to gap acceptance is to consider the interactions between traffic streams as situations in which queuing takes place. The mathematical theory of queuing can then be applied to indicate the capacity of a traffic facility and the delays incurred in using it. The results obtained from queuing theory are similar to those from gap acceptance, though not necessarily identical as the basic theoretical assumptions may not be the same. Queuing theory may provide a more versatile framework for traffic analysis in many situations.

Generally, the prime objective in designing for the efficient movement of vehicles through a traffic facility is to ensure that the capacity of the facility can accommodate the average traffic demand to use it. Transient queues may form as part of the normal traffic performance of the facility (e.g. during the red period at a signalised junction, or at a stop line). If the demand is less than the capacity of the facility, these queues will clear over time (e.g. during the green phase, or when long gaps occur in the major stream). Standing queues (which exist for protracted time periods) will be formed if the demand exceeds the capacity of a facility. Mathematical theory may be used to provide a description of these phenomena. The main performance measures of interest are queue length, delay, and proportion of units (customers) delayed. A complete description of a system requires the specification of:

- the distribution of customer arrival times;
- the distribution of service times;
- whether or not the level of demand is finite;
- the number of servers (or 'channels') available (e.g. single server, multiple server) and the service discipline (e.g. service in parallel, or in series), and
- the queue discipline, i.e. the means for determining the order in which customers will be served (e.g. 'first-in, first-out').

The system is said to be in state 'n' if there are n items in it, including those waiting in the queue and those being serviced, and the utilisation factor (or traffic intensity) (ρ). This is the ratio of the average arrival rate (q_a) to the average service rate (q_b). It measures the degree of saturation of the system ($\rho = q_a/q_b$). If $\rho \geq 1$, the system is overloaded and the queue length will increase over time, as long as the excess demand exists. If $\rho < 1$, the queue is stable, i.e. there is a fixed probability of the queue being in a particular state.

4.2.1 Arrival pattern

The arrival pattern or distribution of arrival times indicates how customers arrive at the point. Arrivals may be in terms of individual units, travelling independently, or in bunches, or following some cyclic pattern. The classical arrival distribution is random arrivals (see Section 3.3.2), and this is the pattern assumed in the subsequent discussion leading to a simple theoretical model of a queue. The parameter of interest in this case is the mean rate of arrivals, q_a. Other arrival patterns of interest include regular arrivals, multiple arrivals (when customers arrive in groups of random size), time dependent arrivals, and system dependent arrivals (e.g. when the presence of a long queue discourages later arrivals from joining that queue).

4.2.2 Service mechanism

The service mechanism consists of three parts. The first is the number of servers available. The second part is the service rate. Classical theory is based on a single server, which provides service according to a negative exponential distribution of service times, at a mean service rate of q_b. This rate is independent of the rate of arrivals. The third part is an operational one: when is the service available, and what restrictions might be imposed to reduce the number of customers who can be served simultaneously, below the capacity of the system? Many real queuing systems will operate in other ways. There may be one or more server, which may speed up or slow down service in response to the level of demand. Service time distributions may follow other patterns, e.g. a constant service time for all customers.

4.2.3 Queue discipline

Queue discipline refers to the manner of selection of the next customer to be served. The most common queue discipline is 'first-in, first-out', but there are others. For instance, there may be priority customers who can bypass the queue. When there are multiple servers operating, customers may form a single queue, or form separate queues for each server.

In multiple channel queues (i.e. where there is more than one server), certain servers may specialise in catering for particular types of customers (e.g. automatic toll booths for 'cars only' or for 'correct toll, no change given', or for electronic cash transfer). This specialisation may lead to particular benefits for some customers and more efficient operation of the facility, such as significantly reduced service times for that class of customer, or a reduced need for a large number of servers.

4.2.4 Traffic intensity

The traffic intensity of a system (ρ) is

$$\rho = \frac{q_a}{q_b} \tag{4.6}$$

A value of $\rho < 1$ means that the queue exists in a stable, equilibrium state. This means that although there may be different numbers of customers in the queue at different times, the mean queue length and mean delay remain constant over time.

The closer the value of ρ is to unity, the greater the amount of delay that will be expected. A traffic intensity greater than or equal to one indicates that the queue is overloaded; customers arrive faster than they can be served and the queue will grow indefinitely, or until the arrival rate falls. Traffic intensity is thus equivalent to the degree of saturation. Equilibrium in a queuing system has to be ascertained by observation over a finite period of time long in comparison with the individual arrival and service times. The concept of equilibrium can now be used to derive a simple queuing model.

4.2.5 Derivation of simple queuing model

A simple but useful queuing model for traffic applications may be found by considering the situation of a queue with random arrivals, one server and random service times. 'Random' means that the probability of an event occurring during a small time interval ($t, t + \delta t$) is $q\delta t$, where q is a constant equal to the average rate at which such an event occurs, and is independent of any other event. This pair of assumptions is identical to that used in formulating the Random Traffic Model (Section 3.3.2). The resulting distributions of arrival times and service times follow the negative exponential distribution defined by equations (3.21) and (3.22). A further consequence of the assumed probability of an event being $q\delta t$ is that the probability of more than one event occurring in the time interval δt is of the order $(\delta t)^2$ and can thus be ignored if δt is sufficiently small.

Let q_a be the arrival rate and q_b be the service rate. Then $q_a \delta t$ and $q_b \delta t$ are the respective probabilities of an arrival and a departure during $(t, t + \delta t)$. Therefore, the average time interval between successive arrivals is $1/q_a$, the mean service time is $1/q_b$, and the traffic intensity $\rho = q_a/q_b$. The probability that there are n customers waiting in the queue at time $t + \delta t$ is equal to the sum of:

(1) the probability that there are n customers in the queue at time t and no one arrives or leaves during $(t, t + \delta t)$;
(2) the probability that there are $(n + 1)$ customers in the queue at time t and one customer leaves during $(t, t + \delta t)$, and
(3) the probability that there are $(n - 1)$ customers in the queue at time t and one customer arrives during $(t, t + \delta t)$. [This condition can only apply for n > 0, whereas (1) and (2) apply for $n \geq 0$.]

Given that $p_n(t)$ is the probability that there are n customers in the queue at time t, including the customer being served, then this summation of component probabilities may be written as the following pair of equations:

$$p_0(t + \delta t) = p_0(t)(1 - q_a \delta t) + p_1(t) q_b \delta t$$
$$p_n(t + \delta t) = p_n(t)(1 - q_a \delta t - q_b \delta t) + p_{n+1}(t) q_b \delta t + p_{n-1}(t) q_a \delta t \quad (4.7)$$

where n takes the values 1, 2, 3, ... in the second expression in equation (4.7). Note that the probability of n customers in the queue at time t and with no arrivals or departures during $(t, t + \delta t)$ is $1 - q_a \delta t - q_b \delta t$, and that all terms of the order $(\delta t)^2$ may be ignored. Transferring $p_0(t)$ and $p_n(t)$ to the left hand side and dividing by δt yields:

$$\frac{p_0(t + \delta t) - p_0(t)}{\delta t} = -q_a p_0(t) + q_b p_1(t)$$

$$\frac{p_n(t + \delta t) - p_n(t)}{\delta t} = -q_a p_n(t) - q_b p_n(t) + q_b p_{n+1}(t) + q_a p_{n-1}(t)$$

Letting δt tend to zero, so that in the limit the left hand sides become the derivatives with respect to time, yields a pair of differential equations:

$$\frac{dp_0(t)}{dt} = -q_a p_0(t)$$

$$\frac{dp_n(t)}{dt} = -q_a p_n(t) - q_b p_n(t) + q_b p_{n+1}(t) + q_a p_{n-1}(t)$$
(4.8)

The definition of the equilibrium state for a queue is that the probabilities p_n are constant over time, which means that the derivatives must vanish. Thus, for a queue in equilibrium, the equation pair (4.8) becomes:

$$q_b p_1 - q_a p_0 = 0$$

$$q_b p_{n+1} - q_b p_n - q_a p_n + q_a p_{n-1} = 0$$
(4.9)

This pair of equations may be further rearranged and simplified. For $n = 0$, it follows that

$$p_1 = \frac{q_a}{q_b} p_0 = \rho\, p_0 \qquad (4.10)$$

while for $n > 0$,

$$p_{n+1} = p_n(1 + \rho) - \rho\, p_{n-1} \qquad (4.11)$$

which has defined a set of recurrence relations between the probabilities p_n. Applying the result from equation (4.10) for p_2 in equation (4.11) yields $p_2 = \rho^2 p_0$, $p_3 = \rho^3 p_0$, $p_4 = \rho^4 p_0$, ..., so the general result is:

$$p_n = \rho^n p_0 \qquad n = 1, 2, 3, \ldots \qquad (4.12)$$

All that remains is to determine the probability p_0. This may be found from the general result that as p_n is a probability distribution,

$$\sum_{n=0}^{\infty} p_n = 1$$

and using equations (4.10) and (4.12), it follows that:

$$P_0 \sum_{n=0}^{\infty} \rho^n = 1$$

Now $\Sigma_n \rho^n = 1/(1-\rho)$ for $\rho < 1$, a well-known result for the sum of this infinite series. Thus,

$$P_0 = 1 - \rho \qquad (4.13)$$

so that equation (4.12) may be rewritten as:

$$P_n = \rho^n (1 - \rho) \qquad (4.14)$$

This distribution of queue sizes is the geometric distribution, which was introduced in Section 3.3.1.

4.2.6 Results of simple queuing theory

The following results emerge from the simple queuing model defined by equations (4.13) and (4.14), based on the assumptions of random arrivals, a single server, first-come first-served discipline, negative exponential service times, and traffic intensity $\rho < 1$.

The mean queue length (n_q) is:

$$n_q = \frac{\rho}{1-\rho} \qquad (4.15)$$

and the variance of the queue length $(\sigma^2(n))$ is:

$$\sigma^2(n) = \frac{\rho^2}{1-\rho} \qquad (4.16)$$

The probability $Pr\{n>N\}$ of observing more than N units in the queue is:

$$Pr\{n > N\} = \rho^{N+1} \qquad (4.17)$$

The mean delay (the average time spent in the queue before being served) is w_q, which is equal to the product of the mean queue length and the mean service time. This result follows from the observation that the average delay involves waiting for the average number of customers to be served and the time taken for each customer will average out to be the mean service time. For the geometric distribution of queue lengths this yields:

$$w_q = \frac{n_q}{q_b} = \frac{\rho}{q_b(1-\rho)} \tag{4.18}$$

A more general result for the mean delay in a queue with random arrivals and a single server is given by the Pollaczek-Khintchine formula,

$$w_q = \frac{\rho(1+v^2)}{2(1-\rho)q_b} \tag{4.19}$$

This formula applies to any system with random arrivals and where service times are independent of queue length and v is the coefficient of variation of the service times. For negative exponential service times, $v = 1$ and the mean in-queue delay from the Pollaczek-Khintchine formula reduces to the form given by equation (4.18), as expected. The Pollaczek-Khintchine formula provides a useful generalisation of the basic model, by indicating the differences in mean delays that result from different service time regimes. For example, if service time is constant (i.e. all customers experience the same service time) then $v = 0$ and the mean delay reduces to half of the value found for random service times. This is indicative of a general outcome, in which, given all other factors the same, the greater the variation present in the system, the greater the possible delays that may be experienced. We will revisit this outcome in our considerations of traffic signals (see Section 4.3.8).

The total mean delay (w_m), including and service time (e.g. the wait for an acceptable gap when at the head of the queue) is given by:

$$w_m = w_q + \frac{1}{q_b} \tag{4.20}$$

i.e. the sum of the mean service time and the average wait in the queue. In terms of the gap acceptance (Adams) delay as discussed in Section 4.1.2, equation (4.20) would be rewritten as:

$$w_m = w_q + w_h \tag{4.21}$$

Note that this expression for the mean total delay will not *necessarily* give identical results to those of equation (4.3) because the assumptions used in the queuing theory are slightly different from those used in the gap acceptance model. In most cases (all but the extremes where the utilisation factor is close to one) the difference will be small, and can be ignored.

4.3 Theory of traffic signal operation

Interrupted flow is a hallmark of urban traffic systems. The major factors affecting urban traffic are those related to intersections in the road network, where sets of traffic streams must compete for limited resources of road space and time. The effects are two-way, for the operation of the intersections affects the nature of the traffic flow, and the traffic flow affects the operation of intersections.

Traffic signals have become the major form of urban traffic control and management, especially for major intersections where large volumes converge. The basis of traffic signals operation is that each of the intersecting traffic streams will be offered a window of time (the 'green time') during which it will have the opportunity to traverse the intersection. The design task is to determine the lengths of these green times so that the intersection can provide efficient operation for all of the intersecting traffic streams.

The cycle of operation of a traffic signal is the total time required before the repetition of the same sequence of the traffic signal *stages* (termed *phases* in North America and Australasia). A stage (phase) is a distinct part of the signal *cycle* in which one or more movements receive right of way (i.e. have the green light). A stage (phase) is identified by at least one movement which gains right of way at the start of it and at least one movement which loses right of way at the end of it.

A *movement* is a separate queue leading to the intersection (e.g. right or left turning traffic, or the separate lanes available for passage through a junction). Figure 4.3 shows an junction plan and simple staging (phasing) diagram for a T-junction. There are six vehicle movements (1-6) and two pedestrian movements (7 and 8) possible. The staging (phasing) diagram shows the movements given right of way in the three phases (A, B and C) of the signal cycle. An important part of the analysis and design of a signal cycle is to determine the set of *critical movements* which determine the capacity and timing requirements at the junction.

Theories of Interrupted Traffic Flow 87

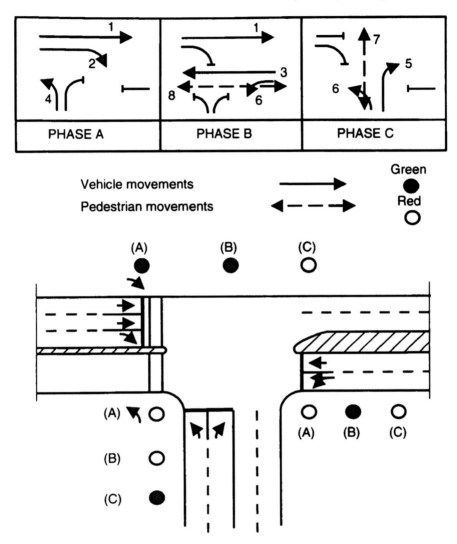

Figure 4.3 Sample junction plan and signal staging

4.3.1 Basic parameters for signal operation

Five main factors influence the capacity of a single movement at a signalised intersection. These are:
- the *saturation flow* (s) for each of the lanes used by the movement;

88 *Understanding Traffic Systems*

- the *number of lanes* available for the movement;
- the *lost time* (*l*) for the movement;
- the *cycle time* (*c*) for the intersection, and
- the *green time* (*g*) for the movement.

Capacity is considered to be the maximum volume of traffic that can be discharged through the intersection over an extended time period (i.e. more than one cycle). The saturation flow is the maximum instantaneous flow rate, and is only possible during the green time for the movement.

4.3.2 Wardrop-Webster model

The Wardrop-Webster model was developed in the UK during the 1950s and is the basis of all signal calculations. Akcelik (1981) provides a useful description of the model. Modern implementations of it are usually implemented as software packages, such as Akcelik's SIDRA package.

The basic model is formulated in the following way. Consider a simple four-arm junction, controlled by two stages (one for east-west traffic, the other for north-south traffic) as in Figure 4.4. This figure shows the flow of traffic over time on the east-west road, under the assumption that this road is carrying a sufficiently high volume such that there is always a queue present (i.e. the east approach is *saturated*). The top flow diagram in Figure 4.4 shows the observed flow at the stop line for one lane of traffic on each approach. Once the signal changes to green on one approach, traffic starts to move through the intersection: the rate of discharge (flow rate) builds up rapidly to a value of about 0.5 veh/s (1800 veh/h, implying an average headway of two seconds between successive vehicles), at which it levels off. Flow at this rate will continue across the stop line of the approach until:

- the queue is cleared, after which time it would fall back to the arrival flow rate of traffic on the approach, or
- the end of the green period, when the flow will reduce to zero again, depending on which event happens first. The pattern is repeated in each cycle - one way to view this periodic phenomenon is as a series of pulses (waves) of traffic released through the signal.

Webster suggested that for saturated flows on an approach to a traffic signal, the observed flow profile as shown in the top diagram of Figure 4.4 could be replaced by a simple rectangular volume-time profile, as indicated in the bottom section of the figure. The area of the rectangle is equated to the area under the observed profile. Now, as the height of the rectangle is determined by the saturation flow, the width of the rectangle is taken as the *effective green time* (g_A) for stage (phase) A.

Theories of Interrupted Traffic Flow 89

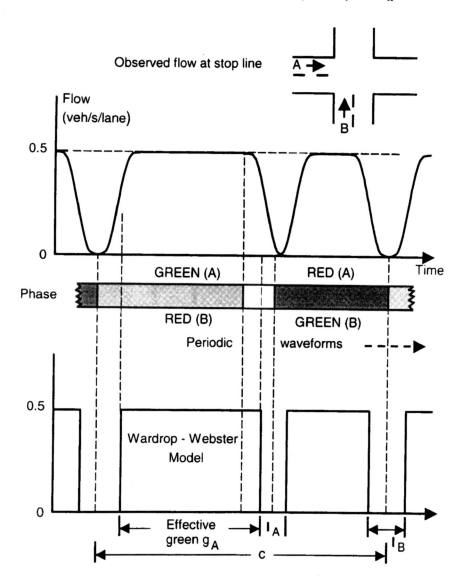

Figure 4.4 Basic Wardrop-Webster model for a junction

There is a second effective green time (g_B) for phase B. But note from the lower part of Figure 4.4 that g_A and g_B do not add up to the cycle time (c). There is an amount of lost time at the end of each phase. In reality,

90 Understanding Traffic Systems

this lost time is useful because it provides a buffer between the major flow movements (that can be used, for example, to allow vehicles on one road to clear the intersection before the cross-traffic movement starts, and/or to allow vehicles making opposed turns to clear the intersection). The amounts of lost time for each phase in Figure 4.4 are l_A and l_B. Total lost time (L) is an important parameter for the intersection. This time is defined as the difference between the cycle time and the sum of the effective green times for all of the stages (phases):

$$L = c - \sum_i g_i = \sum_i l_i \qquad (4.22)$$

The amount of lost time depends on the number of stages (phases) and the intersection geometry.

The Wardrop-Webster model has gained world-wide acceptance and provides the basis of all traffic signal calculations. There are, however, some deficiencies in it, such as:
- opposed turn lanes with no separate turning phase allowed;
- a shared lane, i.e. one used by more than one movement (e.g. a lane shared by through traffic and turning traffic), and
- parking in the vicinity of an intersection, where there may be short lengths of kerbside lane available for traffic flow near the stop lines, but these lanes are then blocked further away from the intersection.

A number of modifications and extensions to the basic theory have been made to overcome these problems (e.g. Akcelik, 1981), whilst many of the software packages for signal analysis extend the model further, to cater for cases involving shared lanes and multiple movement phases.

4.3.3 Capacity of one movement

The capacity of a movement is the maximum number of vehicles that can be discharged through that movement over an extended period of time. The maximum number of vehicles than can be discharged for one movement per cycle (of length c) is equal to the area of the flow rectangle (e.g. $s_A g_A$ for phase A in Figure 4.4 is the total number of vehicles that can be discharged from approach A during one cycle). Capacity is normally expressed as a flow rate (in veh/s or veh/h) so that, in general, the capacity (C_i) of movement i is given as:

$$C_i = \frac{s_i g_i}{c} = s_i u_i \qquad (4.23)$$

The ratio $u_i = g/c$ is the proportion of time that movement i has the green signal. Now consider the relationship between the arrival rate of traffic (the traffic demand) on a movement and the departure rate. If the arrival flow rate for the movement is q_i, then the number of vehicles arriving per cycle is $q_i c$. For the signals to function satisfactorily (i.e. be able to cope with the demand), adequate capacity must be provided to clear the vehicles arriving in one cycle, i.e.

$$C_i = s_i g_i \geq q_i c \qquad (4.24)$$

and it is relation (4.24) that provides the first clue as to what the green time should be.

4.3.4 Capacity of entire intersection

The overall capacity of the intersection is found by determining the capacities of the critical movement phases in a full cycle. First consider the simple two-phase system at a cross-road, as seen in Figure 4.4, looking at each of the approach roads (or movements) that operate in the signal cycle (Figure 4.5).

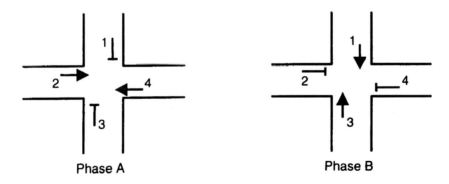

Figure 4.5 Simple phasing arrangement at a cross intersection

92 Understanding Traffic Systems

Let q_i be the flow on approach (movement) i. The green time for Phase A is g_{24}, and the green time for Phase B is g_{13}. From relation (4.24) we can write the following set of inequalities that must be satisfied if the intersection is to have sufficient capacity to handle the traffic demand:

Phase A:

$$s_2 g_{24} \geq q_2 c$$
$$s_4 g_{24} \geq q_4 c \tag{4.25}$$

Phase B:

$$s_1 g_{13} \geq q_1 c$$
$$s_3 g_{13} \geq q_3 c \tag{4.26}$$

subject to the constraint that:

$$c = g_{13} + g_{24} + L \tag{4.27}$$

where L is the total lost time. Note that there are two inequalities to be satisfied in each of the phases A and B. Within each phase (stage), one inequality will dominate, and the other will be redundant. The dominant inequality for a stage defines the critical movement for that stage. This result will apply in the general case, where there will be N movements in any one stage. The critical movement will be the one with the dominant inequality, and there will be $(N - 1)$ redundant inequalities. Now define new variables: (1) (the 'y-value') for each movement, where $y_i = q_i/s_i$, and (2) (the 'u-value') for each movement, where $u_i = g_i/c$. Then we can rewrite relation (4.25) for phase A as:

$$u_A = \frac{g_{24}}{c} \geq y_2$$
$$u_A = \frac{g_{24}}{c} \geq y_4 \tag{4.28}$$

and relation (4.26) for phase B as:

$$u_A = \frac{g_{24}}{c} \geq y_2$$
$$u_A = \frac{g_{24}}{c} \geq y_4 \qquad (4.29)$$

Then we select the largest y-value for each phase as the representative value for the phase, and this identifies the critical movement, i.e. $y_A = \max\{y_2, y_4\}$ and $y_B = \max\{y_1, y_3\}$. Providing adequate capacity for each critical movement will also provide enough capacity for all of the other movements in each stage. Note that the critical movement is not necessarily the one with the heaviest traffic flow. It is the ratio of traffic flow to saturation flow that is important. The constraint equation (4.27) may now be rewritten as:

$$c = g_{13} + g_{24} + L \geq L + c(y_A + y_B)$$

which may be rearranged into:

$$c \geq \frac{L}{1-(y_A+y_B)}$$

In general, when there are n stages (phases) in a cycle, the *practical cycle time* is given by

$$c \geq \frac{L}{1-\sum_i y_i} = \frac{L}{1-Y} \qquad (4.30)$$

where $Y = \sum y_i$. Now, $Y \leq 1$, otherwise it will not be possible to find signal settings that provide sufficient capacity for the intersection. Rearranging relation (4.30), we find that

$$Y \leq 1 - \frac{L}{c}$$

which suggests that Y will always be less than one, for L is always positive, never zero. In practice $Y \leq 0.75$, otherwise operation of the intersection will be unsatisfactory. The higher the value of Y, the higher the degree of

saturation, and hence the greater the delays expected. A number of measures might be considered to reduce an excessively high Y-value, by reducing the y-value(s) on one (or more) movements, for example:
- providing extra lanes (or greater road width) for a through movement;
- providing separate turn phases;
- banning parking near the intersection;
- banning one or more of the turning movements, especially those turns to be made against an opposing traffic stream, and
- providing slip lanes for unopposed turns, or allow such turns during the red period when safe to do so.

4.3.5 Capacity analysis

The capacity of one movement is given by sg/c. Thus we can increase the capacity of that movement by:
- increasing the g/c ratio. This will boost the capacity of the movement but may take capacity away from other movements;
- increasing cycle time (c) whilst keeping the g/c ratio constant. The total effective green time (G) = Σg_i and $G = c - L$. As lost time (L) is a constant for a given signal timing design (one complete cycle), fewer cycles per hour will mean less lost time per hour. However, increased cycle times will mean increased delays for all movements, and
- increasing the saturation flow (s), e.g. with more but narrower lanes.

4.3.6 Degree of saturation

The degree of saturation of a movement (x_i) and of the intersection (x_p) are the fundamental design parameters in traffic signals design. The movement degree of saturation is defined as the ratio of the traffic demand on the movement (the number of vehicles arriving a cycle) to the capacity of the movement (the maximum number of vehicles that could be discharged during the green period in the cycle):

$$x_i = \frac{q_i}{C_i} = \frac{q_i c}{s_i g_i} = \frac{y_i}{u_i} \qquad (4.31)$$

and the intersection degree of saturation x_p = max$\{x_i\}$, i.e. the maximum of the movement degrees of saturation. For design, a maximum value of 0.9 is often put on x_p.

4.3.7 Saturation flows

The saturation flow is the maximum instantaneous flow rate (possible during green time only). The success of the signal design process depends on the correct choice of saturation flow and lost time. The primary factors affecting saturation flow include the environment class of the intersection, lane type, lane width, gradient and traffic composition.

UK practice for estimating saturation flows requires the transformation of vehicle flows into equivalent flows of passenger car units (pcu), using the following steps:
- assign pcu equivalent factors to compute saturation flows in pcu/h. The pcu equivalents are 1.0 for light vehicles, 1.5 for medium commercial vehicles (defined as vehicles with two axles but more than four wheels), 2.3 for heavy commercial vehicles (those with more than two axles), 2.0 for buses and coaches, 0.4 for motorcycles, and 0.2 for bicycles. The basic saturation flow (s_h) is taken to be 2080 pcu/h (for a lane 3.2 m wide, non-nearside);
- lane saturation flow depends on the proportion of turning traffic (f), radius of turn (r metres), gradient (G, δ_G = 1/0 (uphill/downhill)), lane position (δ_n = 1/0 (nearside/other)), and lane width (w metres). Modifying factors are applied for each of these. The saturation flow of an approach is the sum of the lane saturation flows;
- apply the modifying factors to give the overall result for the saturation flow of a single unopposed lane as:

$$s(r,f,n,G,w) = \frac{2080 - 140\delta_n - 42\delta_G G + 100(w - 3.25)}{1 + 1.5\frac{f}{r}} \quad (4.32)$$

- for opposed movements in a lane, the saturation flow $s = s_g + s_c$ where s_g is for the departure of vehicles during the effective green time and s_c is for departures immediately after the end of the effective green time. These component saturation flows are defined by equation (4.33):

$$s_g = \frac{s_o - 230}{1 + f(T-1)}$$

$$s_c = P(1 + N_s)(fX_o)^{0.2} \frac{3600}{uc}$$

(4.33)

where $T = 1 + 1.5/r + t_1/t_2$, $t_1 = 12X_0^2/(1 + 0.6(1 - f)N_s)$, $t_2 = 1 - b(1 - f)N_s$. X_o is the traffic intensity in the opposing direction (= $q_o c/s_o g$), s_o is the saturation flow for the opposing direction, f is the proportion of right-turning traffic, N_s is the number of storage spaces within the intersection that right-turners can use without blocking through traffic, c (seconds) is the cycle time, u is the g/c ratio, and P is a conversion factor (for transforming veh/h into pcu/h). P is defined as:

$$P = 1 + \sum_i p_i(\alpha_i - 1)$$

where p_i is the proportion of type i vehicles and α_i is the corresponding pcu equivalent. A full description of the UK method is given by Kimber, McDonald and Hounsell (1986).

Australian practice for estimating saturation flow uses the following steps:
- for each lane allocated to a given movement, choose a base saturation flow (s_b) value on the basis of environment class and lane type (see Table 4.1). This base saturation flow will be in units of *through car units per hour* (tcu/h);
- adjust this base saturation flow to allow for the various factors that impinge on the particular movement (these will include effects of intersection geometry, gradient and traffic composition), to obtain an estimate of lane saturation flow in units of *vehicle per hour* (veh/h), and
- add the lane saturation flows to determine the total saturation flow of the movement.

Table 4.1 Base lane saturation flows (s_b tcu/h) for Australian conditions

Environment class	Lane type		
	Type 1	Type 2	Type 3
A	1 850	1 810	1 700
B	1 700	1 670	1 570
C	1 580	1 550	1 270

Table 4.1 shows the values of base saturation flows. The two dimensions of the table are the *intersection environment class* and the *lane type*. Three classes of intersection environment are defined, as follows.
- *Class A*: *ideal* or near ideal conditions for free movement of vehicles on both approach and departures sides of the movements, good visibility, little interference from pedestrians or parked vehicles. This environment is typically that occurring in a suburban residential or parkland setting.
- *Class B*: *average* conditions, which means adequate intersection geometry, small to moderate numbers of pedestrians, some interference by parked vehicles or goods vehicles. These conditions might be expected in an industrial or shopping area.
- *Class C*: *poor* conditions, involving large numbers of pedestrians and considerable interference from parked vehicles, goods deliveries and perhaps buses. Restrictions on visibility could also be expected. Such conditions are typical of central city precincts.

Lane type is also defined as three categories:
- *Type 1*: *through lane*, containing only through vehicles
- *Type 2*: *turning lane*, containing any type of turning traffic (exclusive left turn or right turn lane, or a shared lane from which vehicles may turn left or right or continue straight through. Adequate turning radius and negligible pedestrian interference to turning vehicles
- *Type 3*: *restricted lane*, as for Type 2, except that turning vehicles face a small turning radius or some pedestrian interference.

The base saturation flows from Table 4.1 need to be adjusted to fit the particular characteristics of a given intersection, to allow for factors such as lane width, gradient and traffic composition. Adjustments are made using adjustment factors for the lane width (f_w), the gradient (f_g) and the traffic composition (f_c). These are defined as follows:

$$f_w = 1 \text{ if } 3.0 \leq w \leq 3.7,$$

$$f_w = 0.55 + 0.14w \text{ if } w < 3.0$$

$$f_w = 0.83 + 0.05w \text{ if } w > 3.7.$$

$$f_G = 1 \pm \frac{G_r}{200}$$

where G_r is the per cent gradient. Use $+G_r$ for downhill gradients, to increase the saturation flow, and $-G_r$ for uphill gradients, to decrease the saturation flow, and

$$f_c = \frac{1}{q} \sum_m e_m q_m$$

where q_m is the flow rate (veh/h) in the movement for turn or vehicle type m, q (veh/h) is the total vehicle flow rate for the movement, i.e. $q = \Sigma q_m$, and e_m is the *through car equivalent* for turn or vehicle type m. The 'unit' for e_m is tcu/veh, i.e. the number of through car units equivalent to a single type m vehicle.

The through-car-equivalent factors are given as:
- *opposed* left turning cars, $e_{LT} = 1.25$;
- *unopposed* left turn, $e_{ULT} = 1.00$;
- *unopposed* right turning cars, $e_{URT} = 1.00$ [except for a one-way street where the value of $e_{URT} = 1.25$, and
- *opposed* right turning cars. This requires knowledge of the opposing through movement and the amount of green time that may be available. The opposed right turn equivalent e_{RT} is given by:

$$e_{RT} = \frac{0.5g}{s_u g_u + n_f}$$

where q is the green time for the movement with the opposed turn, s_u is the opposed turn saturation flow (see Akcelik (1981) for methods of estimating s_u), and g_u is the *unsaturated* part of the opposing movement green time. This time is given by:

$$g_u = \frac{sg - qc}{s - q}$$

The term n_f represents the number of turning vehicles from the movement who depart the movement *after* the green time, and
- trucks and buses, $e_t = 2.0$ cars/truck, except that $e_t = e_{RT} + 1$ for an opposed right turn.

The adjusted saturation flow for a lane or movement (s veh/h or veh/s) is then given by:

$$s = \frac{f_w f_G}{f_c} s_b \qquad (4.34)$$

The saturation flows for each lane in a movement are then summed to yield the total saturation flow.

4.3.8 Measures of performance

A number of measures of performance can be defined to help assess the worth of a particular signal design. These fall into two broad classes:
- the performance of traffic using the intersection (e.g. the delays experienced by the traffic), and
- the performance of the traffic control system and the junction itself (e.g. the amount of idle time, such as time that any one movement has the green but there is no traffic to take advantage of it).

An assessment of performance in terms of traffic behaviour may include considerations of: delay time (e.g. stopped delay, time in queue, and maximum delay), number of stops, queue length, pedestrian delay, fuel usage and pollutant emissions, and 'overflow' queuing (the presence of a residual queue at the end of a *green* period). Sometimes, composite measures of performance are used, for example SIDRA provides a *Performance Index* combining delays, number of stops, fuel consumption and queue length.

In terms of junction performance, important measures include: the degree of saturation (x_p), intersection flow ratio (Y), intersection green time ratio ($U = \Sigma u_i = G/c$, where $G = \Sigma g_i$ is the total amount of green time for the intersection), and cycle failure rate (the proportion of cycles that do not clear queues). Queue length is an important performance measure, e.g. if queues build up so far that they begin to block neighbouring junctions. Queue management is an increasingly important concern in urban traffic control.

Equation (4.30) defined a practical cycle time, by indicating the minimum cycle time required to provide adequate capacity for the intersection. If optimum performance of an isolated intersection is sought, e.g. for minimum delay, fuel consumption or vehicle operating cost, then Akcelik (1981) indicates that the *optimal cycle time* (c_o) is given by

$$c_o = \frac{(1.4 + k)L + 6}{1 - Y} \qquad (4.35)$$

where $k = 0.4$ for minimum fuel consumption, $k = 0.2$ for minimum operating cost, and $k = 0$ for minimum delay. Note that a different optimal cycle time can exist for each of these measures of performance. A further operational constraint on cycle time is that c should not be too long. Excessive cycle times (e.g. 150 seconds or longer) may cause drivers to suspect a fault with the signals, and then to start to disregard them!

The *formation of queues* (and thus the *imposition of delay* on an least some of the traffic) is an integral part of signalisation. Each cycle must contain some time when each movement is prevented from travelling through the intersection, although traffic may continue to arrive to make that movement. Thus queues build up. Further, once the signal changes to green, only the vehicles at the head of the queue can start moving. A finite amount of time that will elapse before vehicles at the back of the queue can move off. New arrivals during the green also have to join the back of the queue, unless this has cleared. The trajectory diagram in Figure 4.6 provides a simple illustration of this process.

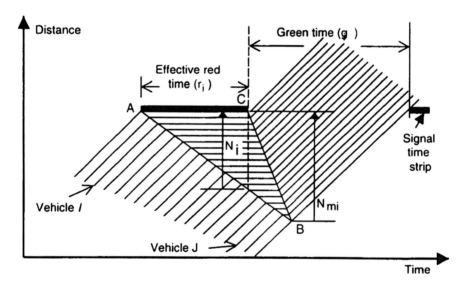

Figure 4.6 Trajectory diagram of arrivals and departures for one movement

Consider the trajectory of vehicle *I* in Figure 4.6. This vehicle drives up to the intersection during the red time (r_i) for the movement, where r_i is indicated by the thick horizontal line AC. Vehicle *I* joins the end of the existing queue. When the signal changes to green, the queue starts to move, with vehicles being discharged at a constant rate (the saturation flow.) Thus vehicle *I* starts to move forward some time after the start of the green phase.

The horizontal line segment in the trajectory for vehicle I is the delay time experienced by that vehicle, and the total area of the *delay triangle* ACB in Figure 4.6 is the total delay experienced by all of the vehicles in the queue. The queue length at the start of the green period is $N_i = r_i q_i$.

Now Figure 4.6 indicates the continued existence of the queue some time after the end of the red period. For instance, vehicle J arrives after the start of the green period, but still has to join the queue. The total number of vehicles that pass through the queue formed as a result of the red period is N_{mi}, which is known as the *back of queue*:

$$N_{mi} = \frac{N_i}{1 - y_i}$$

An alternative representation of the queuing process at a signal is given in Figure 4.7. The top part of this figure shows the flow rate of arrivals and departures over time, the bottom part shows the cumulative numbers of arrivals and departures. The queue length $N(t)$ at time t is given by the difference between the cumulative number of arrivals and the cumulative number of departures. Thus N_i is the vertical distance between the two cumulative flow lines at the start of the green period. Notice the step form of the departure flow in Figure 4.7. This indicates that the movement is undersaturated ($x_i < 1$). Discharge at the saturation flow rate only continues until the queue has cleared. Once the queue has cleared, the subsequent arrivals to the movement during the remainder of the green proceed straight through the intersection, and the departure rate for that residual green period is equal to the arrival rate q_i. In the limit (as $x_i \to 1$), the time at which the queue clears approaches the finish of the green period. The point of time during the green interval at which the queue will clear can be obtained by equating the areas of rectangles OABC and FEDB in Figure 4.7. The back of queue is given by

$$N_{mi} = N_i + \Delta N$$

Total delay time over the time interval 0 to t incurred by vehicles on the movement is given by

$$D = \int_0^t N(z) dz$$

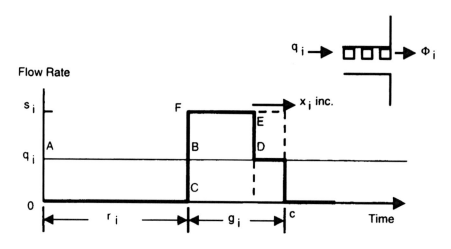

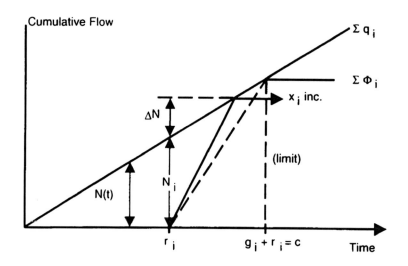

N(t) = queue length at time t

N_i = queue length at start of green

Figure 4.7 Queue formation by comparison of arrival and discharge rates

The mean uniform delay per vehicle (for uniform arrivals and the Wardrop-Webster model) is then

$$d_u = \frac{r_i^2}{2c(1 - y_i)} \qquad (4.36)$$

This average delay term is an *under-estimate* of the total real world delay as it does not account for random fluctuations in the arrival rate.

A component of 'overflow' delay (d_f) to account for these fluctuations is thus usually added to d_u, to yield a total mean delay (d):

$$d = d_u + d_f \qquad (4.37)$$

An important consideration in determining d_f is the time period (T_f) over which the traffic demand (q) persists. Average delays are expected to be higher, the longer that high volume demand occurs at an intersection. This results from the differences in volume between peak and off-peak conditions, and the greater opportunity for overflow queues to occur as time period (and hence number of cycles) increases. Akcelik (1981) gave the following expression for d_f:

$$d_f = \frac{1}{4} T_f \left\{ (x - 1) + \sqrt{(x - 1)^2 + \frac{12}{QT_f}(x - 0.67 - s\frac{g}{100})} \right\} \qquad (4.38)$$

where s is in veh/h and g is in seconds. Equation (4.38) applies for $x \geq 0.7$. A number of similar expressions for d_f have been proposed, and Burrow (1989) showed that these are all special cases of a general equation

$$d_f = \frac{1}{4} T_f x^n \left\{ (x - 1) + \alpha + \sqrt{(x - 1)^2 + \frac{m}{QT_f}(x + \beta)} \right\} \qquad (4.39)$$

with parameters n, α, m and β defined by Table 4.2.

The average *number of stops* per vehicle for a movement, including both vehicles that join the queue or are able to pass unimpeded, is given by h_i where

$$h_i = \frac{N_{mi}}{c\,q_i} = \frac{1 - u_i}{1 - y_i}$$

Now this is a theoretical result and we know that, in practice, many drivers will avoid coming to a full stop by decelerating well before reaching the back of the queue, so that they never quite come to rest. Thus Akcelik (1981) suggested applying a correction factor to cope with these 'partial stops'. This correction is:

$$h_i = \frac{0.9(1 - u_i)}{1 - y_i} \qquad (4.40)$$

Table 4.2   Burrow's (1989) comparison of traffic signal overflow delay expressions

Expression	n	m	α	β
US HCM	2	4	0	0
Australian	0	12	0	-(0.67+sg/100)
Canadian	0	4	0	0
Transyt-8	-1	4	0	0
Alternative HCM	0	8	0	-0.5
TRRL	0	4χ	-2χ/QT$_f$	1+χ/QT$_f$

Note: 'Alternative HCM' is that proposed by Akcelik (1988), whilst the parameter χ in the TRRL expression takes a value dependent on the arrival and departure flow patterns at the junction (Burrow, 1989).

4.4 Link congestion functions

An important application of interrupted flow theory is to describe the overall traffic performance of an element (e.g. a route, link or junction) in a traffic network. The relationship between the amount of traffic using a network element and the travel time and delay incurred on that element is important. In Chapter 5 we will see how such relationships are used, in models of traffic network performance, in impact assessment, and in estimating the effects of tolls and charges. The discussions of waiting times and delays based on gap acceptance, queuing theory and traffic signals operations presented earlier in this chapter provide the basis for these considerations. The total travel time to traverse a network element is directly related to the traffic volume using that element. As volume increases, so delay, and hence travel time, increases. The rate of increase in travel time accelerates as volume approaches the capacity of the element. Previous discussion has focussed on the junction as the network element. Now we need to consider the network link.

Traffic movement along a link in a network may be seen as consisting of two components. The first component is cruising, with traffic moving along the link largely uninterrupted (except for the possibility of side friction, say due to vehicle parking manoeuvres). Travel along the link may also be punctuated by points of interruption, such as pedestrian crossings, bus stops and, most importantly, road junctions. For example, the junction at the downstream end of the link may dictate the traffic progression along the link. Movement through the interruption points can be handled using the methods for intersection analysis and queuing theory described previously. What is also needed for many transport planning and traffic impact analyses, particularly for urban areas or other places where congestion is expected, is a composite relationship that can include the two components simultaneously. A *congestion function* (or *speed-flow relation*) may be used to describe the relationship between link flow and speed or travel time on a network link or road section containing a set of network elements, as in Figure 4.8.

The most convenient way to represent a congestion function is in terms of the travel time on a link, e.g. $t = f(q,\beta)$ where t is the travel time on the link when it is carrying traffic at a flow rate of q, and the vector β represents a set of parameters that describe the characteristics of the link. The function starts with a finite travel time (t_o) at zero flow, and the actual travel time then increases with volume. The rate of increase is small for low volumes, but accelerates once volumes build up towards the capacity of the link. The excess travel time for finite volumes above t_o may be taken as a measure of the 'system delay' (see Section 11.2) on the link, and reflects the state of congestion on the link. A typical congestion function is that developed by Davidson (1978):

$$t = t_0\left(1 + \frac{Jq}{C-q}\right) \quad (4.41)$$

in which J is an environmental parameter that reflects road type, design standard and abutting land use development, and C is the absolute capacity for the link.

106 *Understanding Traffic Systems*

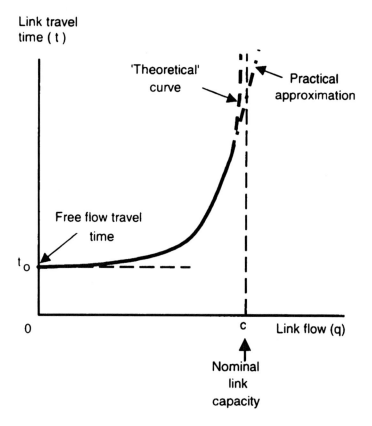

Figure 4.8   **Typical form of a link congestion function**

The Davidson function has proved popular in economic analysis and travel demand modelling for road networks, mainly because of its flexibility, and its ability to suit a wide range of traffic conditions and road environments. However, the original form (equation (4.41)) has one serious flaw; it cannot define a travel time for link volumes which exceed the capacity (*C*). This can provide computational problems in (say) a traffic network model which determines link volumes in an iterative manner, and may consequently occasionally overload some links in computing its intermediate solutions. A modification involving the addition of a linear extension term as a second component to the function has thus been proposed and used in transport planning practice (Tisato, 1991). The modified Davidson function is then:

$$t = t_0\left(1 + \frac{Jq}{C-q}\right) \qquad q \leq \mu C$$

(4.42)

$$t = t_0\left(1 + \frac{J\mu}{1-\mu} + \frac{J(q-\mu C)}{C(1-\mu)^2}\right) \qquad q > \mu C$$

where μ is a user-selected proportion, usually in the range (0.85, 0.95), as discussed by Taylor (1984). This proportion sets a value of q after which the travel time increases as a linear function of q, and this removes the computational difficulties associated with the original function. It should be noted that all of the above congestion functions are 'steady-state' functions, in that they are based on an assumption that the flow q will persist indefinitely.

More recently Akcelik (1991) compared the Davidson function to the delay equations found in traffic signals analysis, and proposed a new link congestion function, better able to model link travel time when intersection delay provides a significant part of the total link travel time. Further, this function also has a 'time-dependent' form as well as the steady state form. The time period over which the volume q (the 'travel demand' to use the link) is maintained has an important bearing on the level of delay experienced. The longer this time period T_f persists, the higher the delays will be. Further, the time-dependent equation is designed to cope with periods of oversaturation, as defined by a degree of saturation (or volume/capacity ratio) $x = q/C$ which is greater than one. In this way oversaturation is regarded as a normal condition, which may exist for a finite time period as it does in the real world. The steady-state form of Akcelik's congestion function is

$$t = t_0\left(1 + \frac{Ax}{Ct_0(1-x)}\right) \qquad (4.43)$$

where A is Akcelik's environmental delay coefficient defined as $A = I\kappa$ where I is a factor representing the intensity of delay elements along the link (e.g. the junction density in intersections per kilometre) and κ is a delay parameter reflecting the level of randomness (or regularity) of the arrival and service processes at the interruption points along the link. Appropriate values for κ might be $\kappa = 0.6$ for isolated traffic signals, $\kappa = 0.3$ for coordinated signals, and $\kappa = 1.0$ for roundabouts or other unsignalised intersections.

The time-dependent form of Akcelik's congestion function is

$$t = t_0 + \frac{1}{4}T_f\left\{(x-1) + \sqrt{(x-1)^2 + \frac{8A}{CT_f}x}\right\}$$

(4.44)

$$t = t_0\left\{1 + \frac{1}{4}r_f\left[(x-1) + \sqrt{(x-1)^2 + \frac{8A}{Ct_0r_f}x}\right]\right\}$$

where r_f is the ratio of the flow period (T_f) to the minimum travel time t_o on the link. Akcelik (1991) provided a set of representative parameter values for use with this model, as shown in Table 4.3. The column t_m/t_o in Table 4.3 is the ratio of link travel time (t_m) for $x = 1$ to the zero-flow travel time t_o. This ratio provides an indication of the additional travel time and delay on the link when it is saturated, and may be taken as one measure of the effects of congestion on that class of road. For instance, the value of 1.587 for a freeway suggests that the travel time at saturation is about 59 per cent higher than the zero-flow travel time, whereas on an arterial road with interruption points (e.g. traffic signals) along it, this percentage increase is about 104 per cent, meaning that the travel time has slightly more than doubled.

Table 4.3   Representative parameters for Akcelik's congestion function

Road class	Description	t_0 (min/km)	C (veh/h/lane)	A	t_m/t_0
1	freeway	0.50	2 000	0.1	1.587
2	arterial (uninterrupted)	0.60	1 800	0.2	1.764
3	arterial (interrupted)	0.75	1 200	0.4	2.041
4	secondary (uninterrupted)	1.00	900	0.8	2.272
5	secondary (high friction)	1.50	600	1.6	2.439

5 Theories of area-wide traffic flow

Major changes to a traffic system, such as the introduction of a new traffic generator or the imposition of a new traffic management scheme, may have effects on traffic movement and traffic impacts at locations far away from the sites where the changes are implemented. The optimum passage of vehicles through a network may be sought by coordinating the traffic signal timings at successive intersections so that vehicles released through one intersection will arrive at the downstream intersection just as its signals turn to green. Thus traffic analysts need to consider the area-wide, or network, aspects of traffic flow and travel demand. This chapter introduces some of the theories that may be applied in traffic impact analysis at the network level. Traffic network analysis is important for three main reasons:
- the need for network considerations in urban traffic control systems;
- the effects of changes in travel demand on network performance, and
- the multimodal nature of the transport systems in most cities.

Urban traffic control must take a network perspective. A control system such as SCOOT (the UK Split, Cycle, and Offset Optimisation Technique, see Hunt *et al* (1981)) or the Sydney Coordinated Adaptive Traffic System (SCATS, see Sims and Dobinson (1979)) attempts to coordinate signal phasings at the intersections along routes through the network, so that traffic progression can be expedited. The specific objective is to allow the smooth passage of platoons of vehicles along major routes with a minimum of stops, and the preservation of the platoons. The capacity of the intersection is a property of the intersection alone (although the cycle time might be set by capacity considerations at some nearby 'critical' intersection). Achieving the best utilisation of the available capacity requires coordination between intersections, so that platoons on intersecting roads arrive in sequence to use the green times on the respective approaches.

Changes to the signal phasing and geometry for different movements at one intersection (i.e. changes to the transport infrastructure *supply*) may alter the pattern of traffic usage (the travel *demand*) of the intersection and the surrounding network. For instance, if a turn is banned at one junction, then those drivers who previously made that turn at the intersection will have to change their routes, if by no more than making their turn at a neighbouring intersection. Perhaps they may undertake a more radical change of route and avoid the area entirely? Likewise, improving the facilities for a turn, such as providing a separate turning phase and/or extra turning lanes, may attract traffic to the intersection from neighbouring junctions.

Urban transport systems are multimodal, and consideration must be given to other modes of transport such as public transport, cyclists and pedestrians. In addition, the destination choices of travellers may be made from amongst a set of alternatives (e.g. where to go shopping). We need to know what people's travel needs and patterns are, how much usage will be made of all the modes, how destination choice will depend on the relative accessibility of alternative destinations, what impact congestion will have on the choice of mode and route and the timing of trips, and what interactions will be observed in different parts of the network. For instance, there may be a need to favour one particular mode in one region (say pedestrians in the central city), or to provide interchange facilities so that people can move easily between modes (e.g. bus-rail interchanges at node points in the suburbs).

The application of Intelligent Transport Systems (ITS) technologies to transport networks, such as electronic road pricing (ERP) schemes or real time incident detection and incident management systems is further reason for taking the network perspective (CTS, 1998).

5.1 Principles of network analysis

Traffic network analysis aims to describe or forecast the distribution of traffic flows over a road network in a given time period or over a set of time periods. Modelling is usually undertaken in one of two main stages: in the *description* of an existing traffic system, where the aim would be to calibrate or evaluate a model, and in the *synthesis* of a future or alternative system (e.g. a proposed traffic management plan), where a calibrated model would be applied to test the performance or impact of the future or proposed system. Analysis and modelling may be applied in a series of steps, with the following five steps and the basic question each seeks to answer being:

(1) *trip generation* - shall I travel?
(2) *trip distribution* - where shall I go?

(3) *trip timing* - when shall I travel?
(4) *modal choice* - how shall I travel?
(5) *trip assignment* - which way shall I go?

The sequence given by steps (1), (2), (4) and (5) forms the traditional 'four-step' (sic) procedure which has been applied in transport planning for many years (usually in the context of daily flows on urban transport networks). Question 3, regarding the timing of trips, was traditionally ignored (hence 'four-steps') but becomes important when flows over time of day are considered, e.g. the relationships between peak period and off-peak travel and the influences of congestion levels on travel behaviour. Although the five steps may be seen as a logical progression of the travel-related decisions to be made by an individual, the true decision making framework is more complicated. The decisions are in fact unlikely to ever follow in the strict order implied by the above sequence and the outcomes will more often involve a complex set of interrelated decisions including whether or not to engage in an activity at another location (e.g. Jones et al (1983), Ettema and Timmermans (1998)). Many model packages have now dispensed with the sequential structure but the five questions still provide a useful indication of the principal dimensions to be considered in an analysis. Further consideration can also be given to models of land use-transport-environment interaction (e.g. Hayashi and Roy, 1996).

5.1.1 Trip generation

Trip generation seeks to define the numbers of trips which start (originate) at one site and finish at another. [In transport planning, a trip is defined as *the one way movement from an origin to a destination* and therefore there are two *trip ends* associated with each trip (one at the origin and one at the destination).] A narrow but practical definition of trip generation useful in traffic impact analysis is that it is the measured level of traffic activity associated with a given site, development or land use. Usually (in practice) the emphasis is on vehicular traffic, but this need not be so. Trip production (P_i) is the number of trips which start (are produced) at a given site (i). Trip attraction (A_j) is the number of trips finishing (attracted to) site j. The total of the productions and attractions for a given site is termed the *trip end total*.

Trip productions from residential areas are usually modelled in terms of average trip rates, or by the use of 'category analysis' tables, which classify the trip rates by household variables such as household size, income, car ownership and access, etc. Trip attractions (and trip productions from non-residential land uses) are usually modelled by regression relationships,

112 *Understanding Traffic Systems*

with the independent variables selected from the physical or economic characteristics of the land use. For example, trips attracted to a shopping centre might be related to the floor area of the centre (see Chapter 13).

5.1.2 Trip distribution

In transport network analysis trip distribution follows trip generation, as it is concerned with forming linkages between the trip productions (P_i) and trip attractions (A_j) to form the trip matrix T_{ij}, the number of trips that will go from i to j. This matrix is known as the origin-destination trip matrix (O-D) matrix or sometimes as the demand matrix.

5.1.3 Trip timing

Modern transport and traffic planning is concerned with the timing of journeys during the hours of the day, for instance in terms of the duration of periods of peak demand, the opportunities for spreading peaks over longer time periods to lessen congestion levels and thus reduce travel costs (e.g. delays, excess fuel usage and emissions), and the influences of congestion levels on the travel behaviour of individuals. In addition, constraints may be placed on an individual's ability to change the times at which trips are made through constraints imposed by non-travel activities undertaken by that person (e.g. working hours) or by the interactions between the members of a household. Trip timing is often undertaken by proportioning of a daily O-D matrix (T_{ij}) between a set of time periods $\{t\}$ to yield a separate O-D matrix T_{ijt} for each time period.

The timing decision is now seen as a consequence of higher order activity scheduling decisions – see, for example, D'Este (1997) and Kitamura and Fujii (1998).

5.1.4 Modal choice

Modal choice analysis is used to predict the numbers of travellers who will use the different modes of transport available in an area, and how the patronage of the different modes will vary with changes to the transport and land use systems in that area. The most reliable models of modal choice attempt to predict the probability that a given person will choose a particular mode for a given journey. This probability will depend on the socio-economic characteristics of the individual, the characteristics of the available transport modes (e.g. fares, travel time, waiting time, reliability, comfort), and the proximity of the mode to the origin and destination of the individual's trip. The economic theory of consumer utility (e.g. Hensher and Johnson, 1981) forms the basis for most modal choice modelling.

Many models make use of the concept of a generalised cost of travel, in which all of the component costs of a journey are combined into a single overall value by applying weighting and conversion factors. Such components commonly include fares, parking charges, tolls, walking time to access the mode, waiting time, in-vehicle time, distance travelled, time taken to walk to destination after leaving the vehicle, vehicle operating costs, fuel costs, and others besides. If C_i is the generalised cost of using mode i for a journey, then

$$C_i = \sum_l \lambda_{ji} c_{ij} \qquad (5.1)$$

where c_{ij} are the components of travel cost for mode i. The λ_{ji} are weighting factors, applied to each of the components. The usual unit for generalised cost is money, so, for instance, there will need to be a weighting factor (money value of travel time) to convert each component of travel time into an equivalent amount of money. Different components of travel time are often distinguished because people tend to weight them differently. For instance, a rule of thumb is that public transport users value access (walking) times and waiting time at about twice the amount of in-vehicle time. The use of a generalised cost function implies that travellers will trade-off between the different components of travel cost when making their travel decisions. One output from modal choice modelling for use in traffic impact analysis would be the O-D matrix of trips T_{ijk} by mode k, with the possibility of determining matrices T_{ijkt} for mode k and time period t.

5.1.5 Trip assignment

Trip assignment modelling is used to determine the route choice of drivers or travellers, and to build up the vehicle or passenger flows on each link in the network. This is accomplished by taking the origin-destination matrices T_{ijkt} for trips by mode k in time interval t and routing these along paths through the network, and so accumulating the flows on each link or on each turning movement in the network. Trip assignment is the 'final' phase of the travel demand modelling process. Assignment models exist for private modes (e.g. cars, pedestrians) and for public modes (e.g. train and bus services).

5.2 The origin-destination matrix

As indicated throughout the above discussion on the principles of network analysis, the origin-destination trip matrix provides fundamental information for transport systems planning and impact analysis. It reveals the spatial pattern of trip making in the study region and, if trip timing information is available, temporal patterns may also emerge. The O-D matrix thus describes the pattern of travel in terms of the trip movements (travel desire) between a set of origin points and a set of destination points on the boundary of the region (the *cordon line*) or inside the region over a given time period. This matrix then defines the travel load to be borne in the study area at that time. Trips with origins and/or destinations inside the area are described as *local trips*, whilst trips with origins and destinations outside the study area (or on the cordon line) are described as *through trips*. The flows on the links of the network inside the area thus result from the routing of the trips in the O-D matrix along the network links. Figure 5.1 depicts through and local trips in a study area, as defined by an external cordon line drawn around the area.

Note that there are three different types of local trip that can occur, as seen in Figure 5.1: *external-internal* local trips, where the trip origin lies outside the cordon, *internal-external* local trips, where the trip destination lies outside the cordon, and *internal-internal* trips, where both the origin and destination of the trip lie inside the study area. Note that all trips except internal-internal local trips cross the cordon line, with through trips crossing the cordon twice. The internal-internal trips will never be observed in any traffic study based solely on observations on the cordon. Figure 5.1 shows a *screen line*. This line consists of a series of observation points inside the study area. Recording trips movements at the screen line will capture some of the internal-internal trips (those that pass through the screenline observation points) and also provides some information on the routes taken by vehicles inside the study area. Table 5.1 shows a typical O-D matrix structure subdivided by trip type (through or local).

The O-D matrix may be determined in three different ways:
- by *direct observation*, using traffic surveys (e.g. a registration plate survey) or questionnaire surveys (see Chapter 12);
- by *synthesis* from observed flows on the links in the network. The mathematical theory for matrix synthesis is introduced in Section 5.2.1. One simple application of the method is the Hauer-Kruithof method for estimating turning movement flows (see Section 11.1), and
- by *modelling*. Wilson's 'gravity' models of trip distribution are widely used for this purpose (Wilson, 1967).

Theories of Area-wide Traffic Flow 115

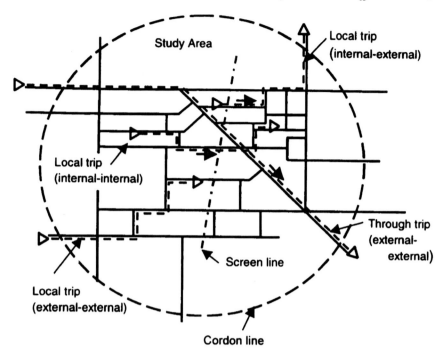

Figure 5.1 A traffic area defined by an external cordon line and showing types of through and local trips

Table 5.1 Typical structure of an O-D matrix for a traffic area

Origin \ Destination	External			Internal		
	1	...	M_1	1	...	M_2
External 1 .. N_1	Through trips (external-external)			Local trips (external-internal)		
Internal 1 .. N_2	Local trips (internal-external)			Local trips (internal-internal)		

Methods for the direct observation of O-D matrices are discussed in Chapter 12. These methods tend to be expensive, limited in scope and error-prone (unless great care is taken in their execution). The other methods allow an analyst to estimate the contents of the matrix from other traffic data, that may be easier and cheaper to collect, such as link volumes and trip generation volumes.

5.2.1 Synthesis of O-D matrices from link counts

Link volume data are relatively easy to collect at the wide scale, and thus offer a useful data source if an O-D matrix estimation procedure is available. There has been extensive interest in this field of analysis, with the work of Van Zuylen and Willumsen (1980) providing the impetus for the use of a method derived from the mathematical theory of information. The procedure may be summarised as follows. Assume that consistent traffic counts Q_e exist for a set of links in a network and that these counts result from the passage through that network of the travel demand defined by an O-D matrix T_{ij}}. The number of possible ways of selecting an O-D matrix $\{T_{ij}\}$ where the total number of trips $T = \Sigma T_{ij}$ is

$$W(\{T_{ij}\}) = \frac{T!}{\prod_{ij} T_{ij}} \qquad (5.2)$$

from the mathematical theory of information, the 'most probable' matrix $\{T_{ij}\}$ is found by maximising the function $W(\{T_{ij}\})$, or some monotonic function of W such as $S = \log_e(W)$, subject to a set of constraint equations that explain the relationships between the known data (the link volumes Q_e) and the desired information (the matrix contents T_{ij}). Using Stirling's approximation for $\log_e(X!)$, which is

$$\log_e(X!) \approx X \log_e(X) - X \qquad (5.3)$$

and noting that the total number of trips T is a constant, we can write

$$\max S = -\sum_{ij} T_{ij} \left(\log_e T_{ij} - 1 \right) \qquad (5.4)$$

subject to the constraints

$$Q_e \sum_{ij} T_{ij} p_{eij} = 0 \qquad (5.5)$$

for all counted links e. In the constraint equation (5.5) p_{eij} is the probability that a trip from i to j will use e. This probability may be found using a traffic assignment model (see Section 5.3). For the simplest case of an 'all-or-nothing' assignment in which all trips from i to j follow a single path through the network, p_{eij} will be a binary variable (0/1).

The mathematical programming problem defined by equations (5.4) and (5.5) is solved using the method of Lagrange multipliers. The solution is

$$T_{ij} = \prod_e X_e^{p_{eij}}$$

which is a multi-proportional problem for which a standard solution method is available to find the factors $\{X_e\}$.

One particular application of the method is where an old O-D matrix $\{t_{ij}\}$ exists, but needs to be updated, say by taking new traffic counts in the network. In this case additional prior information is available, in the form of the old matrix. Equation (5.2) can then be replaced by

$$W(\{T_{ij}\}) = \frac{T! \prod_{ij} \left(t_{ij} / \prod_{ij} T_{ij} \right)}{\prod_{ij} T_{ij}} \qquad (5.6)$$

from which the method of Lagrange multipliers indicates that

$$T_{ij} = t_{ij} \prod_e X_e^{p_{eij}} \qquad (5.7)$$

The form of the solution offered by equation (5.7) is the one usually applied; if an old O-D matrix $\{t_{ij}\}$ is not available then the t_{ij} are set to one in the equation. The following algorithm solves the problem:

Step 0: set the iteration counter n equal to one
Step 1: for each available link e set X_e^n equal to one
Step 2: set an initial value for the normalising factor Z^1, where

$$Z^n = \frac{\sum_e Q_e}{\sum_{eij} P_{eij} t_{eij}}$$

Step 3: for each available link e, calculate a value for the proportioning factor X_e^{n+1},

$$V_e = \sum_{ij} P_{eij} Z^n t_{ij} \prod_f (X_f^n)^{p_{fij}}$$

$$Y_e = Q_e / V_e$$

$$X_e^{n+1} = X_e^n Y_e$$

Step 4: Calculate a new value for the normalising factor,

$$Z^{n+1} = \frac{Z^n \sum_{ij} t_{ij} \prod_e (X_e^n)^{p_{eij}}}{\sum_{ij} t_{ij}}$$

Step 5: Check for convergence, by comparing the values of $\{X_e^{n+1}\}$ and $\{X_e^n\}$. If the process has converged, then go to step 6. Otherwise, return to step 3.

Step 6: Calculate the new trip matrix

$$T_{ij} = t_{ij} \prod_e (X_e^n)^{p_{eij}}$$

and finish.

5.2.2 Modelling of O-D matrices for known trip end totals

The general form of the gravity model of trip distribution is

$$T_{ij} = K P_i A_j f(c_{ij}) \qquad (5.8)$$

where K is a calibration constant, and $f(c_{ij})$ is an impedance function indicating the separation between i and j. The following alternative forms of the impedance function are in use:

- $f(c_{ij}) = c_{ij}^{-n}$, for which the choice of $n = 2$ gives an equation similar to Newton's law of gravitation;
- $f(c_{ij}) = exp(-\beta c_{ij})$, which is the functional form found from Information Theory (this gives the most probable O-D matrix). The value of the coefficient β is found by calibration from an observed trip matrix, and
- $f(c_{ij}) = c_{ij}^{-n} exp(-\beta c_{ij})$, which is a composite impedance function that can degenerate to each of the simpler forms of $f(c_{ij})$ by a suitable choice of the parameters n and β.

Two particular forms of the gravity model of trip distribution are in common use. These stem from different considerations on the available knowledge of the trip productions and attractions. For example, if only the trip productions P_i are known, then the singly-constrained gravity model can be used. For instance, this model can be applied to shopping travel. We know how many shoppers (shopping trips) will be made, but not necessarily which shopping centres the trips will go to (the shopping centres compete for the available shopping trade and the market shares of the centres are part of the model output).

The total number of trips T will be given by $T = \Sigma P_i$. We can use a proxy for the trip attraction of each shopping centre, say the size (floor area, F_j) of the centre, and perhaps assume that $A_j = F_j^{\gamma}$ where γ is a constant (typically $\gamma \approx 1.5$). Then we can rewrite equation (5.8) as

$$T_{ij} = a_i P_i A_j f(c_{ij})$$

which assumes that the calibration constant a_i depends only on the origin i.

Noting that $\Sigma_j A_j = T$, it follows that the distribution of shopping trips from origin zone i to each retail centre j is:

$$T_{ij} = \frac{P_i F_j^\gamma f(c_{ij})}{\sum_j P_i F_j^\gamma f(c_{ij})} \qquad (5.9)$$

which is a 'share' model (as it proportions each P_i between the alternative shopping centres).

In the case of journeys to work, we know both the numbers of workers (P_i) in each origin zone and the number of jobs (A_j) in each destination zone. Given a one-to-one correspondence between the total number of workers and the total number of jobs (any surplus workers are unemployed, any surplus jobs are vacancies), we can say that $\Sigma_i P_i = \Sigma_j A_j$ and $P_i = \Sigma_j T_{ij}$. Then the (doubly-constrained) gravity model of trip distribution is given by

$$T_{ij} = a_i b_j P_i A_j f(c_{ij})$$

where $\{a_i\}$ and $\{b_j\}$ are sets of origin-specific and destination-specific constants respectively. These constants can be shown to be

$$a_i = 1/\sum_j b_j A_j f(c_{ij})$$
$$b_j = 1/\sum_i a_i P_i f(c_{ij})$$

Although the algebraic formulation of this model is straightforward, its application is not necessarily simple. The reason may be seen in the two expressions for a_i and b_j given above: each depends on the other. Both sets of calibration constants $\{a_i\}$ and $\{b_j\}$ have to be found by iteration, and this may prove to involve significant computation.

5.3 Network flow modelling

Given one or more O-D matrices describing the travel demand pattern in an area, the next task is to load the trips contained in the matrices on to the network, to build up the levels of flows on the network links, from which travel times, delays, congestion levels and travel costs may be derived. In a network with congestion, there will be an interaction between the travel time and travel costs associated with traversing a network element and the traffic volume on that element. Part of this interaction was described in Section 4.4, which considered link congestion functions for estimating travel times from link volumes. The other part of the interaction is that drivers will attempt to seek routes which minimise some factor related to their travel time or travel costs, so that as volume builds up on a link and hence travel times increase on it, the drivers may become less inclined to use that link. Thus a feedback loop between volume and travel time (or cost) is established. The common strategy for the traffic analyst is to use models that seek an equilibrium point in this feedback loop. Equilibrium assignment involves the solution of a mathematical programming problem, in which the objective function is non-linear but the constraints are linear. The constraints are concerned with conservation of flows in the network, and are set to ensure that the travel demand described by the O-D matrices is satisfied. The objective function represents the strategy adopted by drivers in their route choices.

In broad terms, a traffic assignment model consists of three separate components:
- a set of *congestion functions* to determine the relationships between traffic volumes and travel times (travel costs) on a network, as described in Section 4.4;
- a strategy for selecting the 'best' route for a journey through the network, and
- an algorithm for finding shortest paths through a network.

5.3.1 Assignment strategies

The three commonly used strategies for assignment modelling are Wardrop's principles (Wardrop, 1952) and Jewell's principle (Jewell, 1967), which may be used to define a family of equilibrium assignment models. Different members of this family will have different objective functions in the mathematical programming formulation for the equilibrium assignment problem.

Under *Wardrop's first principle* the journey times on all of the routes used for travel between an origin and a destination will be equal at the equilibrium point, and will be less than those times which would be experienced on any other route. No individual driver can gain an advantage by a unilateral change of route. This strategy implies that drivers seek a route which minimises their individual travel times given that all drivers are attempting this strategy for themselves. The strategy is one of individual travel time optimisation, and it implies competition between drivers, who are all seeking the best outcomes for themselves independently of each other. However, the resulting pattern of flows on the network represents a stable equilibrium, for at the equilibrium solution no driver can gain from a unilateral deviation from that solution. To do so will incur a longer travel time, so no advantage is gained for the individual. The equilibrium assignment problem for Wardrop's first principle is known as *user travel time minimisation.*

Wardrop's second principle considers the overall minimisation of the travel task represented by the total travel time (vehicle-hours of travel, VHT) in the network. In this case drivers will select their routes to produce the minimum VHT which is necessary for the travel demand to be satisfied, i.e. for all of the trips in the O-D matrices to reach their destinations. This model is described as *system travel time minimisation.* The solution to this problem implies a degree of cooperation between drivers to attain this result. Although the total VHT will be less than that arising in the user travel time minimisation, some individual drivers will encounter much longer travel times than the minimum available to them. Should such drivers decide to improve their own situations, then the system-wide optimum solution will be lost, and there is no incentive (other than the ideal of cooperation for the overall gain of the community) for them not to do so. (The means of providing such an incentive is by the introduction of congestion pricing, as described in Section 5.5.3.) Without external intervention this solution is unstable. It does, however, define a datum in terms of the best distribution of flows that could occur if the overall minimisation of 'travel effort' (e.g. VHT) were to be achieved, and other solutions (e.g. for user travel time minimisation) may be compared to it on those grounds.

Jewell's principle is a generalisation of the two Wardrop principles, each of which can be seen as a special case of Jewell's principle. Jewell's principle is that the assigned flow pattern should optimise some overall economic objective for the network. This objective may be the minimisation of travel time, either by individuals (Wardrop's first principle) or for the system as a whole (Wardrop's second principle). Other definitions of economic objective can be chosen, such as minimum fuel consumption, vehicle operating costs, generalised cost of travel, or pollutant emissions. The optimisation problem for any of these objectives may be defined as either a user minimisation problem or a system minimisation problem.

5.3.2 Mathematical formulation of equilibrium assignment

The basic equilibrium assignment model for fixed (inelastic) travel demand is an expression of Wardrop's first principle. This model formulation provides a useful macroscopic simulation of travel on an urban network. It may be written as the following non-linear optimisation problem, for which a convergent solution may be found (as indicated, for example, in Taylor (1984)):

$$Z = \min \left\{ \sum_e \int_0^{q(e)} c_e(x) dx \right\} \qquad (5.10)$$

subject to the continuity of flow constraints

$$T_{ij} = \sum_r X_{rij} \qquad \forall i, j \qquad (5.11)$$

and

$$q(e) = \sum_{ijr} \delta_{eijr} X_{rij} \qquad \forall i, j \qquad (5.12)$$

where δ_{eijr} = 1   *if and only if e is in path r from i to j,*
= 0   otherwise

X_{rij} is the number of trips using path r between i and j, and the function $c_e(q)$ is the congestion function for link e.

The equivalent system-wide travel time minimisation problem (i.e. the assignment flow pattern corresponding to Wardrop's second principle) may be written as a similar optimisation problem, with objective function

$$Z = \min \left\{ \sum_e q(e) c_e(q(e)) \right\} \qquad (5.13)$$

with the same conservation of flow constraints.

The Wardrop principles may be treated as meeting different economic objectives for network travel, if travel time is taken as one possible alternative measure of travel cost. Thus they may be seen as particular cases of Jewell's assignment principle that the ultimate pattern of flow in a network will satisfy some explicit economic objective, for instance minimum generalised travel cost or minimum fuel consumption

124 Understanding Traffic Systems

(both either individual or system-wide). For instance, generalised cost functions including travel time, fuel consumption, pollutant emissions, tolls and road user charges, vehicle operating costs etc could also used in equilibrium assignment. Possible models for generalised cost and vehicle operating costs are introduced in Section 5.5.

Given a distribution of link flows across a network, consideration can be given to area-wide traffic control, especially for networks containing sets of signalised junctions, as a useful means of regulating traffic flows and minimising delays, fuel consumption and emissions.

5.4 Area-wide traffic control

Contemporary traffic control systems attempt to provide for free-flowing traffic (or at least minimum disruption) on major traffic routes. Linking of the signals at successive junctions (i.e. so that the start of the green periods at successive signals are coordinated to allow for the free movement of platoons) is one way to minimise delays and stop-start driving. The purpose of a coordinated signal system is to allow the maximum volume of traffic to pass without stopping, while catering for the demands of cross street traffic. Further, coordination can be used to prevent queues from extending back to interfere with upstream intersection movements.

The basics of signal coordination are shown in **Figure 5.2**. Platoons of vehicles are aimed at each 'green window', giving rise to the expression 'the green wave' for through traffic flow. Signals at intersection number 2 turn green ϕ_{12} seconds after those at intersection number 1. This time is known as the *offset* between the signals. There is constant (critical) cycle time for the system (the *common cycle time*), which is set to provide sufficient capacity at the *critical intersection* in the group. The green times at all intersections are set to provide adequate capacity and to give opportunity for free passage of the platoons. A coordination cruise speed ($tan\theta$ in Figure 5.2) is determined as a design parameter. The essence of coordination is to keep platoons together but the nature of the driving task is for platoons to break up.

5.4.1 Platoon dispersion

A platoon is a cluster of vehicles, travelling at short headways and moving at about the same speed. Platoons form naturally in the process of vehicles moving away from a standing queue, as happens at a traffic signal.

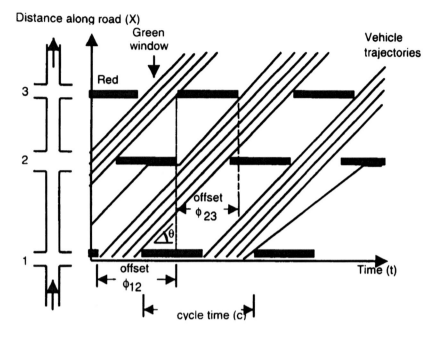

Figure 5.2 Signal coordination principles and the 'green wave'

Once formed, the natural tendency of the platoon is to disperse. The headways between vehicles in the platoon will tend to increase as the speed of the platoon increases, so that the size of the platoon[1] grows. In addition, some drivers and vehicles may be unable or unwilling to maintain speed at the level set by the platoon leaders. Thus as the platoon moves down the road towards the next set of signals it will occupy an increasing physical length of road and an increasing time band. This is the process of platoon dispersion. Dispersion makes the task of setting signal offsets for traffic control more difficult. Methods are needed to estimate the size of the platoon (i.e. the physical space it occupies) as it disperses so that offsets and green times at downstream signals can be designed to accommodate the bulk of, if not all of, the platoon.

The most common method for modelling platoon dispersion is Robertson's recurrence formula. This is an integral part of the TRANSYT computer model (Robertson, 1969) for designing network-wide signal

---

[1] The size of a platoon is the length of road that it occupies.

settings. The method is based on an assumed relationship between the flow passing a point in one time interval and a previous time interval. This relationship is

$$q_2(i + BT) = F q_1(i) + (1 - F) q_2(i + BT - 1) \qquad (5.14)$$

where $q_2(i)$ is the predicted flow downstream in the ith time interval, $q_1(i)$ is the flow of the initial platoon in the ith time interval, B is the 'travel time factor' (the ratio of the platoon leader travel time to the average travel time of the entire platoon), T is the average travel time of the entire platoon, over the distance for which the platoon dispersion is being calculated, and F is a smoothing factor. The value of F is given by

$$F = 1 / [A + BT] \qquad (5.15)$$

where A is a dispersion factor to account for the degree of platoon dispersion. A and B are empirical parameters, ranging from 0 to 1. If $A = 0$, then no dispersion occurs. If B is constant and A increases from zero, the level of dispersion increases. Values of A and B may be determined from field data, but this is often infeasible. Typical values that may be used as defaults are $A = 0.5$ and $B = 0.8$.

Other models for platoon dispersion exist, and the interested reader is referred to Seddon (1972), Tracz (1975), Young, Taylor and Gipps (1989) and Stamatiadis and Gartner (1999). Consideration of platoon dispersion in terms of a relatively simple model such as that described above can then be used to formulate analytical models of traffic flow under signal coordination.

5.5 Congestion

The management of congestion is an important issue in traffic planning. Much of the work of traffic engineers, planners and analysts focuses on how to ameliorate the effects of congestion, or indeed on how to use congestion to regulate traffic movement through an area. The impacts of new developments or traffic arrangements on existing levels of congestion are always important issues in traffic impact assessment. Congestion is an integral part of a transport system, but its specific definition and identification are not immediately obvious. Various definitions of traffic

congestion and the observed phenomena associated with it were reviewed by Taylor (1992, 1999). Three recurrent ideas were found:
- congestion involves the imposition of additional travel costs on all users of a transport facility by each user of that facility;
- transport facilities (e.g. road links, intersections, lanes) have finite capacities to handle traffic, and congestion occurs when the demand to use a facility approaches or exceeds the capacity, and
- congestion occurs on a regular, cyclic basis, reflecting the levels and scheduling of social and economic activities in a given area.

The following definition of congestion was proposed for use in traffic studies: 'traffic congestion is the phenomenon of increased disruption of traffic movement on an element of the transport system, observed in terms of delays and queuing, that is generated by the interactions amongst the flow units in a traffic stream or in intersecting traffic streams. The phenomenon is most visible when the level of demand for movement approaches or exceeds the present capacity of the element and the best indicator of the occurrence of congestion is the presence of queues' (Taylor, 1999). This definition extension recognises that the capacity of a traffic systems element may vary over time, e.g. when traffic incidents occur. Congestion may always be present in a transport system, but that the level of congestion may have to exceed some threshold value to be recognised.

5.5.1 Measuring the level of congestion

The investigation of any traffic planning or traffic management strategy requires the determination and possible subsequent monitoring of the level of congestion. Thus there is a need to collect and analyse data on congestion. Several measures can be used, and although the definition of traffic congestion would suggest that delay time and queue length are essential parameters, they are almost certainly not sufficient measures. The set of factors reflecting the level of congestion includes:
- *delay*, possibly disaggregated to consider delays to different road users (e.g. private vehicles, public transport, pedestrians, etc) or delays on different roads (major arterial roads, local roads and streets, etc);
- the *equitable distribution of delays* between competing traffic streams;
- the *reliability of travel times and travel costs*. Delays reflect an overall (or average) level of congestion experienced by travellers. Under congested conditions individual travellers may experience considerable

variation in travel times which have significant effects on travel choices;
- *queue management*, which is of importance in urban traffic network control, in the attempt to prevent queuing and congestion at one point in a network from moving upstream to block other intersections. Queue management is necessary in congested road networks to maintain the overall capacity of the network;
- *incident detection*, which is important for traffic flow control on limited access facilities such as freeways;
- *excess energy consumption* caused by delays and queuing;
- *additional emissions of gaseous and noise pollution* caused by delays and queuing, and
- the possible increase in *accident potential* due to reduction in manoeuvring space and increased frustration and anxiety of driving in congested conditions.

5.5.2 Generalised cost of travel

The concept of a generalised cost of travel has been widely used in transport planning, especially in relation to the analysis of modal choice (see Section 5.1.4). One particular formulation of generalised cost was proposed by Wigan (1976) for network studies of the interaction of road-based components of travel cost. This formulation is

$$C = \left(A + Bu + \frac{E}{u^2}\right)X + m$$

where X is link length, u is the unit travel time on the link (time taken per unit distance), m is road toll, congestion charge or other direct out-of-pocket money cost incurred on the link, and A, B and E are parameters characterising the network and traveller behaviour. Parameter A represents the operating cost of the vehicle per unit distance while B can be taken as the traveller's valuation of travel time. Both A and B can be split into components: $A = A_t + A_c$ and $B = B_m + B_t + B_c$ from which a further breakdown of the cost function into policy related components can be made:
- $A_t + B_t u$ indicates the cost per unit distance due to taxation on fuel, tyres and oil;

- $A_c + B_c u$ indicates the operating cost per unit distance net of all taxation, and
- $B_m u$ is the money cost equivalent of travel time per unit distance

The term Eu^{-2} is associated with the fuel consumption and emissions of vehicles and can be taken as an approximation of the more detailed models for fuel and emissions to be described in Section 5.6.

5.5.3 Congestion pricing

Modern transport planning practice increasingly seeks to allow travellers to make 'more informed' choices through the provision of current information on travel conditions and through the provision of 'pricing signals' by which trip-makers make direct payments for their travel decisions. In road networks this may occur through the use of toll facilities on bridges and high capacity roads, through road user charges (especially for commercial vehicles), and through congestion pricing, with ERP technology now allowing charges to be automatically deducted from individual vehicles whilst moving at highway speeds.

Congestion of itself provides a natural but partial restraining mechanism on travel demand. The additional costs (delays, queuing and inconvenience) resulting from congested conditions can act as a deterrent to the generation of further travel demand. However, there is widespread belief amongst transport economists that the congestion 'price' of itself is inefficient as a demand management tool. Individual drivers may not be fully aware of the true costs that they impose on other travellers and the transport system on the basis of congestion delays alone. Some other pricing signal is required to this end. Assuming that travellers will respond to a composite generalised cost (i.e. a total travel cost containing components from travel time, travel distance, out-of-pocket expenses, fuel cost, wear and tear, etc.) by trading-off the different cost components in their travel decision making, the further step is to impose a congestion tax, toll or road pricing charge on travellers in an intelligent, selective fashion (e.g. for travel on some parts of a network at some times of day). There is a resurgence of interest in this topic, largely as a result of the new technological capabilities for vehicle identification and monitoring, e.g. May et al (1996).

The economist's conceptual model for a congestion tax or price is that shown in Figure 5.3. This shows the average travel cost curve, which may be equated to the congestion function curve shown in Figure 4.8. The marginal cost curve is also shown in Figure 5.3; this curve indicates the additional travel cost imposed by each new driver using the facility.

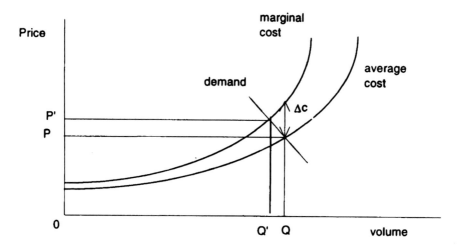

Figure 5.3   **The economic price of congestion and travel demand**

Average cost and marginal cost are similar for small traffic volumes, but marginal cost increases more quickly than average cost. The unified definition of traffic congestion says that congestion is any additional travel cost above the minimum cost to traverse the system element. The congestion may not be apparent at small flows, a threshold value may have to be reached first. The growing disparity between marginal cost and average cost might be used to indicate the threshold. Figure 5.3 also shows an economic demand curve, which provides the relationship between the travel demand to use the facility and the cost of doing so. The intersection of the demand curve and the average cost curve (at (Q, P)) may be taken as the actual level of flow on the facility. Now the average cost (P) is less than the marginal cost, so the motorists are not meeting their full marginal cost. A congestion charge imposed on motorists would see a decrease in traffic volume to Q' with the cost of travel at P' (including a congestion charge of ΔC).

Now, the relationships between the curves of Figure 5.3 indicate that the actual congestion charge to be imposed is not easily determined. It requires detailed knowledge of the characteristics of the facility and its average and marginal cost curves, and the level of traffic flow. If a mathematical relationship (such as the Davidson or Akcelik congestion functions, see Section 4.4) is available, then the marginal cost can be found from the derivative of this function. If the marginal travel cost on link e is c_{me} where

$$c_{me} = \frac{\partial C_{Te}}{\partial q_e}$$

and C_{Te} is the total travel cost on the link, given by $C_{Te} = c_e q_e$, then, for the Davidson function defined by equation (4.42), the marginal cost of travel is given by:

$$c_m = c_0 \left\{ 1 + J \frac{\mu(2-\mu)}{(1-\mu)^2} \right\} \qquad \mu < \rho$$

$$c_m = c_0 \left\{ 1 + J \frac{\rho}{1-\rho} + J \frac{(2\mu - \rho)}{(1-\rho)^2} \right\} \qquad \mu \geq \rho$$

(5.16)

Application of congestion pricing (in its theoretically pure form) requires considerable data, including some continuous monitoring over the time that the congestion charge is being imposed — a significant task. The technological capability to perform this task now exists, for example see May et al (1996). The further result of new technology may be the ability to provide drivers with information that will assist them in making their travel choices. This might be the factor that wins community acceptance of a congestion pricing scheme.

5.6 Fuel consumption and emissions

The family of models of fuel consumption and emissions from traffic streams proposed by Biggs and Akcelik (1986) provides specific models to cover a wide range of traffic circumstances, from the performance of an individual vehicle in traffic to a model for a total door-to-door trip. The models are:

(a) an *instantaneous model*, indicating the rate of fuel usage or pollutant emission of an individual vehicle continuously over time;

(b) an *elemental model*, relating fuel use or pollutant emission to traffic variables such as deceleration, acceleration, idling and

132 *Understanding Traffic Systems*

cruising, etc. over a short road distance (e.g. the approach to an intersection);

(c) a *running speed model*, yielding emissions or fuel consumption for vehicles travelling over an extended length of road (perhaps representing a network link), and

(d) an *average speed model*, estimating the level of emissions or fuel consumption over an entire journey.

The instantaneous model is the basic (and most detailed) model. The other models are aggregations of it, and require less and less information but are also increasingly less accurate. The elemental model provides a useful model of fuel consumption and emissions at an intersection, and is the most suitable model for traffic engineering applications. The running speed model is particularly suitable for application in strategic networks, for it can be used at the network link level. The average speed model is useful for sketch planning applications. Taylor and Young (1996) provides a full description of each of these models.

5.7 Exposure measures for accident analysis

Traffic accident analysis often requires the assessment of accident rates, involving the determination of the incidence of accidents in terms of numbers of accidents per unit exposure (see Chapter 14). The basic underlying theory behind this analysis is based on the assumption that the expected number of accidents at a given site is a function of exposure and accident propensity. That is, the expected number of accidents, E(A) is given by the relation

$$E(A) = f(q)\,g(q)$$

where $f(q)$ is defined as the exposure variable and $g(q)$ as the propensity.

Exposure is defined as the number of opportunities for accidents of a given type in a given area. Propensity is defined as the conditional probability that an accident occurs given the opportunity for one. Thus the expected number of accidents is defined as the product of exposure and propensity. While there are excellent references for appropriate exposure measures for various accident types (e.g. Layfield et al, 1996), the same cannot be said of propensity. A review of 30 studies by Satterthwaite (1981) found inconsistency in about half of them. In general the propensity for accidents at a given location depends on traffic volume and a number of driver, vehicle and environmental factors including: traffic control devices, light and weather conditions, road surface conditions, road design standard, and vehicle and driver performance.

The usual measure of exposure applied to traffic on a link is the vehicle-kilometres of travel (VKT) on the link, possibly disaggregated by vehicle type. VKT is a measure of the total 'traffic effort' on the link, being the product of traffic volume and link length. The resulting accident exposure index is number of accidents per unit VKT.

The interactions between road crashes and traffic at junctions can be studied in two ways. The first uses a function of flows using the intersection as the exposure and compares the observed crashes with this exposure. The other approach uses empirical accident and flow data for many locations to find a relationship between crashes and traffic flow. This approach assumes that the risk at each location is the same. These approaches plus actual vehicular movement observations have led to four exposure measures being suggested for crashes at intersections: the total traffic using the intersection, product of the cross flows on conflicting paths, square root of product of the cross flows and the observed number of conflicts at a location.

Part C

DATA CAPTURE

6 Principles of survey planning and management

So far we have considered the planning process and various theories of traffic movement. These provide basic foundations for the consideration of traffic studies and for impact analysis based on data collected in those studies. Our attention now turns to the means for the collection and examination of traffic data to apply with the theoretical models as part of the planning process. Data collection is expensive, time consuming and not always straightforward, so great care must be exercised in its planning, design and conduct. This chapter considers the necessary steps in survey planning and management.

Traffic analysts may, on occasion, have access to a previously collected database containing all the variables needed for a given analysis. Anticipating such circumstances, and with a vague knowledge that more and more data is being collected automatically, some analysts might feel that they need not trouble themselves with the details of traffic survey planning and data collection methods. They would be foolish to take such a view! Not only would it seriously restrict the analyses they could conduct but also, without a knowledge of the survey procedures they could misinterpret the meaning of variables and, perhaps most serious of all, their understanding of traffic systems would be the poorer for not having been involved in the data collection process. Furthermore as we shall see, there is reason to believe that in certain circumstances, such as research studies, the optimum survey design is only achieved if the analyst is directly involved in the survey planning.

6.1 Elements of the survey planning and management process

Figure 6.1 is an idealised representation of a traffic data collection exercise starting with the specification of objectives of the exercise and running through to the archiving of results. Note the existence within Figure 6.1 of various feedback loops indicating that survey design is not a purely sequential process; for example, analysts must be prepared to modify their survey instruments and sample frames in the light of the outcome of the pilot survey.

138 *Understanding Traffic Systems*

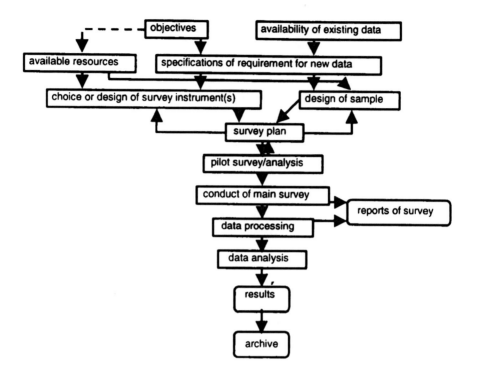

Figure 6.1   Stages in the design and conduct of a traffic survey

6.1.1 Objectives

Any data collection exercise must, of course, start with a full appreciation of the objectives of the exercise; for example is the survey required as part of an ongoing monitoring process or an ad hoc investigation? Are the results supposed to relate to a specific place or are general results sought? What hypotheses are to be tested? What level of disaggregation/detail is required? See Chapter 2 for further discussion of the objectives of data collection exercises.

6.1.2 Availability of existing data

Once the objectives are established, consideration must be given to the extent to which relevant data may already exist or may be being collected elsewhere. Data previously collected, whether by governmental bodies or their agents or by research teams, may remove or reduce the need to collect further data. Data being collected as part of an on-line system or a continuous monitoring process can also prove invaluable provided that access can be obtained (see section 6.4). Even where they cannot obviate the need for a new survey, data from such sources may help to define the data items (e.g. in the interests of maintaining comparability) or may yield information about the extent of variances within the population and so help in the calculation of required sample sizes (see Chapter 7).

6.1.3 Specification of the requirement for new data

In the light of the objectives of the exercise and the availability of pre-existing data it may be possible to specify the data to be collected with considerable precision; which parameters are to be measured, at what frequency, at which locations and over what time scale? In some circumstances, however, particularly if the objectives are associated with exploratory research, there may be some latitude in the specification; some items may be essential and precisely specified while others may be optional and less tightly defined. It is useful at this stage to make a distinction between what is essential and what might be a useful bonus and to resist the temptation to include a lot of redundant items in the specification 'just in case' they turn out to be useful. In the light of this specification, and of the resources available, the survey planner can now select the most appropriate survey instrument and design a sampling strategy.

6.1.4 Available resources

The availability of resources, whether of time, people or money, is usually a real constraint on the specification of a survey and compromises often have to be made between what the analyst would ideally want and what can actually be afforded.

6.1.5 Choice of survey instrument

The choice of survey instrument may be fairly straightforward if there is only one procedure or piece of equipment that can do the specified task (e.g. on-line analysis of exhaust emissions to a given precision) but it is much more common to find that several alternative procedures or items of equipment could be used. In such circumstances a careful appraisal of their relative strengths and weaknesses is needed - the information provided in Chapters 8-14 is designed to facilitate this. Some survey instruments will allow several items of data to be collected simultaneously at little or no extra cost and the utility of such additional data may help to tip the balance in favour of one instrument rather than another. The survey organiser may make such judgement in the light of the specification provided by the analyst but the optimal decision is most likely to be made if the analyst is fully involved in the decision and most perfectly if the analyst *is* the survey organiser!

6.1.6 Design of sample

The selection of survey instrument(s) and the design of the survey sample are interrelated. A given instrument may have particular sampling requirements (e.g. number of observations required to overcome any measurement error inherent in the method) and a given sample plan may have implications for the choice of instrument (e.g. if the sample requires observations every fraction of a second, manual techniques may be ruled out). Chapter 7 gives further information on sample design.

6.1.7 Survey plan

The initial survey plan will be based on the decisions taken on survey instrument and sample but will also include operational/procedural aspects of the exercise such as the recruitment of staff, acquisition of equipment and the schedule of key events. The schedule should be drawn up as a critical path flow chart showing the interdependence of each element of the survey. An example of such a chart is included as Figure 6.2. No two surveys will have the same schedule of events and a large part of the skill in survey planning is to work out an appropriate schedule for the specific circumstances, fitting all the required steps around any fixed dates and other constraints. Fixed dates and constraints will typically include the latest date by which results are required and the window of opportunity for the survey itself - often reflecting seasonal factors (e.g. the desire to survey traffic in a neutral month) and constraints on the availability of staff and equipment. Any given survey may have additional constraints such as the need to take measurements before completion of a network change.

The prudent survey planner will try to make allowance for unexpected delays by building in a certain amount of slack but such luxuries are often casualties of the compromise between the ideal survey plan and the one which provides data at the time when the analyst 'must' have it.

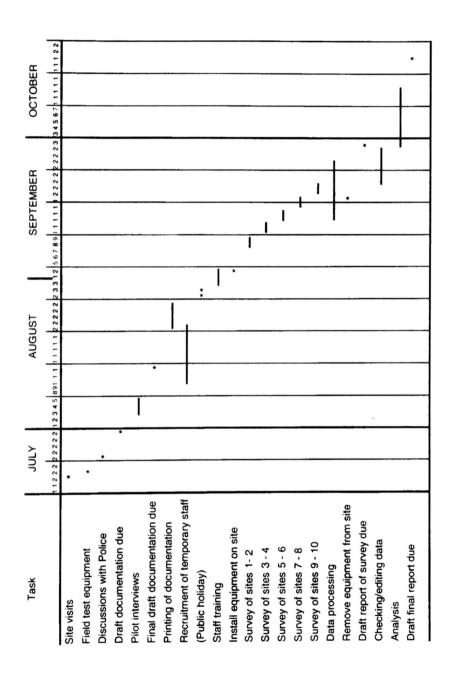

Figure 6.2 Typical survey schedule

6.1.8 Pilot survey

An important element in the survey plan, which is omitted only at great risk to the success of the exercise, is the pilot survey. A full pilot survey will include testing not only of the survey instrument itself but also of associated procedures (e.g. to determine whether the instructions to coders are adequate). Piloting may be reduced if a survey is using standard procedures and equipment but is clearly vital when innovations are being introduced. It may sometimes be appropriate in such circumstances to allow for a 'pre-pilot' during which the design of an innovative instrument (e.g. a new piece of equipment or a new design of form) or procedure (e.g. having one supervisor responsible for several simultaneously operating sites rather than just one) can be tested and refined. Whatever level of piloting is included, the survey plan must allow adequate time and resources for any revision or redesign which is shown to be necessary. If all has gone well then there may be no need to redesign the survey instrument or modify the proposed procedures. The results of the pilot survey may then simply be used to derive estimates of variance in the data and so perhaps suggest a change in the sample requirement. In extreme cases, however, the pilot survey may warrant abandonment of the entire survey program or major adjustment to the survey plan.

6.1.9 Conduct of main survey, data processing and archiving

Once the techniques and procedures have been shown to work, the main survey can be conducted and followed up by such data processing as is required. It is vitally important at this stage that the procedures adopted during the main survey and data processing stages are written up in full and archived along with any information relevant to the conduct of the survey, e.g. a note of ambient weather conditions or other factors that might have affected the data.

The archived report of survey will thus include copies of the survey schedules and all instructions to staff, together with technical descriptions of any equipment or software, site plans, copies of questionnaires, coding sheets, coding instructions and so on. Good practice is to archive the raw data itself, particularly if the processing was potentially subject to error or if any element of subjective judgement was involved in the processing. In practice, however, the indefinite maintenance of a comprehensive archive including all the data in its raw form can become quite costly and many practitioners will now put a limit on the length of time for which they will store raw data. No such limit should be put on the other elements of the report of survey since they are essential not only to any further analysis of the data but also to any interpretation of results derived from it.

Principles of Survey Planning and Management 143

Chapter 15 indicates in more detail how the data from a survey are transformed into useful information and integrated with other information in a database.

6.2 What constitutes a good survey instrument?

While Chapters 8-14 discuss the advantages and disadvantages of individual survey instruments, it is useful to set out some guiding principles in this introductory chapter. The survey instrument must obviously be fit for its purpose, which implies that it can produce the expected data items with:
- minimum but known expense;
- maximum but known reliability;
- maximum but known accuracy;
- minimum but known requirement for highly specialist staff;
- minimum but known hazard to survey staff;
- minimum but known interference with the phenomenon being measured;
- minimum but known disruption to third parties, and
- minimum but known requirement for post processing.

Most surveys will be conducted using relatively well known techniques, not least because the survey managers feel comfortable with them and are in a position to estimate their cost, reliability, accuracy etc with relative ease. However, changing technologies and changing data requirements produce new opportunities and new challenges for the survey manager, who should be ready to use new techniques when appropriate.

Where off-the-shelf equipment is involved, the manufacturer's specification will be the major source of information but should be read in the light of that manufacturer's reputation for quality and its (or its agent's) reputation for delivery on time and for after sales service. Such information may best be obtained by contacting colleagues or fellow professionals who are likely to have used the equipment in question. Where equipment is to be purpose built, modified, or put to a purpose other than that for which it was designed, particular attention will need to be paid to preparation time, ease of use, robustness, reliability, and provision for maintenance and/or replacement. Again the advice of colleagues should be sought in estimating these unknowns, bearing in mind, of course, that the relevant knowledge may not necessarily be found among traffic engineers but perhaps among mechanical, electrical and computing engineers or others involved in fields such as industrial process monitoring, control engineering or remote sensing. Literature searches and inquiries on relevant electronic bulletin boards often prove an effective route to people with the relevant expertise.

Where data are to be derived by enumerators keeping records on a form, the design of the form should be assessed with respect to ease of use in the field and ease of the ease with which the required data can subsequently be extracted. Ergonomic considerations dictate factors such as that forms should fit onto clipboards without need for excessive need for page turning, that boxes should not be too small for rapid identification in field conditions and that they should be easy to use by left-handed as well as right-handed enumerators. It is, unfortunately, quite rare to find the deficiencies of survey forms documented in the literature and it is often only the staff who have to use them who can comment sensibly on their design. It is for this reason that the most useful comments on the design of forms are to be had from members of survey teams employed by highway authorities or their agents. Whether the survey instrument involves items of equipment or forms, survey managers should familiarise themselves with their use in field conditions before venturing to instruct others on how to use them.

6.3 Good practice in survey administration

No matter how well designed a survey may be, and how well chosen the survey instrument and sample size, all can be lost if the administration is lax. The following basic guidelines should therefore be adhered to (with allowances made for the scale and circumstances of the survey in question):
- prepare a realistic survey schedule (see details in Section 6.1 above);
- conduct a pilot survey and modify procedures as necessary;
- prepare, and archive, written copies of all instructions to staff involved in conduct of the survey and processing of the data;
- make provision for spare copies of forms, back-up equipment (including batteries etc) and back-up staff, and plan how they are to be delivered to the necessary spot and when they will be required;
- if a survey, unavoidably includes use of equipment whose reliability is not guaranteed, it may be wise to make plans for a fall-back procedure (e.g. keeping written records of the registration plate numbers of a sample of passing vehicles might be the fall back procedure if the reliability of audio tape recorders, on to which it was planned to speak the registration plate numbers of all vehicles, is in doubt);
- plan realistic work rotas such that enumerators do not get bored, cold, tired or hungry. Depending on the task involved, a two hour work shift may be a realistic maximum, build in sufficient slack to allow for unexpected delays in the arrival of staff;
- plan to pay enough to attract staff of the required quality;

- provide for adequate training and supervision of field staff and inculcate in them a commitment to the success of the survey;
- keep detailed records of site conditions on the day, and
- make back-up copies of raw data and archive them along with the relevant documentation.

6.4 Data capture without surveys

As will be apparent from chapters 8-14, two developments are combining to change the context in which traffic and travel surveys are conducted.

The first is the increased use of permanently installed monitoring equipment as part of system-monitoring and on-line control initiatives made possible by the reduced cost and increased capability of the relevant detection equipment and data storage media. Some of the resulting data is routinely stored for future use but much of it is discarded after it has fulfilled its primary on-line function.

The second development is the spread of intelligent transport system (ITS) applications which manage or control components of the transport system in a way that involves on-line monitoring of network conditions or logging of individual traffic events. (Consider, for example, the detection of queues by adaptive traffic control systems, the detection of speeding vehicles by police speed enforcement cameras or the way that an on-line parking management system might log visits by season ticket holders to a particular carpark). The managers of some of these ITS applications are already providing data for use by third parties – usually on payment of a fee, while others have yet to overcome commercial or privacy sensitivities.

The potential volume of data from such sources is enormous. A number of highway authorities are now providing real-time traffic condition data via Internet sites and are making detailed record available for research purposes. If current trends continue, it will soon be unnecessary for analysts to conduct new surveys except where they have a particular requirements which cannot be met from one of the continuously monitored sites or ITS sources. Examples would include the need for site-specific data or the simultaneous recording of data on an unusual combination of variables. As such sources become increasingly important, analysts will find that they need to supplement their skills as survey designers with skills as data miners – an issue which was raised in section 2.5.

7 Experimental design and sample theory

Data are collected for many purposes. They may be required to provide a description of a particular occurrence (e.g. the distribution of parking durations); they may be collected to test a particular hypothesis (e.g. do the speeds of vehicles on a road change after the introduction of a public transport priority scheme?); or they may be collected to learn something that was previously unknown (e.g. what are the factors influencing the speed of traffic on a residential street?). This chapter addresses the question of setting up experiments to investigate these issues. It discusses appropriate experimental design before turning to the choice of sample design and the determination of appropriate sample sizes.

7.1 An introduction to experimental design

The process of experimentation is an iterative one. Traditionally, it starts with an hypothesis; data are then collected and analysed. If the analysis does not provide a solution to the problem, it may be necessary to reformulate the hypothesis and repeat the process until an appropriate answer is found. This process of moving from an initial hypothesis through a series of experiments may not provide a unique route to the answer. Each experimental design is influenced by noise or errors in the data, which may be inexact or false and therefore misleading. Thus it is necessary to continually update the hypothesis and the data, to deduce the true nature of the system. Because of this inexactitude, two equally competent analysts may take quite different routes to the same solution. What is important is convergence to a unique solution. Convergence will be quicker if:
- efficient methods of experimental design are used to reduce the effect of experimental error; and
- sensitive data analysis is used to indicate legitimate conclusions about the hypotheses and suggest new ideas that could be investigated.

Of these two resources, the experimental design is the most important. If the experimental design is poorly chosen, the resultant data will not contain the required information, and no matter how sophisticated the analysis procedures, correct answers cannot be found. If, however, the experimental design is wisely chosen, a great deal of information is available, and sensitive data analysis will extract a valid answer. Experimental design and statistical analysis are, however, intertwined. This intertwining will be illustrated in this chapter by reference to appropriate statistical tests for particular designs.

Chapter 6 outlined the recommended elements of a survey process. It highlighted the setting of objectives, the determination of the criteria to be used in assessing the impacts, and the collection of data through sampling. Another important aspect of the survey process is the determination of representative data for the variables to be investigated. This aspect is usually referred to as experimental design and is of fundamental importance. It begins with the selection of the variables to be studied.

7.2 Choice of variables

The collection of information on as many variables as possible is likely to be very expensive and often counter productive. It is more logical to collect data on a small set of variables while minimising the variation in all variables. For instance, in a 'before and after' study of the effects of street improvements on travel time, it is essential to collect data on the travel times. Associated with the improvement may be a change in traffic flow, so this variable should also be measured. The composition of traffic may also change, so data on this should also be collected. The list of variables could go on and on. Eventually, however, there comes a point where the collection of more data becomes too onerous. In such a case the variables not being measured should, if possible, held constant.

The choice of variables depends on the adopted hypothesis and should explore those parts of knowledge that are not totally clear. However, it may be more efficient in terms of cost and labour to estimate the effects of several variables simultaneously and, in such circumstances, it may be desirable to incorporate a number of variables (or effects) into one experiment. It often occurs, however, that the initial set of data does not enable certain hypotheses to be fully analysed. In such cases it may be necessary to collect a new set of data for a new set of variables. Sometimes there is a need to include the same variables in a number of successive hypotheses. Experimental design should, therefore, be seen as a moveable window through which aspects of the true nature of the system can be viewed.

An initial step in setting up an experiment is the determination of a 'reference set' or 'reference distribution' of variable values. This set describes the set of values that are of interest. It is, for instance, pointless to measure the speeds of vehicles on a racing track when considering the effect of traffic control devices on speeds of vehicles in residential streets. Similarly, it is pointless to investigate the impact of road width on speed if the width of road cannot be changed.

The reference set should be determined so as to include the characteristics of outcomes which could occur if the modification was to be entirely without effect. In the transport field this is usually carried out by collecting data on the existing situation, or by determining what would happen if nothing were done. Thus, when determining the road needs at a future point in time, the flow of vehicles expected at that date could be loaded onto the existing system to see what effect this would have on travel times and delays.

The variable range should also cover all the values needed in the investigation. This is often a problem in real world experiments since either the range of values is not present, or some of the variables are correlated. The lack of variation in a variable of interest will result in the variable appearing to have little impact, while correlation will result in one of the correlated variables contributing little to the explanation offered by the data. Attempts should always be made to overcome these problems. If they are not, it is likely that the final result will be biased.

7.3 Alternative experimental designs

Experimental designs can be constructed to compare simple quantities, such as the mean value of certain variables, or to investigate complex interactions. The previous section discussed the selection of variables and the range of values appropriate to each variable. We will now consider, in turn, experimental designs that are appropriate for comparison of two alternatives, of more than two alternatives, and of interactions between variables.

7.3.1 Comparison of one variable over two alternatives

The simplest set of experiments involves the comparison of one variable over two situations. This can involve studies over time or space. A common example is the before and after study. Experiments of this type can use random data or data that are grouped in some manner.

Since the total population conceptually contains all values that can occur from a given observation, any set of observations collected from the population is considered a sample. An important statistical idea is that of *random data*: in certain circumstances, a set of observations may be regarded as randomly drawn from the total population, i.e. there is an equal chance of drawing each element of the sample from the population. This basic assumption, which is adopted in many experiments, makes analysis straightforward, perhaps involving something as simple as the comparison of means.

The assumption of randomness is not always correct because the variable being studied may be systematically influenced by extraneous factors. For example, if we wished to study the effect on car speeds of a new road geometry associated with a public transport priority scheme, simple comparison of speeds before and after introduction of the new layout could be misleading if the change had been associated with reduced flow of vehicles (perhaps due to the attractiveness of the newly prioritised public transport mode). Such problems can be overcome by grouping the alternatives and subjecting them to the same set of base conditions. This is termed *blocking*.

A simple form of blocking is the 'paired comparison' design which compares two alternatives by subjecting them to the same set of conditions. For example, the comparison of the reflectivity of two types of roadside reflector could be carried out by subjecting both to the same set of laboratory experiments, or, if a field test was thought desirable, by mounting the two reflectors side by side such that both are subject to the same environmental conditions. Data could then be collected on the performance of both reflectors simultaneously, ensuring that the ambient conditions (such as weather and light levels) are controlled to be the same for each. Taking the difference in the performance of each reflector then enables the external conditions to be removed, and comparison can be made. A paired comparison t-test may be an appropriate analysis technique. The paired comparison technique is a useful method that may result in efficiency of experiments in some situations.

Both the paired and random procedures are valid experimental designs and can take into account the problems associated with unavoidable sources of variability. A good experimenter will identify important extraneous factors ahead of time and eliminate these effects where possible, and thus *combine blocked and random procedures*. Representative variations between blocks should also be encouraged. In the reflector example, it would be advantageous to vary the road conditions, car light intensity, drivers sight distance and light conditions as much as possible so as to get a representative set of conditions.

7.3.2 Comparison of one variable over a number of alternatives

The comparison of one variable over a number of situations is an extension of the procedures discussed above. There are, however, refinements in the procedures used to analyse the data and the general approach to collecting the data. This section discusses the collection of random data, randomised block data and latin squares experiments.

Random data are collected as described in the preceding section, but for more than one alternative. Examples of situations appropriate to this category might include studies with a number of before and after surveys, or comparisons of travel times over a number of routes. The analysis of these data is usually carried out using analysis of variance techniques.

The concept of pairing the observations, discussed in the preceding section, is a special case of blocked designs. As stated earlier, a block is a portion of the experimental material that is expected to be more homogeneous than the aggregate. In a blocked design two kinds of effects are contemplated: firstly those of the alternatives which are of major concern to the experimenter; and secondly those of blocks whose contribution we wish to reduce. A *randomised block design* has the advantages of providing the opportunity to eliminate block to block variations while providing a wide inductive space for experiments run with uniform raw material several alternatives can be run with several different blockings.

Blocking may involve running experiments close in time or in space. For example, in a study of the effectiveness of accident countermeasures, spatial blocking could be used to remove seasonal effects. In the following example the blocking design for a test of the robustness of different brands of tyre makes use of the fact that a test car can be equipped with a different brand of tyre on each wheel. Determination of the amount of wear on each of four brands of tyre (A D) could be investigated by carrying out a series of tests with each brand of tyre fitted successively to each wheel of the car (see Table 7.1).

Analysis of randomised block designs can be carried out using analysis of variance. In some situations it may not be possible to get the number of tests equalling the number of alternatives. For instance, it may only be possible to obtain three tests for each blocking when there are four alternatives. Table 7.2 presents a possible blocking for this situation. Once again the random design reduces potential bias due to interaction effects due does not altogether eliminate it.

This interaction bias could be eliminated by including all permutations. Using the example of Table 7.2 this would require 24 (= 4! = 4×3×2×1) tests, but clearly this 'full factorial' design is an expensive option. An alternative, cheaper, approach is based on the so-called *latin square*. In mathematical terms a latin square is a data matrix containing n×n cells. Each cell contains one of n specific numbers so that they appear once and only once on each row and column.

Table 7.1 Possible blocking situations when number of tests equals the number of alternatives

Wheel	Non random test test no				Randomised test test no			
	1	2	3	4	1	2	3	4
Front left	A	B	C	D	A	B	C	D
Front right	B	C	D	A	B	D	A	C
Rear left	C	D	A	B	C	A	D	B
Rear right	D	A	B	C	D	C	B	A

Table 7.2 Possible blocking situations when number of tests is less than the number of alternatives

Wheel	Non random test test no			Randomised test test no		
	1	2	3	1	2	3
Front left	A	B	C	A	B	C
Front right	B	C	D	B	D	A
Rear left	C	D	A	C	A	D
Rear right	D	A	B	D	C	B

To illustrate the use of latin squares, consider the determination of the effect that different petrol additives have on the emissions of oxides of nitrogen from vehicles. The experiment is set up to test the effect of four additives (A, B, C and D) and uses four drivers and four cars. Now, even if the cars are the same models, slight differences in the performance of the car will exist. A similar situation will exist with the drivers. It is the aim of the latin squares design to eliminate the car to car and driver to driver variations. The latin squares arrangement shown in Table 7.3 allows this to be carried out. This design assumes that there is no interaction between the drivers and cars and that their effects are approximately additive. The different additives - A, B, C and D are arranged in columns and rows so that they appear only once in each. The statistical analysis develops a relationship between the additives and the emission. Analysis of variance techniques are used to analyse data collected in this manner and more complex latin square designs can be developed to study more complex situations (Box, Hunter and Hunter, 1978).

Table 7.3 Latin squares design for car emissions using 4 drivers and 4 cars

Driver	Car			
	1	2	3	4
I	A	B	C	D
II	C	D	A	B
III	B	C	D	A
IV	D	A	B	C

7.3.3 Analysis of situations with multiple variables

The preceding section considered the comparison of one variable over a number of alternatives. Other experiments may wish to investigate the effect of two or more variables. In testing these effects, it is necessary to have an idea of how these variables interact. Consider an experimental study of the effects of alcohol and coffee on the reaction time of a driver operating a simulator. If the stimulants were considered separately, it may be found that coffee improves reaction time by 0.20 seconds per cup and that each shot of rum decreases reaction time by 0.45 seconds per shot. However, if the effect of coffee and rum combinations were considered, a combined effect may be found. It would be convenient if the effects were linear and additive. If this were the case, two shots of rum and three cups of coffee would reduce reaction time by only 0.30 seconds. However, if ten shots of rum and 23 cups of coffee were taken, this would only reduce reaction time by 0.10 seconds. It is much more likely that the effect of the rum is related to the number of shots drunk previously and is, therefore, non-linear with an interaction between the effect of the rum and the coffee. An experimental design that is able to collect data in such a way that both linear and additive effects can be investigated is termed a *factorial design*.

A factorial design has no blocking, and each of the variables considered is of equal interest since it is possible that the factors interact. To perform a general factorial design, an investigator must define the number of values for each of the variables to be considered. Each test should include each value for each. This design requires a large number of runs of the experiment to be made, in order to include all combinations. For instance, if we had to examine the emissions from two types of fuel additives in three types of cars under at five different speeds, we would need 2×3×5 = 30 tests. Factorial designs are important since they:
- allow thorough study of a local problem;

- provide the opportunity to investigate a wide range of variables and, therefore, increase efficiency in the early stages of the experiment;
- allow designs to build on each other and thus build up the degree of complexity of the analysis, and
- allow the interpretation of the observations produced by the design to proceed by the use of common sense and elementary arithmetic.

A 3×3 factorial design for use in the investigation of speed on residential streets is shown in Table 7.4. This factorial design can be replicated a number of times. It can be analysed using regression or maximum likelihood methods.

Table 7.4 3×3 factorial experiment investigating vehicle speeds on residential streets

Traffic flow	Road width (m)		
(veh/h)	5.0	7.0	9.0
200	40	45	50
400	35	40	45
600	30	35	40

Factorial designs can also be used to investigate the impact of qualitative variables. Table 7.5 shows a 2×2×2 factorial experiment in which there are two quantitative variables (road width and traffic flow) and a single qualitative variable (built environment). The response is vehicle speed. Regression techniques may be used to analyse this design.

Table 7.5 2×2×2 factorial experiment of vehicle speeds in residential streets

Traffic flow	Central business district Road width (m)		Suburban environment Road width (m)	
(veh/h)	5.0	7.0	5.0	7.0
200	35	45	40	50
600	30	40	35	45

7.4 Sampling methods

Almost all traffic surveys involve observing a number of members of a target population in order to infer something about the characteristics of that population. In this sense they are sample surveys. The effectiveness of the survey is dependent on having chosen an appropriate sample. Sample design is thus a fundamental part of overall experimental design. It includes the following elements:
- definition of target population;
- definition of sampling unit;
- selection of sampling frame;
- choice of sample method;
- consideration of likely sampling errors and biases; and
- determination of sample size.

7.4.1 Target population

Changes to a traffic system are usually location specific and aimed at a particular class of traffic or road user. Their performance needs to be assessed with respect to that area and group of users together but should ideally be extended to cover other groups or areas which might be affected (and perhaps a similar group or area who will not be affected but who may act as a 'control' sample). The particular group of road users or sites that the survey is intended to cover is the target population. Clear definition of the target population will subsequently reduce the level of uncertainty associated with decision making, and lead to optimisation of the resources needed to complete the survey.

Although, as explained above, the target population should include all impacted groups, it is important to differentiate between the effects on different subgroups of that population; failure to do so might cause important effects to be missed. For example, in a before and after study of delays due to a street closure, a small decrease in delays for the whole population might hide very large increases for a particular subgroup. Analysis of data from a pilot survey can help to define the important subgroups within a population.

7.4.2 Definition of sampling unit

A population consists of individual elements (e.g. individual vehicles in a stream of traffic) but it may not be practical, or necessary, to survey at the level of these individuals. For example, although in some studies we may be interested in individual vehicles, in others interest may be focussed on types of vehicles, vehicles making particular manoeuvres, or vehicles travelling at a particular speed. The definition of the sampling unit depends primarily on the nature and purpose of the study but may also be constrained by the practical considerations involved in collecting the required data. Typical sampling units for traffic surveys would include such entities as: individuals; vehicles; companies and organisations (including households); geographic areas (traffic generating sites, car parks, zones, towns, regions); junctions, and road links.

7.4.3 Selection of sampling frame

The sampling frame is the 'register' of the target population. It is some list or definition which defines and contains the members (units) of the target population. A complete definition of the sampling frame is required for sampling to proceed, and for any inferences to be made about the target population from the sample results. The actual choice of the sampling frame depends on the needs and constraints of the particular study. For instance, the sampling frame might consist of the list of vehicles registered as being garaged in a certain area, or it could be all those vehicles using a particular road or junction during a specified time period. The main requirement is that the sampling frame can be quantifiably identified so that the actual size of the target population may be determined.

7.4.4 Choice of sampling method

Two main methods exist for selecting samples from a target population. These are judgement sampling and random sampling. In random sampling, all members of the target population have a chance of being selected in the sample, whereas judgement sampling uses personal knowledge, expertise and opinion to identify sample members. Judgement samples have a certain convenience. They may have a particular role, by way of 'case studies' of particular phenomena or behaviours. The difficulty is that judgement samples have no statistical meaning; they cannot represent the target population. Statistical techniques cannot be applied to them to produce useful results, for they are almost certainly biased. There is a particular role for judgement sampling in exploratory or pilot surveys where the intention is to examine the possible extremes of outcomes with minimal resources. However, in order to go beyond such an exploration, the investigator cannot attempt to select 'typical' members or exclude 'atypical' members of a population, or to seek sampling by convenience or desire (choosing sample members on the basis of ease or pleasure of observation). Rather, a random sampling scheme should be sought, to ensure that the sample taken is statistically representative. Random samples may be taken by one of four basic methods (Cochran, 1977): simple random sampling; systematic sampling; stratified random sampling, and cluster sampling.

Simple random sampling allows each possible sample to have an equal probability of being chosen, and each unit in the target population has an equal probability of being included in any one sample. Sampling may be either *with replacement* (i.e. any member may be selected more than once in any sample draw) or *without replacement* (i.e. after selection in one sample, that unit is removed from the sampling frame for the remainder of the draw for that sample). Selection of the sample is by way of a 'raffle'. A number is assigned to each unit in the sampling frame, and repeated random draws are made until a sample of desired size is obtained. For most applications, the use of a table of random digits is the most convenient means of drawing a random sample. Most statistical text books contain such tables.

Systematic sampling is a simple and convenient method of selecting a pseudo-random sample. It has two advantages over simple random sampling. Firstly, it is quick and demands only limited resources once the sampling frame exists. Secondly, it can easily be applied by unskilled workers, which may be particularly convenient in some surveys. It provides a definite rule for on-site sampling, for example the method used by Almond (1963) for sampling vehicle free speeds. The method involves selecting an arbitrary starting point in the sampling frame, then reading off every nth entry in the frame. For example, a ten per cent systematic sample is obtained by selecting an arbitrary starting point in the first ten members of the sampling frame and then select every tenth member.

Systematic sampling differs from simple random sampling in that, although (initially) each element has the same probability of inclusion in the sample, each sample does not have an equal probability of being selected. Once the first element is selected, the sample is completely defined. If the nth element in the sampling frame is in the sample, then the (n+1)th element has no possibility of selection (unless the 'sample' was 100 per cent of the sampling frame!). Systematic sampling is often employed because of its convenience. For example, it provides a convenient way for the selection of locations for traffic counting sites along a highway. There are potential problems, however, due to the non-random selection procedure. In particular, biases will be introduced if some surveyed parameters correlate with the order of elements in the sampling frame (e.g. age of motor vehicles and registration plates).

Stratified sampling requires division of the target population into relatively homogeneous groups, or strata. Two methods may be used to draw a stratified random sample: Either draw a simple random sample of a specified number of elements from each stratum, corresponding to the proportion of that stratum in the target population, or draw equal numbers at random from each stratum and weight the results according to that stratum's proportion in the target population.

Both approaches guarantee that every element in the population has a chance of selection while ensuring that 'rare' strata will be represented in the overall sample. This overcomes a possible difficulty in simple random sampling where 'interesting' members of the population may not be found in the resulting sample. For example, commuter cyclists might be missed in a random sampling of arterial road users by simple random sampling, but their presence in the sample is guaranteed by stratified random sampling. When properly defined, stratified samples can more accurately and efficiently reflect the character of the target population than other kinds of samples (Cochran, 1977). The method is particularly relevant when it is important to ensure that 'minority' groups (of whatever kind) are included in the sample.

Cluster sampling requires division of the target population into groups (clusters) and then selecting a random sample of the clusters. For example: in a trip generation survey, we might select a random set of streets in a municipality and then survey each household in that set of streets; or, in a registration plate survey, we might consider all vehicles whose plates end with a selected digit. One advantage of a *geographical cluster*, such as the street cluster example mentioned above, is that the distribution of questionnaires, etc is easier than if an equivalent number of households were selected by simple random sampling. Another advantage is that examination of the cluster may reveal important interactions within it. The difficulty lies in assuming the absence of bias in selecting all households in some streets to represent the municipality (e.g. would there be a bias due to misrepresentation of socio-economic groups, age of housing, etc?).

Both stratified random sampling and cluster sampling require the division of the target population into well-defined groups. Stratified sampling should be used when each group has small variations within itself, but there is a wide variation between groups. Cluster sampling should be used when there is considerable variation within groups, but the groups have essentially the same characteristics.

7.4.5 The basis of statistical inference

Systematic sampling, stratified random sampling and cluster sampling have been developed for their precision, economy or ease of application. However, it is the principle of simple random sampling that lies behind the mathematical theories of statistical inference, i.e. the process of drawing inferences about a population from information about samples drawn from that population. We now consider the methods of statistical inference, based on the simple random sample, but applicable also to the other sampling methods described above.

7.4.6 Sampling error and bias

There is always a possibility that the sample does not adequately reflect the nature of the parent population. Random fluctuations ('errors') which are inherent in the sampling process are not serious because they can be quantified and allowed for using statistical methods. However, if due to poor experimental design or survey execution, there is a systematic pattern to the errors, this will introduce bias into the data and, unless it can be detected, this will distort the analysis. A principal objective of statistical theory is to infer valid conclusions about a population from unbiased sample data, bearing in mind the inherent variability introduced by sampling. The following summary provides a guide to the design of samples for traffic data collection.

A distribution of all the possible means of samples drawn from a target population is known as a sampling distribution. It can be partially described by its mean and standard deviation. The standard deviation of the sampling distribution is known as the 'standard error'. It takes account of the anticipated amount of random variation inherent in the sampling process and can therefore be use to determine the precision of a given estimate of a population parameter from the sample.

7.4.7 Sample size determination

Traffic surveys for specific investigations (as opposed to general monitoring of the traffic system) will probably attempt to provide data for the estimation of particular population parameters, or to test statistical hypotheses about a population. In either case the size of the sample selected will be an important element, and the reliability of the estimate will increase as sample size increases. On the other hand, the cost of gathering the data will also increase with increased sample size, and this is an important consideration in sample design. A trade-off may occur, and the additional returns from an increase in sample size will need to be evaluated against the additional costs incurred.

If the target population is infinite, then the standard error (s_x) of variable X is given by

$$s_{\bar{x}} = \frac{s}{\sqrt{n}} \qquad (7.1)$$

where s is the estimated standard deviation of the population and n is the sample size, assuming that the sampling distribution is normal. Even when the sampling distribution is not normal, this method may still apply because of the Central Limits Theorem which states that the mean of n random variables from the same distribution will, in the limit as n approaches infinity, have a normal distribution even if the parent distribution is not normal. The standard deviation of the mean is inversely proportional to √n. The implication of equation (7.1) is that as sample size increases, standard error decreases in proportion to the square root of n. Here is an important result. The extra precision of a larger sample should be traded off against the cost of collecting that amount of data. To double the precision of an estimate will require the collection of four times as much data.

A further qualification applies to equation (7.1) because it based on the assumption that the sample is drawn from an infinite population. Such populations do not actually exist in our world, although in many instances we can assume that a target population is effectively infinite (e.g. the number of motor vehicles garaged in a large metropolitan area is an effectively infinite population for most purposes in traffic surveys). A modified form of equation (7.1) applies for a finite population of size N_p:

$$s_{\bar{x}} = \sqrt{\frac{N_p - n}{N_p / \sqrt{n}}} \frac{s}{\sqrt{n}} \qquad (7.2)$$

The proportion n/N_p is termed the sampling fraction. If the target population is reasonably large (say $N_p > 1,000$) and the sampling fraction is reasonably small (say < 0.05), then there is little difference between the standard errors predicted by equations (7.1) and (7.2). Consequently, it is the absolute size of the sample that determines sampling precision, rather than the fraction of the target population. Contrast this with common usage which tends to refer to an 'x per cent' sample size.

These results provide preliminary information on how to estimate a desirable sample size. For example, if we wish to estimate a mean speed at a site, with an accuracy of 0.5 km/h, and we know that the standard deviation of the speed distribution is 6.7 km/h, then by applying equation (7.1) we find that n = $(6.7/0.5)^2 \approx 180$ (rounding up from 179.6).

More refined methods also exist for determining the required sample size. For instance, if a sample estimate of a specified percentile value (u, say the 50th percentile, 85th percentile or 90th percentile) is sought from a population assumed to follow a normal distribution, to a given confidence level (α, e.g. 95 per cent or 99 per cent confidence level), then the minimum sample size required (n) may be estimated as:

$$n = \frac{\alpha^2 s^2 (2 + u^2)}{2 d^2} \qquad (7.3)$$

where s is the standard deviation of the sample and d is the permissible error in the estimate. Note, again, that this equation requires advance knowledge or estimation of s. For convenience, values of $\alpha^2(2 + u^2)$ are given in Table 7.6. For example, if the permissible error in the estimate of the median speed (50th percentile) is ±3 km/h, and speeds are distributed according to the normal distribution, and a 95 per cent confidence level is required for the median estimate then the minimum required sample size is $(7.7 s^2) / (2\times32) = 0.428 s^2$ (where s is the standard deviation of the speeds in km/h).

Table 7.6 Values of $\alpha^2(2 + u^2)$ for use in sample size determination, when determining percentile estimates with level of confidence α per cent

Percentile to be estimated	Desired confidence level (percentage)		
	90	95	99
15	8.3	11.8	20.5
50	5.4	7.7	13.3
85	8.3	11.8	20.5

There are other factors which impinge on the determination of a suitable sample size. Firstly, how do we obtain a value for the population standard deviation(s)? Secondly, for what purpose is the sample required -

especially when testing statistical hypotheses? These issues are dealt with at length in subsequent chapters.

7.4.8 Requirements for sample design

Traffic phenomena arise by chance from combinations of factors, all of which have high degrees of variability. In consequence, we need to treat events in traffic systems as random variables, and to use statistical theory to investigate the impacts of the interacting factors. The success of a statistical investigation depends largely on the size and validity of the traffic survey used to collect the data. For useful results, the investigator needs to:
- define the precise aims of the survey;
- define the target population;
- avoid introducing unquantified bias into the data;
- specify the parameters to be estimated and the desired level of accuracy associated with them, and
- ensure the sample is of sufficient size and truly represents the target population.

8 Vehicle counting and classification surveys

Vehicle flow data are the fundamental source of information about the state of the traffic system and the loads that it is carrying. They form an important input for all phases of the planning, design, operation and maintenance of traffic networks. The flows carried at different times and in different parts of a network provide basic information about the relative importance of different links in a system and about the pattern of demand which it is expected to carry. When taken together with data on capacities, flow data can point to the critical times or links within the network and when taken together with data on accidents it can indicate levels of exposure. Flow data are clearly major inputs to any investigation of network performance and are often used as the proxy for, or predictor of, environmental effects. Changes in flow are often used as measures of traffic impacts.

For some of the above purposes the interest will be in the two-way flow while in others it will be useful to distinguish by direction of movement and by class of vehicle.

8.1 Measures of flow

Traffic volume is usually expressed as a flow per unit time; the flow may be expressed in terms of vehicles (perhaps disaggregated by type), axles or passenger car units (pcus) and the time unit may be anything from a minute to a year depending on the purpose of the study. Some commonly used measures are: *pcus per minute* (or per cycle-time) - which is often adopted in junction performance studies; *vehicles per hour* - which is perhaps the most commonly used descriptor of link flow and *daily (24 hour) flow* - which is of particular value in evaluation studies.

Flows at a given site will, of course, vary according to the time of day, the day of the week and the season of the year. The resulting patterns are known as *'flow profiles'* (e.g. see Figure 8.1). The profiles to be expected at any given site will depend on climatic, economic and land use factors and are likely to be different for different types of vehicle. These profiles must be taken into account when planning a survey and analysing the data.

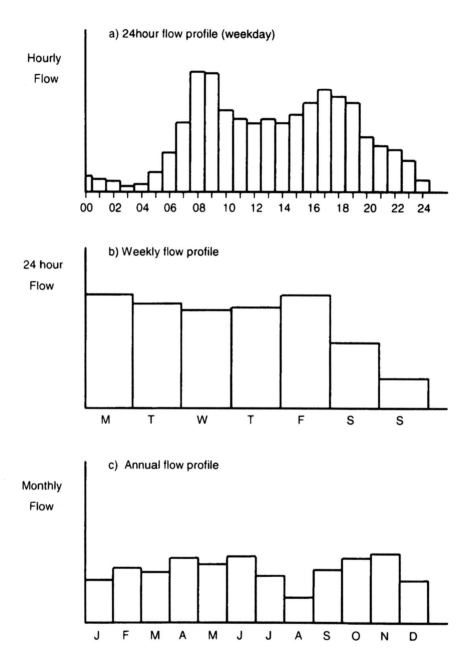

Figure 8.1 Typical flow profiles on an urban radial road in the UK

Particular interest is usually paid to the flow during the busiest hour of the day (the *'peak hour flow'*) since this is the period during which the system will be under greatest stress and, hence, on which most design effort will need to be focused. The *'peak hour factor'* expresses the ratio between the peak hour flow and that of the whole day - typical values being in the range eight to twelve per cent. In some analyses it will be appropriate to identify not only the highest hourly flow but also any subsidiary peaks. Many analysts seek to establish the busiest period of the day (the *busiest n hours*) - where n may be any number below 24 depending on the purpose of the study and the characteristics of the site. Some commonly adopted values for n are 2, 10, 11, 12, 16 and 18. Concentration of the survey effort within these periods (once identified) will clearly lead to savings in survey resources; a 12 hour count will typically account for about 85 per cent of daily flow in an urban area or perhaps 80 per cent on an interurban route.

In addition to the more-or-less predictable variation which is encapsulated in flow profiles, flows will vary from day to day in a fairly random way due to local factors. For this reason it is not be wise to base a design or evaluation on flows measured for one day only. It is therefore usual to seek annual averages or other long-term measures. The most widely used are: the *annual average daily traffic* (AADT) which is total annual flow divided by number of days in the year; the *annual average weekday traffic* (AAWT) which is similar, but for Monday through Friday only, and the *design hour volume* (DHV). AADT and AAWT are used primarily for network and maintenance planning, and evaluation, while, as its name suggests, DHV is used for design work. Although some authorities will design to the average peak hour flow, most will want to cater for the more demanding circumstances which occur relatively infrequently. Hence an interest in the *nth highest hourly volume* (nth HHV) - which is the flow in the nth busiest hour in the year (e.g. if n equals 30 then there will be 29 hours in the year with higher flows). Although there is interest in the very highest hour in the year (i.e. where n equals 1) it is not common practice to design for this flow (which may be a freak outliner) and there is greater interest in hours taken from further down the distribution e.g. the 30th highest, the 80th highest or the 100th highest hour.

8.1.1 Classifications of flow

Traffic can be classified in various ways. The most widely used are: type (and perhaps occupancy) of vehicle, speed and manoeuvre.

Type of vehicle may be required for evaluation purposes but is also required in traffic analyses where the different vehicle types put different loads on the system either because they have different space and/or manoeuvrability characteristics (as will be reflected in their different pcu values) or because they cause different amounts of damage to the road surface. A distinction is accordingly made between motorcycles, cars and light commercial vehicles, buses and coaches, and heavy goods vehicles (HGVs). Given the much greater damage done to the road surface by heavy vehicles it is common practice to require a survey to show the percentage of HGVs (*'percentage heavy'*) in the flow. For some purposes the HGV class is split into sub classes on the basis of number and spacing of axles, vehicle weight and length. Some authorities thereby distinguish up to ten types of HGV. There is no worldwide standard classification. Countries typically adopt a system identifying classes of vehicle appropriate to their needs. The Australian (AUSTROADS) system recognises 12 types of vehicle putting motorcycles, cars and light vans in the same group but identifying two sizes of road train. The US FHWA uses a 13 class system to distinguish vehicles ranging from motorcycles to vehicles with seven or more axles, while the UK Transport Research Laboratory has proposed a more detailed system identifying 25 classes of vehicle covering the same range. The definitions adopted in practice are typically a compromise between what is ideally required and what is achievable with the survey methods and resources available. Different survey methods have different capabilities in this respect.

Vehicle occupancy (i.e. the number of occupants in a vehicle) is rarely required in standard traffic analyses but may be important in the context of transportation planning studies where, for example a 'true' picture of modal split is required.

The number or proportion of vehicles travelling at each of a series of *speed ranges* (sometimes known as *'speed bins'*) may be required particularly if safety or enforcement are issues. As will be seen later, many automatic traffic detectors can produce these data quite readily.

If a count is being made at a junction it will clearly essential to classify vehicles according to *type of manoeuvres* they are making. It is common practice to identify each movement separately (e.g. from north entry to east exit, from south entry to east exit, etc) and then to aggregate them subsequently where appropriate. In the case of mid link counts it is often desirable to be able to separately identify the two directions of flow.

8.1.2 Choice of survey technique

The choice of survey technique is constrained by the resources available and the site conditions and will depend on the precise data requirement - notably on the length and disaggregation of the required count. A useful distinction can be made between methods based on field observation by survey staff, those using video images and those involving automatic detectors.

8.2 Manual counting methods

The trend is clearly away from the use of human enumerators as a consequence of rises in the costs of labour on the one hand and of the accuracy and cost effectiveness of automatic methods on the other. Nevertheless, field observation by survey staff is still a very effective method particularly for relatively short and ad hoc surveys. Its enduring strength lies in its relatively low set up costs and the ability of humans to distinguish between vehicles according to an almost infinite range of parameters - type, occupancy and the manoeuvre being made are the most commonly required.

The basic idea of this type of survey, which is often referred to as a 'manual classified count', is that survey staff located at the roadside, or other vantage point, record the number of vehicles passing the survey site or making the designated manoeuvre. Depending on the purpose of the study, they might distinguish between different categories of vehicle to a greater or lesser extent. The survey period will typically be split into a number of intervals (e.g. a four hour survey might be composed of 16 quarter hour intervals) for each of which sub totals will be required.

Manual classified counts differ according to the method of recording being used. The simplest, and cheapest, recording medium is a sheet of paper on a clipboard. Specially designed forms can be used to facilitate separate recording of different types of vehicles or of vehicles making different manoeuvres. Figure 8.2 shows a design of form widely used for mid link counts; it provides for different types of vehicle and requires the surveyors to tick off the next box each time they see a vehicle of the specified type. While this form allows for very quick transfer of results onto a summary sheet, it is sometimes criticised as having excessively small boxes and requiring the surveyors to take their eyes off the traffic each time they record a vehicle. Figure 8.3 shows an alternative design which allows for vehicles to be recorded more rapidly using free format strokes or the 'five bar gate' method (whereby strokes are grouped into fives to facilitate subsequent summing up). Figure 8.4 shows an alternative design disaggregated into separate turning movements. Some designs of form have a separate sheet for each designated time interval while others fit several intervals onto one sheet. In either case provision should be made for recording the interval totals at appropriate places on the form.

One of the most common problems with manual surveys is that the enumerators forget about the time intervals and either record several intervals together or have an inaccurately placed time boundary between successive intervals. This problem can be reduced with training or the use of watches which sound on alarm at the designated intervals.

168 *Understanding Traffic Systems*

UNIVERSITY OF SOUTH AUSTRALIA

Transport Systems Centre

Link Traffic Count

Road section: _____ Date: __ / __ / __

Observer: _____

UBD ref: _____ Sheet ___ of ___

Time __ __ : __ ending __ __ : __	FLOW DIRECTION:	FLOW DIRECTION: ....
Bicycles 🚲	1 2 3 4 5 6 7 8 9 10 11 12 13 14 15 16 17 18 19 20 21 22 23 24 25 26 27 28 29 30 31 32 33 34 35 36 37 38 39 40 41 42 43 44 45 46 47 48 49 50 51 52 53 54 55 56 57 58 59 60	1 2 3 4 5 6 7 8 9 10 11 12 13 14 15 16 17 18 19 20 21 22 23 24 25 26 27 28 29 30 31 32 33 34 35 36 37 38 39 40 41 42 43 44 45 46 47 48 49 50 51 52 53 54 55 56 57 58 59 60
Motorcycle 🏍	1 2 3 4 5 6 7 8 9 10 11 12 13 14 15 16 17 18 19 20 21 22 23 24 25 26 27 28 29 30 31 32 33 34 35 36 37 38 39 40 41 42 43 44 45 46 47 48 49 50 51 52 53 54 55 56 57 58 59 60	1 2 3 4 5 6 7 8 9 10 11 12 13 14 15 16 17 18 19 20 21 22 23 24 25 26 27 28 29 30 31 32 33 34 35 36 37 38 39 40 41 42 43 44 45 46 47 48 49 50 51 52 53 54 55 56 57 58 59 60
Cars, taxis and vans	1 2 3 4 5 6 7 8 9 10 11 12 13 14 15 16 17 18 19 20 21 22 23 24 25 26 27 28 29 30 31 32 33 34 35 36 37 38 39 40 41 42 43 44 45 46 47 48 49 50 51 52 53 54 55 56 57 58 59 60 61 62 63 64 65 66 67 68 69 70 71 72 73 74 75 76 77 78 79 80 81 82 83 84 85 86 87 88 89 90 91 92 93 94 95 96 97 98 99 100 101 102 103 104 105 106 107 108 109 110 111 112 113 114 115 116 117 118 119 120 121 122 123 124 125 126 127 128 129 130 131 132 133 134 135 136 137 138 139 140 141 142 143 144 145 146 147 148 149 150 151 152 153 154 155 156 157 158 159 160 161 162 163 164 165 166 167 168 169 170 171 172 173 174 175 176 177 178 179 180 181 182 183 184 185 186 187 188 189 190 191 192 193 194 195 196 197 198 199 200	1 2 3 4 5 6 7 8 9 10 11 12 13 14 15 16 17 18 19 20 21 22 23 24 25 26 27 28 29 30 31 32 33 34 35 36 37 38 39 40 41 42 43 44 45 46 47 48 49 50 51 52 53 54 55 56 57 58 59 60 61 62 63 64 65 66 67 68 69 70 71 72 73 74 75 76 77 78 79 80 81 82 83 84 85 86 87 88 89 90 91 92 93 94 95 96 97 98 99 100 101 102 103 104 105 106 107 108 109 110 111 112 113 114 115 116 117 118 119 120 121 122 123 124 125 126 127 128 129 130 131 132 133 134 135 136 137 138 139 140 141 142 143 144 145 146 147 148 149 150 151 152 153 154 155 156 157 158 159 160 161 162 163 164 165 166 167 168 169 170 171 172 173 174 175 176 177 178 179 180 181 182 183 184 185 186 187 188 189 190 191 192 193 194 195 196 197 198 199 200
Buses and coaches 🚌	1 2 3 4 5 6 7 8 9 10 11 12 13 14 15 16 17 18 19 20 21 22 23 24 25 26 27 28 29 30 31 32 33 34 35 36 37 38 39 40 41 42 43 44 45 46 47 48 49 50	1 2 3 4 5 6 7 8 9 10 11 12 13 14 15 16 17 18 19 20 21 22 23 24 25 26 27 28 29 30 31 32 33 34 35 36 37 38 39 40 41 42 43 44 45 46 47 48 49 50
Trucks (rigid) 🚚	1 2 3 4 5 6 7 8 9 10 11 12 13 14 15 16 17 18 19 20 21 22 23 24 25 26 27 28 29 30 31 32 33 34 35 36 37 38 39 40 41 42 43 44 45 46 47 48 49 50	1 2 3 4 5 6 7 8 9 10 11 12 13 14 15 16 17 18 19 20 21 22 23 24 25 26 27 28 29 30 31 32 33 34 35 36 37 38 39 40 41 42 43 44 45 46 47 48 49 50
Trucks (articulated) 🚛	1 2 3 4 5 6 7 8 9 10 11 12 13 14 15 16 17 18 19 20 21 22 23 24 25 26 27 28 29 30 31 32 33 34 35 36 37 38 39 40 41 42 43 44 45 46 47 48 49 50	1 2 3 4 5 6 7 8 9 10 11 12 13 14 15 16 17 18 19 20 21 22 23 24 25 26 27 28 29 30 31 32 33 34 35 36 37 38 39 40 41 42 43 44 45 46 47 48 49 50

Figure 8.2   Midlink classified traffic count form (reduced from size A4)

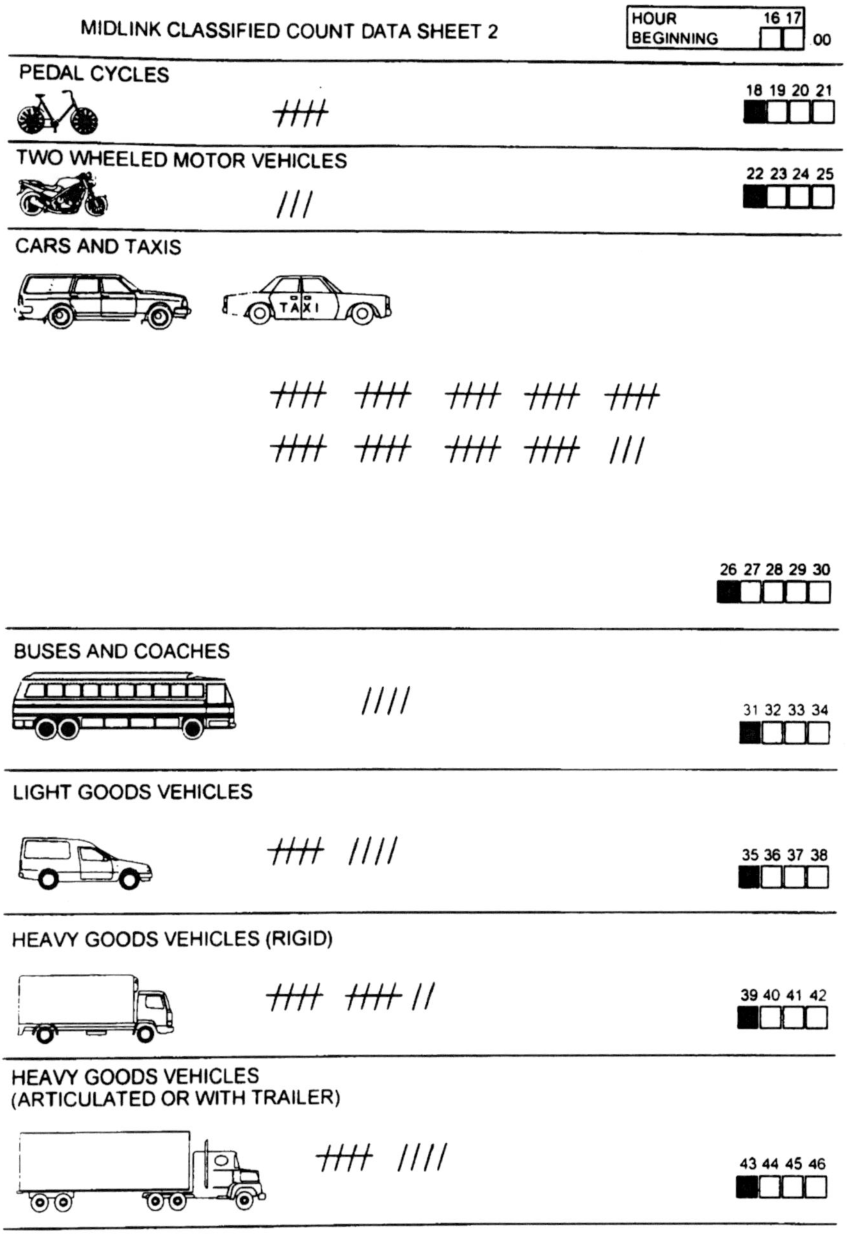

Vehicle Counting and Classification Surveys 169

| MIDLINK CLASSIFIED COUNT DATA SHEET 2 | HOUR BEGINNING | 16 17 □□ 00 |

PEDAL CYCLES

‡‡‡

18 19 20 21
■□□□

TWO WHEELED MOTOR VEHICLES

///

22 23 24 25
■□□□

CARS AND TAXIS

‡‡‡ ‡‡‡ ‡‡‡ ‡‡‡ ‡‡‡
‡‡‡ ‡‡‡ ‡‡‡ ‡‡‡ ///

26 27 28 29 30
■□□□□

BUSES AND COACHES

////

31 32 33 34
■□□□

LIGHT GOODS VEHICLES

‡‡‡ ////

35 36 37 38
■□□□

HEAVY GOODS VEHICLES (RIGID)

‡‡‡ ‡‡‡ //

39 40 41 42
■□□□

HEAVY GOODS VEHICLES (ARTICULATED OR WITH TRAILER)

‡‡‡ ////

43 44 45 46
■□□□

Figure 8.3 Midlink classified traffic count form using 'five bar gate'

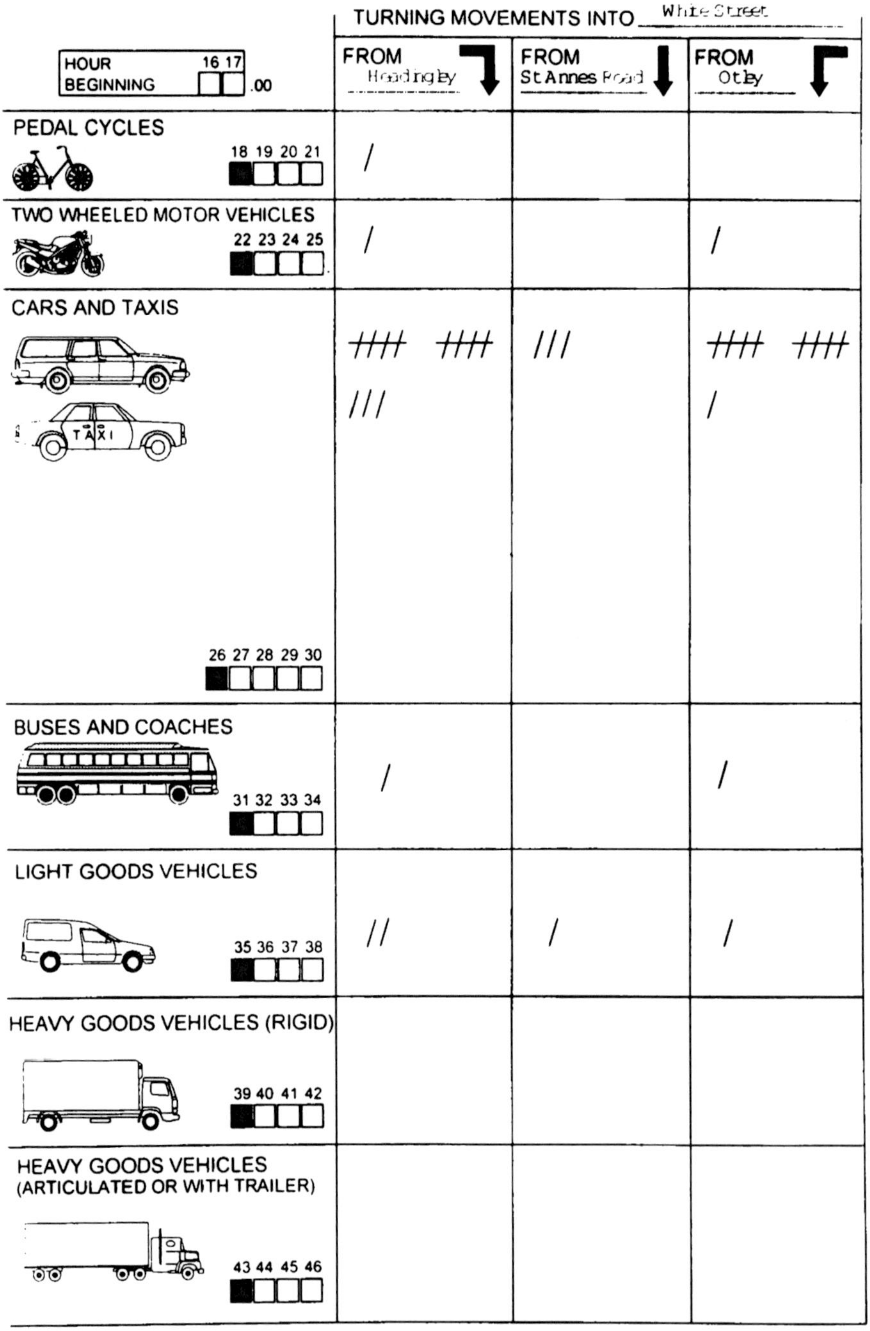

170 *Understanding Traffic Systems*

HOUR BEGINNING 16 17 .00	TURNING MOVEMENTS INTO White Street			
	FROM Headingley ↓	FROM St Annes Road ↓	FROM Otley ↓
PEDAL CYCLES 18 19 20 21	/		
TWO WHEELED MOTOR VEHICLES 22 23 24 25	/		/
CARS AND TAXIS 26 27 28 29 30	//// //// ///	///	//// //// /
BUSES AND COACHES 31 32 33 34	/		/
LIGHT GOODS VEHICLES 35 36 37 38	//	/	/
HEAVY GOODS VEHICLES (RIGID) 39 40 41 42			
HEAVY GOODS VEHICLES (ARTICULATED OR WITH TRAILER) 43 44 45 46			

Figure 8.4 Survey form for 'five bar gate' turning movement study

It is good practice to record the sub totals at the end of each time interval unless to do so would risk failure to record traffic in the next time interval.

Tally counters provide a useful alternative to paper and clipboard methods. Since each vehicle is simply recorded by pressing a key, and since an experienced enumerator can do this, even with a multi-key counter, while still watching the traffic, it is possible to record higher flow rates with tally counters than can be achieved with paper and clipboard. Assuming that the traffic is to be disaggregated into three vehicle categories a flow rate of up to 3000 vehicles per hour can be achieved with tally counters whereas, even with an optimally designed form, an enumerator would have difficulties with flow rates above 2500 vehicles per hour. Although an experienced enumerator can operate up to ten or more tally counters simultaneously (provided that they are of the multi-key variety or firmly fixed to a clipboard), most enumerators begin to have problems with more than three or four keys. A useful hybrid method is to provide the enumerators with tally counters to record the busiest flow groups and to ask them to record the low flow groups directly onto paper. It is clearly essential, when using tally counters, to record current count totals at the end of each interval. In the interests of simplicity, it is better not to zeroise the counters after each intermediate time interval but instead to allow the counter to cumulate right through to the end of the survey.

In the mid 1980s survey staff began to be equipped with handheld computers, (purpose built or customised) as a replacement for mechanical tally counters. The main advantages of these dataloggers are that they can have an automatic time base (thus freeing the enumerator from having to remember or to record sub totals at the ends of time intervals) and that they can be linked directly to a host computer or printer so as to produce tabulated results within minutes of the end of the survey. Another potential advantage of dataloggers is that they keep a record of the time at which the data was entered and this makes it difficult for lazy enumerators to 'invent' their data. The most sophisticated data collection packages allow the survey to be prescheduled by a supervisor following which the machine will go to sleep until a few minutes before the survey is due, at which time it prompts the surveyor to get into position and gives a countdown for the beginning of the observation period. The main disadvantage of dataloggers is the cost; even a relatively modest capital investment may not be warranted if relevant surveys are conducted very infrequently. In the early days some users formed negative views about these devices because some of the early examples were not very ergonomically designed and some were insufficiently robust for field use.

Most dataloggers have keyboard input ideally with large keys arranged in columns or a grid, but some devices have employed a bar-code reader or touch-sensitive screen.

The attraction of the bar-code method is that it allows for great flexibility in the design of the data sheet - e.g. it might represent a plan of a junction with codes placed appropriately to represent different manoeuvres (see Figure 8.5) but inexperienced staff often have problems in getting the bar-code reader to read the codes at first swipe and, if the traffic flow is high, they may tend to panic and produce even more failed reads. Touch sensitive screens offer flexibility advantages similar to those associated with bar-codes but, as yet, the devices have not been sufficiently robust for field conditions and the procedures required to set up a customised screen have been rather time consuming. Most users of these devices now agree that the theoretical advantages of extra flexibility provided by bar-codes and touch sensitive screens have been illusionary and that perfectly acceptable results are achieved with a grid layout keyboard input with customised templates to designate each key's function. The most widely used devices are those which combine a robust and ergonomic design with flexibility achieved through use of customised templates and a simple set-up procedure which enables the keys' functions to be designated and the survey period defined. Such devices can be cost effective for any organisation which undertakes manual classified count surveys more than two or three times a year. For a more complete review of the use of dataloggers in traffic surveys, see Bonsall et al (1989).

Depending on the complexity of the task and the level of flow expected, the enumerator may be required to collect data for several streams of traffic simultaneously or the task may be shared between several enumerators. In allocating streams of traffic to different enumerators it is obviously best to achieve a fairly equitable distribution of work and to avoid very peaked workloads for any individual. An example of the kind of point to bear in mind when doing this is that, at signalised junctions, flows out via a given exit will be less peaked than flows in via a given entry. The productivity and commitment of the field staff will be improved if they can be provided with somewhere to sit, preferably under cover. Many surveys are conducted from strategically parked vehicles but accuracy can be compromised if the vehicle is parked so as to obstruct the flow being measured.

8.3 Counting methods involving video

If a conveniently located Closed Circuit Television (CCTV) installation, perhaps part of a real time traffic surveillance and monitoring system, is available for the duration of a traffic count survey, it may be used to make a video record of the scene. The advantages of this method of data are:
- reduced requirement for field staff;

- survey is less weather dependent;
- existence of permanent record of the scene;
- possibility of extracting very detailed data on simultaneous or interacting events (e.g. vehicle trajectories through a junction, complex turning movements, junction performance studies, gap acceptance behaviour, traffic conflicts), and
- permanent record can be replayed by the analyst if the data contain unusual or interesting features which require some skill to unravel.

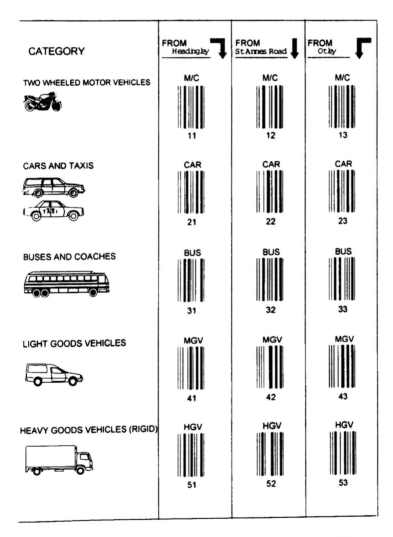

Figure 8.5   A typical bar code sheet (plastic coated for protection)

Generally, of course, there will not be a conveniently located CCTV camera and, if a video record is required, a video camera will have to be positioned for the purpose. The expense and resources involved in this mean that it will rarely be justified for relatively simple counts but may be appropriate when several simultaneous items of data are required or where the required data is particularly complex.

Selection of a site for the camera will be constrained not only by the requirement for a good view but also by access, security and, if a long survey is envisaged, availability of a power supply. The best viewpoint is often high up because obscuration and parallax are reduced while the achievable depth of field is increased. If there is no convenient building, or if permission to film from it is not granted then a telescopic mast can prove useful. Various suppliers and specialist survey firms can provide the necessary equipment which usually comprises a survey vehicle or trailer with its own power source, hydraulic telescopic mast with remotely controlled video camera and ground level monitor.

Whether the images were captured directly on a video tape at the site or transmitted to a base from a CCTV camera, the images will probably be stored on video tape (the falling costs of video disks may make this a competitive alternative in due course). Analysis of the tapes may be done entirely manually, with computer assistance or entirely automatically.

The manual technique involves an enumerator, or team of enumerators, watching the tape and noting the data onto forms or directly into a computer or datalogger. If the tape has a time stamp and the phenomena of interest are intermittent or inherently slow, then all or part of the tape may be played at faster than real time without loss of the basic time base. Although it is theoretically possible to process a tape faster than real time, it often proves necessary to stop the tape from time to time or to rewind it so as to note the full details of a complex interaction or to replay it to capture another set of data (e.g. on the first pass through an enumerator might note one set of manoeuvres and note another set on a subsequent pass). For these reasons it is quite common for the transcription process to take considerably more time than did the original field recording. The resulting laboratory staff time may more than offset the saving in field staff but the likelihood is that a more accurate and detailed database will result.

Computer assisted transcription has been standard practice for several years (e.g. Wootton and Potter, 1981). It is particularly helpful when the required data are too complex to be transcribed in one pass through the tape. The computer is used to build up a log of events which have been transcribed in each pass and alerts the transcriber (usually via an audible cue) to the fact that a given event has been logged in a previous pass. The task is complete when a complete pass can be made without any uncued events. Given an ability to superimpose graphics over a video image, computer assistance can now provide a facility for drawing screen lines across an image, and displaying an appropriate visual cue, in addition to the audible one, on an appropriate part on the playback image.

Automatic analysis of video records is now possible. Specialist software to recognise and log events such as the movement of vehicles across a designated screenline is available. The software works with 'pixels' (elements of the matrix which make up each video image) and records changes in the status of individual pixels, e.g. from light to dark. By looking for patterns in the changing pixels the software can 'recognise' the characteristics of moving shapes as belonging to vehicles moving across the image and can even classify vehicles according to gross type (car, bus) with some accuracy. This technology is related to that used to 'read' vehicle registration plates and to detect incidents (see Chapter 9).

8.4 Automatic detection of vehicles

Many methods are available for detecting the presence of vehicles, and hence for counting them. Different methods predominate in different countries, partly as a reflection of the different ambient conditions (climate, traffic mix and road conditions) but also as a result of commercial decisions and historical accident. Each installation will comprise a detector unit, a decoder and a storage device. Table 8.1 lists some of the most widely used detector methods and summarises some of their characteristics.

8.4.1 Pneumatic tube detectors

Pneumatic tubes count vehicle axles. They are thick-walled rubber tubes mounted on the surface of the road and send a pulse of air to an air switch when squashed by the passage of wheel. The tube must be firmly fixed to the road to avoid the tube being whipped up by passing vehicles and thus triggering multiple signals each time the tube hits the road. They should be located away from locations where children may be tempted to stamp on the tube, and away from locations where vehicles might park on it and so prevent the passage of the air pulse.

The air switch should be calibrated so that it is sensitive enough to detect all vehicles irrespective of speed but not so sensitive as to pick up bounce-signals in the tube. It can sometimes be difficult to adjust the switch so as to reliably include bicycles.

Table 8.1   Automatic detectors

Type	Description (see also figure 8.8)	Method	Mounting
Pneumatic Tube	rubber tube laid across the road surface	each time a vehicle's wheels cross the tube they squeeze it and so send a pulse to an air switch. Air switch is calibrated to detect main pulse but not any pulses due to tube 'bounce' after passage of vehicle.	on road surface
Switch tape/ contact strip/ treadle switch	pair of steel strips, one above the other and held apart by rubber spacers, laid across the road surface	each time a vehicle's wheels cross the detector the two strips are pressed together thus completing an electrical circuit which is detected at the roadside.	on road surface or held in groove
Tribo-electric cable	cable laid across the road surface with coaxial conductors separated by dielectric material	each time a vehicle's wheels cross the cable they cause the conductive material to rub the surface of the dielectric material causing a charge to accumulate and be transmitted to a sensor at the roadside	on road surface or held in special casing anchored in slot
Piezo-electric cable	coaxial conductors separated by polarised ceramic powder are encased in a cable laid across the road surface	each time a vehicle's wheels cross the cable they squash the ceramic powder and thereby induce a charge which is conducted to a sensor at the roadside	on road surface or held in special casing anchored in slot
Induction loop	wire coil buried as a loop just below the road surface or affixed to road surface	alternating current through the coil generates electro-magnetic field, passage of large metal object temporarily induces additional currents which are detected in the roadside sensor	in slot or taped to road surface
Photoelectric beam	any interruption of beam shone across carriageway is detected	beam is broken by passing vehicle (note that reflectors can send beam back across carriageway so that all electronics can be on the same side of the road)	at roadside

Cont\...

Table 8.1 continued

Type	Description (see also figure 8.8)	Method	Mounting
Infrared beam	signal is transmitted onto carriageway and any change in reflected signal is detected	near infrared signal is focussed onto a 'detection zone' on the carriageway, reflected signal is picked up at sensor. Vehicles entering the detection zone will disturb this reflection and so be detected.	oblique or vertically downward from overhead or roadside gantry
Microwave beam	beam is transmitted along the carriageway and any reflected signal is detected and analysed	if the beam strikes a moving vehicle, the doppler effect can be used to analyse the reflected signal to determine the direction of movement and the speed	on overhead gantry or at roadside
Video imaging	digital video image is analysed automatically via patterns at the level of individual pixels	change in pixel values indicates that an object has moved in that part of the image. Method 1 examines patterns in the pixels to recognise vehicle profiles. Method 2 defines 'virtual sensors' across the carriageway in the field of view and objects moving across these sensors are logged.	on overhead gantry, or elevated location at roadside

Pneumatic tubes are very suitable for temporary surveys (24 hours to one month), particularly on low flow roads. The installation will require regular checking to ensure that the tube has not come loose or been holed. Tubes may need replacement after the passage of 250 000 or so vehicles on a good road surface or considerably less than that on a poor road surface or in an extreme climate. Exposure to sun will eventually denature the rubber and high humidity can prevent effective operation. The tubes may be prematurely damaged by hard run wheels or animal traffic and so may be unsuitable for use in some developing countries.

8.4.2 Switch tapes

Switch tapes or contact strips comprise a pair of metal contacts mounted in rubber and stuck to the road surface at right angles to the direction of flow.

They work by completing a circuit whenever pressed together by the weight of a vehicle's wheel. The rules for location, installation and maintenance are essentially the same as for pneumatic tubes. Switch tapes used to have a reputation for poor reliability but this problem has been largely overcome with improved design and manufacture. Even so they have a shorter life than pneumatic tubes (typically needing replacement after about 50,000 vehicles). Their extra cost is justified by their greater reliability (while new), their greater tolerance of climatic extremes and their relative ease of installation. They are often used for limited duration surveys and on low volume roads.

8.4.3 Multicore cables

Various types of multicore cable have been developed as an alternative to pneumatic tubes and contact strips laid across the road surface. The most commonly used, known as a tribo-electric cable, contains cores which induce a charge when they rub together when the cable is temporarily distorted by the passage of a vehicle.

An increasingly popular variant on this is the piezo-electric cable which contains piezo-electric material such as polarised ceramic powder which creates a charged in co-axial conductors when squashed. Top grade piezo- electric cable is relatively expensive and needs careful installation and calibration but, since the change in signal is proportional to the load applied, it can be used for classifying vehicles according to their weight.

8.4.4 Summary on axle detectors

All the above methods involve the use of detectors lying across the carriageway. They are all triggered by the passage of vehicle wheels and thus produce a count of axles rather than vehicles An appropriate conversion factor must therefore be applied. They must be installed at right angles to the flow (otherwise they might detect two wheels on one axle as if they were separate axles) and this effectively prevents their use for counting turning movements. If the detectors are installed in pairs at least 200 mm or so apart, the resulting signals can be analysed to deduce direction of flow and, if they are two to five metres apart, vehicle speed and axle spacing (see Section 9.2.2).

8.4.5 Inductive loop detectors

Induction loops comprise fine wires carrying an alternating electric current. They are usually buried in slots cut in the road surface, but can be stuck onto the road with special tape and bonding agent. The loops detect vehicles via the change in induction created by the presence of metal objects. This technology,

Vehicle Counting and Classification Surveys 179

widely used in real-time traffic control systems and at automatic barriers in car parks, can be used to count vehicles and, because different types of vehicle have different distributions of metal and so produce different induction 'signatures', to categorise them (see Figure 8.6).

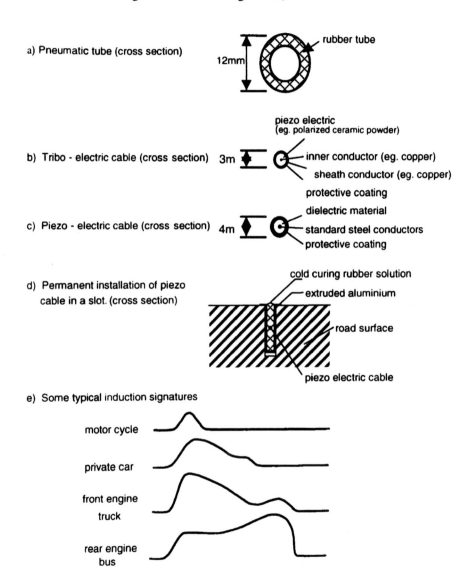

Figure 8.6   Selected vehicle detection technologies

The loops can be configured in a variety of patterns (rectangles, diamonds, chevrons etc) depending on the intended purpose. Careful calibration of the detection equipment is needed. If too sensitive, the loop may be 'side-fired' by vehicles in adjacent lanes or even by a very wet road surface, whereas if too insensitive it might not detect bicycles or motorcycles. Once calibrated, a permanent loop installation can yield a stream of high quality data almost indefinitely with no more than infrequent checking and recalibration. The threat to the loop comes from road maintenance or resurfacing work.

8.4.6 Magnetic imaging sensors

A relatively recent innovation is the development of magnetic imaging sensors which can detect interference to the ambient magnetic field caused by large ferrous objects entering a target area of about two square metres. Portable sensors housed in pads the size of a small book and can be fixed to the road surface or, for permanent installation, buried in it. After acclimatising themselves to the ambient magnetic field, a process which takes about 30 minutes, they can detect any changes to this and, given appropriate software, can count vehicles and classify them according to their length and speed. Since a separate device is required for each traffic lane this technology is likely to be an expensive approach to monitoring flows at multi-lane sites. Also there is, as yet, little corroborated evidence as to the accuracy achievable with these devices. Nevertheless they appear to offer a convenient approach in certain circumstances.

The choice between surface mounting and slot (or groove) mounting for tubes, cables, loops and magnetic sensors will generally be determined by the intended duration of the survey and the local conditions. Surface mounting is quicker and therefore cheaper and less disruptive to the traffic but it leaves the sensor exposed to wear and tear from traffic. If the road surface is poor, or the traffic heavy, or if there is a significant amount of severe acceleration/deceleration in the vicinity of the site, a surface mounted detector may need refixing at frequent intervals. Permanent or semipermanent sensors will therefore invariably be slot or groove mounted rather than surface mounted. Good practice, when resurfacing a road, is to give some thought to whether a counting station might be needed (even temporarily) on that road, and if so, to cut slots and install sensors in anticipation of that need before reopening the road to traffic.

8.4.7 Electromagnetic beams

All the methods so far described involve installation of detectors in or on the road surface and this is inevitably involves some disruption to traffic and some danger to the installers. A number of alternative methods of detection loosely described as electromagnetic beams, are available which do not suffer from this drawback.

The most basic of such methods are based on the principle of a beam, (typically laser or photo electric) transmitted across the road, being interrupted by the passage of vehicles. The count is then derived from the number of times the beam is broken. A rough classification of vehicles can be achieved by having beams at various heights, such that the higher ones are broken only by the largest vehicles or at specified distances apart so that vehicle speed and length can be determined. The standard installation includes a transmitter/receiver on one side of the road and a reflector on the other. When installed in this way the method is clearly of limited use in busy multilane traffic but is very popular in certain circumstances. Multilane applications have been developed, albeit at some cost, using vertically directed beams pointing down from a bridge or overhead gantry and reflected back from a detector on the road surface.

Another group of techniques is based on the use of signals which are reflected back to the transmitter/receiver unit by the vehicles themselves. By measuring changes in the reflected signal the presence, and in some cases the position and speed, of moving vehicles can be deduced. A moving vehicle will cause a change in the frequency of the transmitted signal and the Doppler effect can be employed to deduce speed (see Chapter 9). The most widely used signals are microwave and infrared. Of the two, microwave is cheaper and is more tolerant to variable environmental conditions but infrared deals more effectively with very slow moving traffic, as might be found in queues. Infrared and microwave have advantages where installation of a surface mounted detector would be difficult (e.g. on a motorway).

8.5 Automatic classification of vehicles

So far we have indicated that certain automatic detectors can distinguish vehicles according to their length (e.g. via pairs of tubes, loops or beams), their speed (e.g. via microwave), their physical bulk (e.g. via automatic analysis of digital video images) and their chassis mass and configuration (e.g. via inductive loops). By arranging for an appropriate array of sensors to be in place at a given site it is therefore possible to produce quite detailed classifications. A site consisting only of inductive loops can distinguish up to six types of vehicle with reasonable accuracy (the accuracy of counts for individual vehicle types will always be less than that for total flow). The addition of a tube or cable can increase the precision of the classification by providing more precise information about wheelbase configurations.

The addition of extra sensors can also help to differentiate large vehicles from closely platooned or tailgating vehicles. There is, however, no reliable way of distinguishing trailer combinations or road trains from close tailgating vehicles. This problem aside, the combination of inductive loops and pairs of piezo-electric cables can produce good estimates for up to 13 vehicle configurations and has become a popular configuration for permanent counting stations in, for example, the UK.

Until relatively recently it has not been possible to produce reliable estimate of vehicle weight without slowing the vehicle virtually to a standstill. The development of *weigh-in-motion* (WIM) technology has therefore been very welcome for studies where estimates of pavement damage are a particular issue. The most sophisticated WIM is based on piezo-electric sensors embedded in mats fixed to the road surface. Accuracy of up to ±15 per cent at 100 km/h is now achievable using this technology. Alternative methods using the instrumentation of road bridges and culverts, so that axle weights can be deduced from measured deflections, are also in widespread use.

8.6 Data capture without surveys

Automatic detectors are becoming increasingly sophisticated and now have the potential to produce disaggregate flow data with considerable reliability, more or less indefinitely. This potential is being realised by some highway authorities through the establishment of permanent counting stations at key locations in the national, regional or local road network; some of these stations are producing data continuously ('permanent sites') while others are activated on a rota ('rotating sites'). The data derived from such sites is used primarily for trend monitoring, perhaps through time series analysis and modelling (see section 17.4), but contains within it the basis for the calculation of more accurate seasonal adjustment factors and confidence limits for counts than has previously been possible.

The development of on-line adaptive traffic control systems (e.g. SCOOT, see Hunt et al (1981), and SCATS, see Sims and Dobinson (1979)) has necessitated the establishment of a network of detectors connected to central control computer in many urban areas. These detectors are continuously sending data about link flows and detector occupancy to the central computers. These data can be archived (after suitable summarisation and compression) and are then potentially available as part of a rich and detailed database of traffic conditions within the control area.

The involvement of computers in the collection of tolls and charges on bridges, in tunnels, on motorways and at parking stations offers another potential source of traffic flow data. It should be possible, at marginal cost, to organise and store what would otherwise have been transitory data, in such a way that it becomes a valuable record of flow profiles on key links. The managers of toll facilities may regard these data as commercially sensitive and so may restrict their availability. Equally, however, they may recognise the value of this information and seek to market it.

There are already several examples of highway authorities making their traffic data available for research purposes, either on-line or off-line, and it is becoming increasingly likely that detailed time series traffic data will be available for access without the need for special surveys. Against this background analysts may need to commission or conduct surveys only when they need data relating to a specific site or with an unusual combination of variables.

9 Traffic condition data

Vehicle speeds and headways along with travel times, delays and incidents are crucially important measures of the performance of a transport system. Collection of the relevant data is therefore a prime concern of the traffic analyst.

9.1 The need for data on traffic conditions

Speed data may relate to point velocities at some location in a network or to overall journey speeds (e.g. door-to-door distance divided by door-to-door time). Point velocities are particularly relevant in studies of driver behaviour and safety while journey speeds, along with travel time data, are more relevant to studies of network performance. Data on headways (vehicle time spacing) and traffic incidents are relevant to both driver behaviour and network performance analyses.

9.1.1 Point velocities

Point velocities, sometimes known as 'spot speeds', are a necessary input to detailed studies of driver behaviour, traffic dynamics and safety. At its simplest, interest may be focused on the distribution of speeds at which a population of drivers pass a particular point or undertake a particular manoeuvre. This will involve logging the point velocities of each of a series of vehicles and then examining the resulting distribution. Individual speeds within the distribution can immediately be identified as being above or below any given value (e.g. the design speed or the posted speed limit) and the value of certain key descriptors of the distribution can be calculated (Chapter 15). Particular interest in safety/enforcement work is paid to the mean and to the 85th percentile speed. Interest in the 85th percentile reflects the belief that if a speed limit is exceeded by more than 15 per cent of vehicles then additional enforcement and/or an increase in the limit should be considered. A measure of dispersion of the distribution, such as the range, standard deviation or coefficient of variation (the ratio of standard deviation to the mean), is also a useful indicator in safety and traffic analyses: other things being equal, a large variation of speeds within a given population will indicate the existence of potentially dangerous conflicts (e.g. overtaking manoeuvres) and will be incompatible with efficient signal offsets in area-wide traffic control.

9.1.2 Vehicle headways

Vehicle headway data form an important input to analyses of traffic safety and network performance. They provide contextual information for studies of vehicle speeds and gap acceptance and contribute to the analysis of network capacity. Interest usually focuses on the headways of traffic passing a particular point but some analyses will require data on changes over time in the headways within a given cohort of traffic. Similarly, most analyses rely heavily on the mean headway although much can be learned about bunching by examining the distribution of headways.

Real time analysis of headway data, and of *traffic density* (vehicles per unit length of road) are used along with point velocities as measures of congestion or system overload. In this application interest will be in threshold values which may be used to trigger alarms in the control centre or call up new signal plans designed to deal with the congestion. These data can also be fed to in-vehicle guidance systems in order to warn subscribers of the presence of congestion on strategic routes in the network.

Detailed examination of the behaviour of, and interaction between, individual drivers in a stream of traffic is made possible by the availability of time series data on either the speed of a vehicle over time (the *speed-time profile*, see Figure 9.1), or the headways and speeds of the vehicles in a traffic stream (*vehicle trajectory data*, see Figure 9.2). Both types of data are very expensive to collect, and so are rarely used outside the context of research studies, but they provide revealing insights into traffic dynamics and driver behaviour. From the trajectories in Figure 9.2 it is possible to deduce the distribution of spot speeds at a given point (by observing the trajectory gradients of all vehicles at that point); the average speed of a cohort of vehicles between two given points; the average headway of vehicle passing a given point; the average headway (or spacing) between any two vehicles etc. The interaction between vehicles and the formation and dispersion of queues can also be studied.

9.1.3 Journey times and speeds

Journey time data and journey speed data (sometimes known as 'space speeds') are particularly useful in studies of network performance where they can provide a particularly valuable input to the overall evaluation.

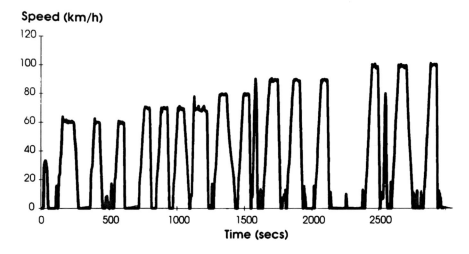

Figure 9.1 Sample vehicle speed-time profile data

Interest may be focused on travel times (or speeds) for whole journeys or for particular sections of journeys (e.g. on a particular type of road or between a particular pair of screen lines). Similarly, some analyses will suggest the exclusion of stopped time (e.g. in queues) while others will want to include the total travel time. Table 9.1 lists some of the commonly used definitions. The focus is often on the mean values of the resulting distribution of travel time (or speed) but there is increasing interest from behaviourally oriented economists in the variability and predictability of travel times and hence in the coefficient of variation and in indicators such as the 90th percentile travel time (i.e. that time which the driver may perceive as likely to be exceeded in only one journey out of ten).

It is seldom possible to estimate or record the speeds of all vehicles within the target group. Thus great care must be taken to ensure that the chosen sample is likely to be representative of the whole group. A classic error in such surveys is simply to record as many vehicles as possible - this results in an under representation of vehicles at the busiest periods during the survey and/or a tendency to over-represent the 'easiest' (e.g. slow moving or distinctive) vehicles. If sampling is necessary, it should be carried out according to carefully chosen criteria and/or in such a way that appropriate weightings can be applied to each observation. Note however, that if a data set contains differentially weighted observations this can complicate the statistical analysis (Section 9.2.1 includes a practical method for sampling speed observations).

188 *Understanding Traffic Systems*

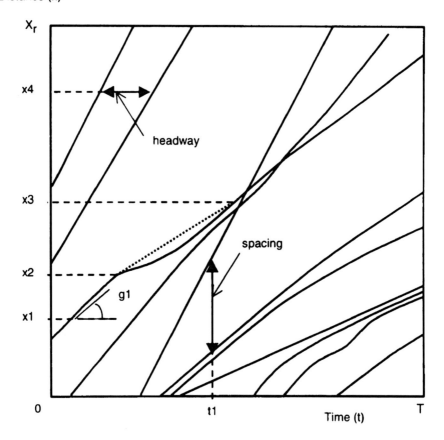

Figure 9.2 Time space diagram showing individual vehicle trajectories

Notes:
(1) Trajectories of nine separate vehicles are shown.
(2) Speeds are shown by the gradients of the trajectories and acceleration/deceleration by rate of change in these gradients.
(3) Spot speeds are deduced from the gradient (g1) of the tangents against each vehicle's trajectory as it passes a specified location (x1).
(4) Journey speeds for individual vehicles are deduced by taking the average gradient of its trajectory between two points (e.g. x2 to x3).
(5) The headway between two vehicles at a given location (x4) is the horizontal (x axis) distance between their trajectories at that point.
(6) The spacing between two vehicle at a given time (t1) is given by the vertical (y axis) distance between the trajectories at that time.

Table 9.1   Various measures of travel time

travel time – the actual (observed) time taken to traverse the survey section

free flow travel time – time taken by an unimpeded vehicle to traverse the survey section

free flow speed – the length of the survey section divided by the free flow time (note that this is often considered to be higher than the speed limit)

delay – the difference between free flow time and observed travel time

stopped time – the total period of time for which a vehicle is stationary within the survey section

running time – the total period of time for which a vehicle is in motion within the survey section (note that travel time is the sum o stopped time and running time)

running speed – the length of the survey section divided by the running time (running speed is likely to be somewhat lower than free flow speed)

9.1.4 Traffic incidents

A traffic incident may be defined as an interruption in the normal flow and/or speed of the traffic on a link. It may be caused by an accident, the erratic behaviour of an individual driver, an abrupt loss in capacity (such as might be caused by a parked vehicle) or by instability due to an excess of demand over supply (Section 4.4). The detection and logging of such incidents can provide a useful measure of system performance and is an increasingly important source of data for traffic studies. The installation of detectors with which the flow can be continuously monitored makes it possible to detect traffic incidents in real time. Software can be used to detect irregularities in the stream of data coming in from traffic counters, occupancy detectors or speed monitoring equipment (e.g. Dia and Rose, 1997). Many highway authorities use such software in on-line management systems in order to alert their staff to potential incidents. There is a particular need for this alert function in control rooms which can receive video data from more surveillance cameras than any individual staff member could be expected to monitor unaided.

190 *Understanding Traffic Systems*

9.2 Methods of collecting data on spot speeds

A number of methods exist for collecting spot speed data, based on the alternative technologies that may be employed.

9.2.1 Manual methods

The traditional way of collecting spot speed data was to define a short stretch of road (30 metres or so) and, with a stopwatch, note the time taken by each vehicle to cover the defined distance. The method relies on quick reactions by the enumerator and can be subject to parallax error. The parallax error can be reduced by the use of painted marks on the road surface, an elevated observation point or the use of mirrors as shown in Figure 9.3.

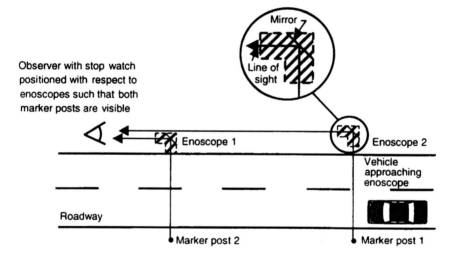

Figure 9.3 An arrangement of mirrors to overcome the parallax problem

Nothing, however, can be done completely to overcome error introduced by slow or variable reaction times on the part of the observer. This error is particularly severe if speeds are high (a 0.1 sec overestimate of time for a vehicle to cover 30 metres will yield an underestimate of speed of 3.2 km/h if the true speed is 60 km/h but an underestimate of only 0.8 km/h if the true speed is 30 km/h). The error can be reduced if the designated stretch of road is lengthened (doubling the length of the designated stretch will halve the error) but when the length of road increases beyond about 40 metres, one is really calculating a vehicle's average speed over a stretch of road rather than its spot speed. Given its susceptibility to human error, manual recording is now used only as a last resort.

Manual recording of vehicle speeds via the stopwatch method is particularly susceptible to sampling bias. The natural tendency to under-represent the busiest periods during the survey is further compounded by the fact that slower vehicles take longer to record and hence there is under representation of the slowest periods during the survey. The simplest solution to this problem, proposed by Almond (1963), is to instruct the surveyor to record the nth vehicle after (the start of) the last observation (where n is the smallest number which can be dealt with at what is expected to be the busiest period during the survey). It is likely of course that an assistant will be required to keep track of the flow and tell the surveyor which vehicles are to be recorded.

9.2.2 Methods involving automatic timing

Manual methods of timing vehicles over a short stretch of road of known distance are susceptible to human error but the simplicity of the concept remains appealing and is widely applied with automatic timing devices. With electronic timing, the time taken for a vehicle to trigger two detectors a known distance apart can be logged with great precision. This enables accurate measurements of speed to be taken even over very short stretches of road, perhaps one to two metres, although some traffic survey professionals prefer to use slightly longer distances between detectors (up to five metres).

Various detectors can be employed as part of such a system, some authorities prefer pairs of pneumatic tubes, switch tapes or overlapping induction loops, others prefer pairs of photo-electric or electro-magnetic beams. Clearly the accuracy of the speed estimate is dependent on the distance between the detectors being known very precisely. Although, it is not possible to install loops, switch tapes, tubes or tapes with absolute precision, provided that they have been laid parallel to one another, it is possible to calibrate the installation using an independent estimate of speed or, at sites where the equipment is also estimating vehicle length, of length. The tendency of tubes to be pushed slightly in the direction of travel puts a theoretical limit on the precision that can be obtained using pairs of tubes, even though if both tubes have been installed with equal and adequate tension this need not be a serious effect. Nonetheless, this problem, together with the relative fuzziness of signals from tubes, causes some authorities to prefer switch tapes rather than tubes for this work. Overlapping induction loops, combined with pairs of tribo-electric or piezo-electric cables, are overwhelmingly preferred at permanent sites. They have proved to be a very accurate source of speed data, so much so that they are now used for enforcement purposes, for example to register speed and trigger 'speed cameras' to photograph the registration plates of speeding vehicles.

Pairs of photoelectric or electromagnetic beams are suitable for temporary sites where it would be difficult to install tubes, cables or switch tapes with sufficient precision. As was mentioned in the previous chapter, the beams can be directed across the traffic stream or, if lane differentiation is required, vertically downwards. The method may be based on the beams being broken or reflected back by the passing vehicle. Magnetic imaging devices can also be configured to detect the speed of individual vehicles passing through their detection zone. A benefit sometimes quoted for the collection of speed data using photoelectric or electromagnetic beams or magnetic imaging devices is that if the transceiver unit (and reflector where necessary) are unobtrusively sited, the installation can be effectively invisible and thus there need be no fear of it affecting driver behaviour. In fact, although tubes and switch tapes, and to a lesser extent loops, will be apparent to drivers, there is little empirical evidence that they have any significant effect on speed unless there is a human attendant present.

Clearly, whichever of the above methods of detection is used, the resulting data can include a traffic count as well as speed and headway. The possibility of collecting classified count data at little extra cost, and so being able to associate speed with flow, traffic mix and vehicle type, has made these methods of collecting speed data particularly popular with traffic analysts.

9.2.3 Methods employing the Doppler effect

The Doppler effect is that property whereby the change in the frequency of a signal is proportional to the speed at which its source is moving towards or away from the receiver. This can be utilised to estimate speed using radar (narrow beam microwave), laser beams, widebeam microwave and infrared signals. A signal of known frequency is transmitted and reflected off a target vehicle back to a sensor where the change in frequency is measured to give an estimate of the speed of the target vehicle relative to the transmitter/receiver.

Radar utilises a narrow beam which can be focused on an identified target vehicle while widebeam microwave and infrared give an average reading for all vehicles within the target area. Radar is therefore appropriate to enforcement applications with a carefully aimed hand held gun while the other methods are more appropriate for monitoring and surveillance from fixed installations directed towards a given stream of traffic.

Radar guns were devised for enforcement but can clearly be used for traffic speed surveys. Provided that they are correctly aligned within a few degrees of approaching or receding vehicles they can give very accurate estimates of an individual vehicles speed. The estimates are displayed on a light emitting diode (LED) or liquid crystal display (LCD) panel and can then be transferred to a record sheet, or datalogger along with other relevant information such as the vehicle type and the current time. A single surveyor can complete both tasks but is likely to miss several vehicles while he or she records details of the previous one. If an assistant is available to record the details the surveyor can concentrate on taking accurate measurements of every nth vehicle (using Almond's method, see Section 9.2.1).

One problem with radar surveys is that the guns have become so associated with enforcement that drivers are likely to slow down when they see them. Some drivers may even have radar detection equipment in their vehicles. This problem can be reduced if the surveyor takes readings from the receding traffic and so obscures the gun from approaching drivers. A major source of inaccuracy in radar surveys is failure to align the gun correctly with respect to the approaching/receding traffic, it can prove difficult to find a place with appropriate alignment and uninterrupted view which is not also very conspicuous. Given these difficulties, and the high cost of mounting a radar gun survey both in terms of equipment and manpower, this method is now generally restricted to very short surveys (of a few dozen vehicles) perhaps in the context of investigation of an accident site, rather than for longer surveys intended to produce a rich database on vehicle speeds.

Widebeam microwave and *infrared* transceivers installed more or less permanently on poles, bridges and gantries are, in contrast to the hand held gun, increasingly used for remote surveillance and monitoring of speeds. The transceivers can be designed and aligned to have a field of view appropriate for high speed traffic with relatively long headways such as might be expected on motorways or for slower speed in denser urban conditions. The transceivers cannot differentiate between separate vehicles within the field of view and are likely to 'see' larger vehicles more easily than small ones so the resulting data cannot be used to ascribe speeds to individual vehicles. The choice between microwave and infrared should reflect their different capabilities (e.g. the relative susceptibility of infrared to environmental interference but its superior accuracy for low speed work) but as often as not it will reflect the relative market position of different manufacturers in different parts of the world.

9.2.4 Methods involving video

It has long been recognised that successive frames of a photographic film or video tape of moving vehicles could be analysed to determine the speed of each vehicle. By slow replay of frames it is possible either to measure the time taken to travel a known distance or to measure the distance covered in a known time. In either case the distance can be judged from calibration marks on the road scene and time can be judged from the interval between successive frames of the film or video tape. Although reasonable results can sometimes be achieved by marking the distance directly onto the monitor screen this method is very susceptible to camera shake and is not recommended. A better method is to put distance markers on the actual roadside or road surface prior to filming. If the traffic is relatively slow moving (below, say, 30 km/h) the easiest procedure is to mark out a distance of about ten metres at the site and then to see how many fields (images) of video are taken to move from one end of the ten metre trap to another. Given that a standard video has 50 or 60 fields-per-second this can give accuracies to within 0.4 km/h at 30 km/h. If greater precision is required then a longer trap should be used. At higher speeds a longer trap would certainly be required but an alternative technique would be to mark out a series of metre lengths at the site and then count how many metres are covered within a set number of frames (e.g. 200 frames) or a set interval of time (using the video's time stamp facility). The accuracy of the method is limited by the number of fields per second, by the precision of the distance markers and by the extent of any parallax problem. Parallax can be reduced by drawing distance marks across the carriageway rather than on the kerb and by placing a camera with zoom lens camera at a considerable distance from the road.

The principles of this method have been established for many years and were at one time applied with cine film. However, the high cost of film restricted its use to situations where no other method was viable. Video has now replaced film except where the higher precision available from high speed film is essential. Current developments in digital image capture and storage are likely to remove even this residual role for conventional film. Although it is sometimes possible to use video derived from permanent urban traffic control (UTC) cameras to derive speeds, such cameras are rarely positioned conveniently for such use. Usually a camera has to be installed for the specific purpose of conducting the survey (see Section 8.3).

Manual processing of the resulting video to derive speeds is usually done by stepping through the frames relating to each vehicle and recording the time (or distance), perhaps together with details of vehicle type, on to paper or direct into a computer or data logger. Automatic processing of digital images (see Section 8.3) now offers the prospect of automating this somewhat tedious task. The conventional method is based on the specification of precisely spaced 'virtual sensors' across the carriageway in the field of view, logging changes in the value of pixels along these sensors (and thereby detecting moving vehicles), then timing the vehicle from one sensor to the other and then calculating its speed.

9.3 Data on vehicle headways

Vehicle headway data are collected by timing the interval between the passage of one vehicle and the next. This measurement has to be taken from the same physical point on each of the vehicles (e.g. front bumper to front bumper, or rear wheels to rear wheels), using similar methods to those used to determine point velocities as described in Sections 9.2.1, 9.2.2 and 9.2.4. The traditional stopwatch method has given way to methods based on automatic timing (pairs of tubes, loops, switch tapes or beams) or the analysis of video images. Automatic timing methods are clearly the most convenient for most applications but may not produce reliable data in dense slow moving traffic. Headways within queues can, however, be deduced from the output of induction loop detectors configured as occupancy detectors or from magnetic imaging devices.

Time series data on individual vehicle speeds and headways are very expensive and rarely used outside the research community. They are obtained by intensive instrumentation of a stretch of road, by analysis of remote video or by instrumentation of target vehicles.

9.3.1 Instrumentation of a stretch of road

By installing a succession of pairs of detectors along a stretch of road, point estimates of speeds and headways of each vehicle in a traffic stream can be made and a profile of the speed and headway of each vehicle can be produced. This method has been used in research studies such as of drivers' behaviour on approach to junctions (Seco, 1991) or other hazards, but the analysis of the resulting data can be very time consuming and subject to error. The main problem being to ensure that the correct readings from each site are matched up to show the passage of each individual vehicle. If there is any overtaking or if any vehicles join or leave the stream of traffic this can be difficult to achieve. Effectively, therefore, the method is limited to sites where the traffic stream contains no joiners, leavers or overtakers. Another potential problem is that, if surface mounted detectors are used, driver behaviour may be influenced by the sight (and feel) of so many cables and/or tubes, although research by Pitcher (1990) suggested that such influences may not be significant.

9.3.2 Analysis of video

It is, of course possible to estimate the speed and headway of each of a succession of vehicles passing through the field of view. The usual procedure is to designate a series of screenlines and to use one of the manual methods described in Section 9.2.4 to estimate each vehicle's speed and headway as it passes each screenline. Accurate estimates will, of course, necessitate marking the road surface with distance markers before taking the video. A remotely sited video will produce less parallax error than a wide-angle lens video sited fairly close. Analysis is simplified if the whole stretch of road appears in one video frame but, in order that accurate measurements can be taken, it may be necessary to have a series of cameras each zoomed in on only two or three screenlines.

Analysis of the video images is extremely time consuming and this might make the method less attractive than the automatic method described in the previous section. Its big advantage, however, is that it is much easier to track each vehicle from one screenline to the next and so it is relatively easy to deal with situations where there is traffic joining, leaving or overtaking. A hybrid method has been tested whereby the video is used to help track vehicles past each of the pairs of automatic sensors but the subsequent editing of records from the automatic detectors proved even more time consuming than taking all measurements from the video.

9.3.3 Use of instrumented vehicles

A quite different approach to the collection of speed profile and headway data for individual vehicles is to instrument a vehicle such that its speed and headway history is automatically monitored. The instrumentation required to derive speed conventionally takes the form of an on-board computer connected to a road wheel rotation counter. Speeds can be logged to an accuracy of ±1 per cent simply by attaching an optical transducer to a speedometer cable, more accurate measurements are theoretically possible but rarely achieved due primarily to variation in tyre inflation and tread ware. Alternative methods of deriving speed from within a moving vehicle are currently under development. One involves employing a transceiver to bounce signals off the road surface and analysing the reflected signal. Another is to use the Doppler effect to deduce speed from rates of change in Global

Positioning Satellite (GPS) signals. Tests have shown (Zito, D'Este and Taylor, 1995) that careful selection of off-the-shelf technology can give speed estimates to within ±2 per cent in rural areas with good satellite coverage and to within ±4 per cent in urban areas with poor satellite coverage (caused by shadowing effects).

The instrumentation to derive headways is usually based on forward, and potentially rearward, facing microwave or infrared. If the Doppler effect is utilised then it is possible to determine the relative speeds of the vehicles immediately to the front and to the rear of the instrumented vehicle. Since the speed of the instrumented vehicle is known it is possible by this method to estimate the speed of three vehicles and the headways of two.

However, the problem with all methods using instrumented vehicles is that the driving style of the driver will directly influence the speed of and headway of that vehicle and may indirectly effect the speed and headway of the following vehicle and, possibly, the speed of the preceding vehicle. This possibility needs to be recognised in the selection of drivers and the interpretation of the results. Some interesting behavioural studies have included, in addition to the automatic estimation of speed and headway, a video of the driver's eye view and of the driver's eyes so that ambient factors influencing choice of speed and headway could be included in the database.

9.4 Data on travel times

Travel time data may be recorded through a wide variety of methods. These variously involve logging the passage of vehicles from selected points along a road section or route, or using moving observation platforms travelling in the traffic stream itself and recording information about their progress. The former group includes: registration plate matching, remote or indirect tracking, and input-output methods. The moving observer methods include the floating car, volunteer driver and probe vehicle methods.

9.4.1 Registration plate matching

This method is used for estimating the travel time (and hence space speeds) between designated points in a network. Records are kept of vehicles passing each of the designated points; each vehicle's registration plate and the precise time at which it passed the survey point is recorded on paper, audio tape or video tape or typed directly into a datalogger. Subsequent analysis, normally using proprietary software, can then use the registration plates to match up the records relating to vehicles observed at both sites. The difference between the times at which a vehicle was recorded at each site is an estimate of that vehicle's travel time. The method can produce very rich data on the distribution of travel times provided that sufficient 'matches' are made.

The traditional method of conducting this survey is to station *enumerators with clipboards and synchronised stopwatches* at each of the designated sites and ask them to record the registration number of each passing vehicle and to record the time to the nearest second at which it passed them. A competent surveyor can usually be expected to record details of about five vehicles per minute which equates to up to a maximum of 300 vehicles per hour depending on the amount of bunching in the traffic. This rate can be increased to perhaps 400 vehicles per hour if, instead of recording complete registration numbers, only partial records are taken. Reliance on partial records does not in practice create a problem provided that there are clear rules on which part of the registration number is required and provided that the risk of 'spurious matches' is not unduly increased.

A spurious match can occur if two vehicles share the same partial registration number (e.g. ABC123 could be confused with DEF123 if only the numerics were being recorded). The chance of a spurious match occurring will depend on the flow and on the distribution of registration numbers within it. UK practice for these surveys is to record four out of seven characters omitting the three alphabetics that identify the area in which the vehicle was first registered. If the 400 vehicles per hour achievable with partial records is still not sufficient to capture all the traffic at a given site it may be possible to increase the number of staff at the site (designating one person as observer/caller and the other as recorder or dividing the task at a multi lane site such that one observer does one lane and the other does the other).

An alternative method which avoids the extra staff cost, is to equip the observers with *audio cassette recorders* into which they can read the registration numbers and speak the time for each passing vehicle. The tape contents can typed directly into a computer by a data processing clerk in the laboratory. Using this method, and partial registrations, up to 700 vehicles per hour can be recorded in ideal conditions but 500 veh/h is a more realistic target. Although cheap recorders and microphones can be used, the extra quality and reliability achieved with more expensive equipment is usually considered worth while. If a voice actuated cassette recorder is used it is possible to reduce the transcription time to a minimum by excluding blank stretches of tape.

The speed of data collection in the field can be further increased (though at the cost of increased transcription time) by dispensing with the time record after each registration number but relying instead on estimates of time derived from the tape revolution counter; during transcription each registration is tagged with the current value of the tape revolution counter and the counter is 'anchored' to real time with time markers spoken onto the original tape at the beginning and end of the survey period. Software can then compute the implied real time for each registration record. Note that the relationship between real time and revolution counter values is not constant and so time markers should be included on the tape every ten minutes or so. Using this method, staff can achieve data rates approaching 1000 veh/h in ideal conditions (with 800 veh/h a more realistic target).

It is sometimes possible to so position a *video camera* that it can record the registration plates of passing vehicles. In practice, however, perfect sight lines are not often available at the sites where the recordings are required; unlike a human, a video camera cannot easily swivel and refocus to capture plates obscured by other vehicles and so the requirements for siting video cameras are much more demanding than those for a human enumerator. It is rarely possible to find suitable sites in urban areas due to the density of traffic but good sites can often be found on motorways as viewed from bridges.

Manual transcription of paper or tape records can be very time consuming. There are two ways to reduce this burden. The first is to key the data directly into *portable computers or dataloggers* at the roadside. The enumerator simply types the registration numbers of passing vehicles directly into the keyboard where they are allocated a precise time stamp by the in-built clock. A variety of machines have been used for this type of survey; ranging from portable computers with conventional QWERTY keyboards to hand held data loggers with very small keys in a grid layout and purpose-built dataloggers designed for easy input of alphanumeric data in field conditions. The best of these allow about 400 veh/h to be recorded while the worst lead to a large error rate due to miskeyed data.

A second way to reduce the transcription burden is to use *automatic transcription technology.* Alternative technologies are available for the automatic recognition of speech (which is relevant if the registration letters have been spoken onto an audio tape) or of visual images (which is relevant if the registration plates can be captured on video tape or didgitally. Speech recognition technologies are improving rapidly but automatic transcription, using voice recognition equipment offers some saving in transcription staff time but, as yet, the costs of the equipment, together with its relatively poor performance when dealing with records made by observers in stressful street conditions (for which 'recognition rates' may be less than 70 per cent) do not make this a practical option.

The image recognition technologies were initially developed primarily for the enforcement market and works by homing in on plates before reading their contents. Accuracies of 90-95 per cent are achievable. The recognition equipment is still beyond the budget of most survey organisations but some agencies have equipped themselves to offer a transcription service to third parties. The most dramatic development however, is in the use of this technology to match registration plates on-line in order to provide drivers and system managers with real-time estimates of travel times along key links in the network.

Whatever methods of data capture and transcription are used the economics of the exercise will normally require the achievement of as high a 'match rate' as possible (i.e. the proportion of upstream observations that can be matched with downstream observations). Many of the above techniques will not manage to capture the registration numbers of all vehicles passing the site. Where this is likely to be the case, a sampling strategy should be planned in advance because if vehicles are randomly missed at each site the matching rate will be much lower than it need be. This is because if, for example, 90 per cent of vehicles are captured at the upstream site, and a random 90 per cent are captured at the downstream site, the matching rate will effectively be limited to 81 per cent (0.9×0.9). The solution is to estimate the likely rate of data capture given the method being used, the skill of the field staff and the amount of bunching in the traffic and then to calculate what proportion this will be of traffic during the busiest part of the survey. This proportion is then used to design a sampling regime. The most commonly adopted technique is to use the final digit to determine whether a given vehicle is 'in' or 'out' of the sample thus if a 50 per cent sample is required one might select all odd numbers, if a 70 per cent sample is required one might select all numbers ending 1 through 7 and so on. An alternative sampling method based on vehicle colour may be considered (Thompson, 1989), but is not generally recommended unless the analyst is satisfied that there is unanimity about what constitutes a given colour and that the possibility of bias due to faster cars being predominantly of a given colour can be discounted.

Appropriate sampling strategy can maximise the match rate for a given flow but it cannot achieve miracles; if a high proportion of the vehicles at the upstream site do not subsequently pass the downstream site or if the downstream site has a large element of flow which did not pass the upstream site then even the best sampling strategy will not be able to produce good match rates. It is for this reason that registration plate matching is normally used within defined corridors of movement rather than across relatively ill defined networks.

When the records from two sites are analysed there will inevitably be some unmatched records. Some of these will be due to sampling errors, some to the fact that the composition of the flow is different at the two sites and some to recording or transcription errors. Some matching software provides the option of matching up 'near misses' from within the pool of unmatched records. This option may be useful to pick up occasional mis-typing (the frequency and occurrence of which is likely to vary according to the layout of the keyboard used) or miss-hearing of spoken records (eg 'B' for 'P') but, if used to excess, it can result in a distortion in the travel time data. Research at the University of Leeds has shown that acceptance of progressively poorer matches eventually leads to an over-representation of travel times equivalent to the offset between the times of maximum flow at the two sites.

9.4.2 Remote or indirect tracking of individual vehicles

Remote tracking can be achieved by making use of vantage points or by accessing vehicle movement logs compiled for a different purpose. Travel times of individual vehicles along relatively short stretches of road can be obtained by monitoring them from a vantage point which has a view of the start and end of the route. Timing can be achieved using stopwatches or dataloggers in the field or by analysis of a video record.

The extra expense involved in using video may be justified by the fact that it enables timings to be obtained for *all* vehicles and not just a possibly biased sample and, if the whole stretch of road is in view, by the relative ease with which it is possible to log intermediate travel times and events such as queues. Good coverage can sometimes be obtained by mounting the video on an airborne platform such as a helicopter, light aircraft, model aircraft or tethered balloon. The development of remote controlled video cameras makes it possible to take advantage of the much lower costs of these latter two. Although the use of video offers the prospect of avoiding biased sampling, the resources required to transcribe the entry and exit times of each and every vehicle can make this advantage illusionary.

A problem inherent in any technique that involves observation of individual vehicles is that the target vehicle may not be typical of its cohort. The error can be reduced by logging the flow patterns surrounding each target vehicle. Under the *input output correction* method, observers are placed at the start and end of the road section. Once the target vehicle enters the test section, the upstream observer counts the number of vehicles passing the starting point during each successive time interval (e.g. each 30 seconds for five minutes). When the target vehicle leaves the study area, the downstream observer counts the number of vehicles passing that point in successive time intervals. The mean travel time is then estimated by equation 9.1,

$$\overline{T} = T + \frac{1}{n_2}\sum_{i=1}^{n_2} t_{2i} - \frac{1}{n_1}\sum_{i=1}^{n_1} t_{1i} \qquad (9.1)$$

where T is the travel time measured by target vehicle, t_{1i} is the time between the entry of the target vehicle and the entry of subsequent vehicles, t_{2i} is the time between the exit of the target vehicle and the exit of subsequent vehicles, n_1 is the number of vehicles counted at entry, and n_2 is the number of vehicles counted at exit. The method is best suited to road sections with no internal entry or exit points, so that all traffic entering the section must eventually emerge at the exit and is ideally repeated for several target vehicles.

An alternative to recording the entry and entry times for a sample of vehicles is provided by the *point-sample* method. This method requires a view of the entire stretch of road and involves taking periodic counts at intervals (Δt) of the number of vehicles (k) on the designated stretch of road while monitoring the number of vehicles coming out at the end (q). The mean travel time can then be estimated using equation 9.2,

$$T = \frac{K \Delta t}{q} \qquad (9.2)$$

where $K = \Sigma k$ (i.e. the total number of vehicles observed on the stretch of road via the periodic counts). The accuracy of the method increases with the number of density counts but deteriorates rapidly at low flow volumes.

The travel times of individual vehicles along a designated stretch of road can sometimes be obtained 'automatically' as a by-product of tolling systems. Many such systems employ time-stamped entry tickets which are surrendered at the exit point. If the time at which the ticket is surrendered is logged, the resultant record can provide a rich source of data on travel times. This data has been used for many years by toll road operators to check for fraud and occasionally by motorway police to support prosecutions for speeding. ERP technology has the capability of providing travel time data without human intervention. If road pricing becomes more widespread, this may become as useful a source of data for traffic analysts and engineers as it already is for toll road operators.

9.4.3 Input-output methods

These methods can be used to estimate mean travel times for cohorts of vehicles travelling between designated survey sites. They are based on the fact that the difference between the means of two sets of observations is equal to the mean of the differences of the two sets. The methods require estimation of the mean arrival time at the upstream site and the mean departure time at the downstream site. The difference between these two estimates is then taken to be the mean journey time between the sites.

The method can only give accurate results in closed systems (such as stretches of motorway between entry/exit points) and requires the target cohort of vehicles to be identified. This can be done by 'injecting' marker vehicles into the stream, one to mark the front of the cohort and one to mark the back or, at lower cost, by nominating distinctive vehicles at the upstream site as the 'start' stop vehicles and phoning or radioing their identities to the downstream site. This technique is, however, not proof against the nominated vehicles being overtaken by (or overtaking) other vehicles and thus destroying the integrity of the cohort. It is therefore important not to bias the composition of the cohort by selecting a vehicle which is moving more slowly than the ambient traffic.

The estimation of the mean time at which cohort members pass the site can be by straightforward recording of each vehicle using a stopwatch and clipboard or, if the flow is too high to allow this to be done accurately, by noting the time of the first and last vehicle at each site and infilling the rest using the disaggregate output from an automatic vehicle detector at each site. An alternative, low tech, method is to note how many vehicles pass the site in each of a series of short intervals. The distribution of vehicles across the intervals can then be used to weight the calculation of the mean. Experience suggests that the intervals should be of about ten seconds duration and so there would be 30 of them during a five minute cohort. Intervals of less than ten seconds give greater potential accuracy but, since they increase the workload on the surveyor, they also invite more error.

9.4.4 Moving observer methods

This group of techniques involves the surveyor ('observer') being in a vehicle in the traffic stream and noting, among other things, the time taken to travel between specified points. These techniques are particularly suitable for relatively long or complex journeys which do not have sufficient through-flow to support registration plate matching or input/output methods. Another advantage of the moving observer methods is that they can yield information about travel times, and traffic conditions, for intermediate stretches of road and so identify the reasons for any abnormality in the overall journey time.

The most *basic moving observer method* simply involves driving a survey car along a particular route and noting the time at various predetermined points along the route. The data are usually recorded by an observer equipped with a stopwatch and clipboard. Some authorities have suggested that, particularly if there are few intermediate points, the observer can be dispensed with and the recordings can be made by the driver onto audio tape. However, practical experience with this method shows that it is prone to error and can compromise safety. If there are a lot of intermediate points, the observer's task can be simplified, and the subsequent need for data transcription eliminated, by using an appropriately programmed datalogger instead of the stopwatch and clipboard.

The obvious problem with the basic moving observer method is that, unless repeated many times, by different drivers, it is not likely to be representative of actual conditions and may be unduly influenced by the driver's driving style. [Note that the problem is not overcome by instructing the driver to tail another vehicle - that would simply replace a bias due to the surveyor's driving style by one due to that of a randomly selected member of the driving population.] It has been suggested that in order to overcome the sampling problem, each route should be driven at least fifteen times with different drivers each time. This is rarely likely to be a practical proposition. The extent of bias due to driving style can be reduced by adopting the *floating car method* whereby the drivers are instructed to attempt to 'float' in the traffic stream, overtaking as many vehicles as overtake them. Overzealous attempts to comply with this instruction can, of course, have safety consequences but even where it is safe to overtake or hang back it may not be obvious whether it is appropriate - it can for example be difficult to judge whether a given vehicle is a part of the main traffic stream in which the driver has to float or whether it is slowing down to join a queue of traffic waiting to exit onto a side road.

A method of correcting for failure to float perfectly was suggested by Wardrop and Charlesworth (1954). They suggested a *correction formula* to calculate the true mean travel time ($\bar{t}_{ab}$) from a to b as shown in equation (9.3):

$$\bar{t}_{ab} = t_{ab} + \frac{O}{q} \qquad (9.3)$$

where t_{ab} is the time taken by a survey car to travel the route from a to b, O is the net number of vehicles overtaken by the survey car vehicle (i.e. vehicles overtaken minus vehicles who overtake) while travelling from a to b, and q is the mean flow rate.

The mean flow rate appears in the formula in order to reflect the fact that overtaking one 'too many' vehicles in a flow of 1000 is less significant than the same thing in a flow of 100. Since q is rarely available from conventional sources (it is likely to differ on different parts of the route) an estimate can be made by having a vehicle travel in the opposite direction ($b{\rightarrow}a$) recording travel time (t_{ba}) and number of vehicles met (m) in the opposing flow (i.e. travelling in the $a{\rightarrow}b$ direction). Then q is given by equation (9.4):

$$q = \frac{m - O}{t_{ab} + t_{ba}} \tag{9.4}$$

The accuracy of the method is improved by taking several runs in the $a{\rightarrow}b$ direction (with matching $b{\rightarrow}a$ runs to estimate q) and it is convenient to utilise pairs of vehicles running back and forth between a and b for this purpose (note that, although common, the use of just one vehicle to provide the $a{\rightarrow}b$ and $b{\rightarrow}a$ times is theoretically flawed since the $a{\rightarrow}b$ and $b{\rightarrow}a$ run should be simultaneous). Note that, if travel times are required for subsections within the $a{\rightarrow}b$ route, it will be necessary to record data to allow separate correction factors to be calculated for each subsection. If executed correctly this method can produce accurate data albeit at some cost. It is however subject to error if conditions fluctuate markedly during the survey.

Information derived from moving observer surveys can be enriched if data is simultaneously recorded on time spent in queues within each segment of the journey. Experience suggests that two observers, particularly if equipped with dataloggers, can quite comfortably record all the data required for t_{ab}, O and time in queues when travelling from a to b, and t_{ba} and m when travelling back from b to a. If the flow is light and if there are few intermediate turning points, it is quite possible to do the whole job with just one observer.

It is technically feasible to instrument survey vehicles to produce 'automatic' moving observer data. The location of the vehicle can be determined by logging distances travelled or by triangulation from terrestrial beacons or GPS. A system clock can be interrogated to derive travel time whenever the vehicle is deemed to have reached a timing point. Time spent in queues could be deduced from a record of distance travelled per unit time while the net overtaking and opposing flow could be derived from an in-vehicle radar. A recent paper by Brackstone et al (1999) indicates the range of data available via this technology.

9.4.5 The use of volunteer drivers and fleets of probe vehicles

The idea here is to use vehicles making 'ordinary' journeys as a fleet of moving observers so that, by sheer force of numbers, the sampling problems inherent in the more conventional moving observer techniques can be overcome. The vehicles may be selected to be 'typical' of the driving population or may be deliberately selected from among high mileage vehicles such as taxis, delivery vehicles and commercial travellers.

The drivers may be asked to keep pencil and paper records or may be supplied with audio tape recorders or dataloggers or may have their vehicles instrumented. Pencil and paper records have widely been used with panels of volunteer drivers to record travel times for their regular journeys. This can be an effective way of building up a time series database but the method is known to suffer from a tendency for drivers to complete their records retrospectively and in so doing to omit outliers and round to the nearest five minutes (e.g. Bonsall and Montgomery, 1984). It is possible to reduce these problems by equipping drivers with voice-actuated tape recorders onto which the driver is asked to speak the time at the beginning and end of the journey, or simple dataloggers with an inbuilt clock, so that the driver can simply press a button at the start of the journey and another at the end. If such devices are used it also becomes feasible to ask them to record the time at which they pass intermediate points on their journey.

Such methods have been used for several years. A more recent development is the use of fleets of instrumented vehicles as probe vehicles. The instrumentation, will typically involve the vehicle periodically reporting its position (and perhaps other data including a recent speed history) back to a base or transmitting a signal which enables its position to be determined externally. The communication will usually be by radio but other media such as infra-red might also be used. The vehicle's position can be estimated using a combination of triangulation from GPS satellites or terrestrial beacons, dead reckoning using road wheel sensors and electronic compass, and map matching software (which corrects a triangulated or dead-reckoned estimate to fit the nearest road on a database in the on-board computer).

These methods have become feasible with the advent of mobile communications and of automatic vehicle location (AVL) and automatic vehicle identification (AVI) as an aid to fleet management and security. Several fleets of vehicles are now appropriately equipped and they are a ripe source of travel time data. Another potential source is vehicles equipped to act as probe vehicles for dynamic route guidance systems such as those piloted in Berlin and Orlando (e.g. see Von Tomkewitsch (1987) and Peters (1993)).

9.5 Data on delays

A basic issue when collecting data on delays is to first determine what constitutes delay. For example; is it time spent stationary or queuing? Or is it the difference between the time taken to traverse the section and some estimate of the time that an undelayed vehicle would take to traverse the same section? Once a definition has been agreed on, there are a number of methods for estimating the amount of delay in a network.

9.5.1 *Estimation of delay*

The delay on a section of road can be defined as the difference between the travel time measured during the study period (i.e. under a specific set of traffic conditions) and a notional undelayed travel time. The definition of the undelayed travel time depends on the specification of an undelayed speed which can act as a reference. The choice of reference speed obviously depends on the purpose of the study but the following alternatives might be considered: posted speed limit; free flow speed, spot speed; or average speed.

Adoption of the *posted speed limit* as the nominal undelayed speed may be justified by the fact that the posted speed limit may be considered as the maximum safe driving speed on the road section.

The *free flow speed* is the average speed on the study route for unimpeded vehicles and is obtained by measuring the speeds of a random set of vehicles travelling through the study area during periods of low demand. Use of the free-flow speed may be justified in preference to the posted speed limit if the posted speed limit is higher than the free-flow speed. Most studies do indeed use the free-flow speed as the reference speed.

The use of *spot speeds* measured at points in the network where there is no immediate interference with traffic flow (i.e. at points as far removed as possible from intersections, pedestrian crossings, etc) may be justified only if estimation of the free flow speed across the entire length of the road is impractical for some reason.

The use of the *average speed* for the population of vehicles may be justified if one wishes to determine the amount of extra delay suffered by a subset of that population (eg those travelling at the height of the peak or those arriving at a particular point in the signal cycle). In the particular case of before and after comparisons, the reference speed may be taken as the average 'before' speed.

Once the reference speed for the section of road has been determined the notional undelayed travel time can be calculated and then the delay can be derived by comparing this with the travel times prevailing during the study period. The travel time can of course be estimated using any of the methods described in section 9.4.

9.5.2 Direct methods of calculating delay

Various methods exist: three of particular note are the path-trace method, the moving-observer method and the point-sample method.

The path-trace method is applicable when data is required on the total time which vehicles spend stationary. It requires a good vantage point providing an uninterrupted view of individual vehicles as they travel along the designated stretch of road. The data may be recorded for a sample of vehicles either 'live' by an observer with a stopwatch or via a videotape record of the scene. Although video cannot cover such a large field of view as an observer on site, it does permit reanalysis of the record either to achieve a higher percentage sample or to seek a consensus between different observers as to the total stopped-time for particular vehicles. The use of the path-trace method is effectively limited by the shortage of good vantage points and the practical difficulties involved in determining the precise instant at which a given vehicle stops or starts.

The moving-observer method offers some advantages over the path-trace method in that the timing of stops and starts of the car in which the observer is travelling can be determined more accurately, particularly if the car is instrumented to record the data automatically. A further advantage is that it becomes practical to record not only the total time spent stationary but also the total time spent travelling below any specified speed. However the problem is that, as discussed in section 9.4.4, it can be difficult to determine whether the test vehicle is being driven in a way that is typical of the total population of vehicles using the road at the time.

The point-sample method of calculating delays is a variant of that described in section 9.4.2 which is particularly used to estimate time spent queuing. It is based on periodic sampling, either of the number of stopped vehicles at the start of the green phase on the approach section of a traffic signal or the number of vehicles stopped when a vehicle departs from a stationary queue. The procedure is most easily adapted to systematic sampling (either fixed intervals or on a per-cycle basis), although it is possible to use other types of sampling (e.g. random sampling). Solomon (1957) calculated the time spent queuing on the approach to a signalised intersection over a specified time period using equation 9.2 (with T representing time spent queuing). The method requires a nearby vantage point for the observer so that accurate counts of the numbers of vehicles in the test section can be made, and thus its use is restricted.

9.6 Off-line use of on-line data

There have been numerous references within this chapter to the growing availability of data from on-line monitoring systems and detectors. If this data is logged and archived it can become a valuable resource for off-line analysis.

For example, use could be made of data originally collected to inform drivers about traffic conditions on key links in the network (such as the spot-speed estimates derived from infra-red detectors fixed to UK motorway bridges and gantries, the travel time data derived from CCD cameras linked to on-line registration plate image matching on other UK roads or the traffic condition data derived from inductive loops along major routes in Australia). Similarly, use could be made of the speed and flow data captured on some French motorways as part of their on-line incident detection system or of travel time data collected as a by-product of road tolling systems. There is no shortage of such data - the analyst's problem is how to negotiate access!

10 Environmental impacts

A central theme of this book is the necessity to take proper account of the full impacts of traffic system developments. This will certainly include consideration of environmental impacts. The performance of traffic systems can have strong environmental impacts, certainly at the local level and most likely at the global level, when individual contributions are aggregated to consider issues such as greenhouse gas emissions. Road traffic systems have significant impacts on fuel consumption by vehicles and environmental degradation in regions. Particular concerns in traffic analysis have arisen with respect to:
- fuel consumption and means of conserving fossil fuels;
- air pollution, including gases and particulates;
- noise and vibration, and
- visual intrusion and physical degradation.

Despite the obvious interest in these special areas, they should always be viewed as symptoms of the broad impacts of traffic systems on land use and the environment. Studies in the areas of fuel consumption and emissions should always be interpreted from this perspective. The systems planning approach to traffic analysis, described in Chapters 2 and 6, provides a useful conceptual framework in which to place surveys of traffic energy use and pollutant emissions. Against this background, we now consider some of the relevant survey management and data collection techniques. The actual data collection and analysis in these specialist areas will often be the province, not of the traffic engineer, but of some authority charged with general environmental protection and/or planning, and may require the expertise of another discipline, such as mechanical engineering or meteorology. Traffic engineers and analysts will be contributing their expertise in traffic systems to a multidisciplinary study team.

This chapter considers some of the means of collecting data on vehicle fuel consumption and emissions, and of assessing the environmental impacts of road traffic. This requires consideration of the performance of individual vehicles in traffic, and of the overall traffic system. In the evaluation of alternative traffic management schemes or the assessment of the impacts of new development proposals, it will also be necessary to consider the relative impacts of alternatives. The starting point is a consideration of the fuel types used in road transport, as it is the by-products of using these fuels that leads to the emission of air-borne pollutants (gases and particulates).

10.1 Transport fuels

Nearly all of the vehicles currently in use are powered by petroleum-based liquid fuels. The reliance on this energy source has been keenly felt since the 'oil shocks' of the 1970s, and governments throughout the world have investigated energy conservation projects. The principal liquid fuel for road transport is petrol (gasoline), which may be found in one of two main variants: leaded petrol (LP) in which small amounts of lead have been used to achieve a high octane rating in the refining process, and unleaded petrol (ULP) where alternative processes have been used. Use is also made of diesel fuel, liquid petroleum gas (LPG), and compressed natural gas (CNG). Each of these fuels has its own properties. The consumption of leaded petrol leads to emissions of the pollutants carbon monoxide (CO), various hydrocarbons (HC) and oxides of nitrogen (NO_x), and particulate lead (Pb) - besides water vapour and carbon dioxide. Diesel fuel is widely used for large vehicles and, to a lesser extent, by passenger cars. The advantages of diesel power are that, for equivalent vehicles, a diesel engine provides greater fuel efficiency (more kilometres travelled per litre) for the same volume of fuel. Diesel fuel offers significant reductions in CO emissions. On the other hand, diesel engine emissions may contain considerably more HC and fine particulates, and may also emit sulphur oxides (SO_x), which are not produced by petrol engines to any great degree. The common measurement of particulate pollution is PM10, the concentration of particulate matter of size 10 microns or less.

LPG may offer a cheap alternative to leaded petrol, for it can be readily used by most petrol-engined vehicles. LPG is 'cheap' because it is an unavoidable by-product of the refining process. Gaseous emissions (of CO, HC and NO_x) from LPG fuel are less than those from petrol. There is also an absence of particulate lead, but fuel consumption (in volumetric terms, i.e. numbers of litres of fuel) is perhaps 25 per cent worse than for petrol. CNG has similar energy and emissions properties to LPG, and can be obtained from the regular sources of natural gas.

Unleaded petrol (ULP) is the other main fuel in common use. This fuel is designed for use with catalytic converter systems in vehicle exhausts, to minimise emissions of the gaseous pollutants CO, HC and NO_x. Lead will poison the rare metal catalysts (e.g. platinum, or palladium) used in the conversion of exhaust gases to harmless emissions. Thus the need for a fuel free of lead.

Alternative fuels to the petroleum-based ones include electricity, methanol and hydrogen but at present none of these fuels is used to any great extent.

10.2 Pollutants from road transport sources

Air pollution in urban areas typically consists of the primary emissions (CO, HC and NO_x), plus oxides of sulphur, fine particulate lead, dust and soot, and derived pollution (e.g. photochemical smog). In addition, road traffic contributes significantly to emissions of the greenhouse gas carbon dioxide (CO_2), which is not strictly a pollutant given that it occurs in large quantities in the natural environment. Studies from cities around the world suggest that road traffic may be responsible for perhaps 80 per cent of CO and NO_x emissions and nearly half of the HC and CO_2 emissions. In addition, LP-powered vehicles form a major *controllable* source of particulate lead emissions.

Carbon monoxide is a dangerous asphyxiant for mammals, as it is absorbed readily into the bloodstream and competes with oxygen in the circulating blood. Lead is a cumulative poison. Hydrocarbons and oxides of nitrogen in the presence of sunlight under stable meteorological conditions, may chemically react over a period of hours, leading to the formation of the derived pollutant, photochemical smog. Ozone (O_3) is one of the many components of smog.

Particulate matter is deleterious to health and there are increasing concerns around the world concerning the effects of airborne particulates on respiratory problems, including asthma. The brown haze common seen as evidence of air pollution in urban areas is mainly caused by fine particulates less than two microns in diameter, which cause scattering or absorption of visible light.

Road traffic is also the most important source of *noise pollution* on urban areas. Studies in the UK indicated that road traffic was the primary source of noise in at least 60 per cent of surveyed sites, and that the next most important sources were ten times less important (Hothersall and Salter 1977).

At the broader level, conflicts between road traffic and land uses manifest themselves as environmental problems. These conflicts are particularly important in residential areas. Buchanan (1963) introduced the concept of *environmental capacity* to indicate the level of possible conflict in a street or an area. Precise definition of environmental capacity has proved elusive over the years, but is normally sought in terms of some level of traffic volume and speed behaviour, e.g. 'the maximum number of vehicles (and associated 50th percentile speed and proportion of heavy vehicles) that may pass along the street in a certain time period and under fixed physical conditions without causing environmental detriment'. Other definitions have attempted to describe the level of sensitivity of a road network to traffic effects, perhaps based on the compilation of inventories of land use, road network and traffic variables, and the assessment of the levels of sensitivity of these variables for the streets in question from which an overall rating of a given street can then be produced. Klungboonkrong and Taylor (1999) provide a review of methods and models employed to assess environmental capacity and environmental sensitivity in urban areas.

Pollution problems may be seen at two distinct levels. In extreme cases the pollutant may be a danger to the physical health and well-being of the individuals subjected to it. Excessive noise and excessive concentrations of some air pollutants (e.g. CO) may inflict immediate damage. Prolonged exposure to large concentrations of other pollutants, or smaller doses of noise and CO, may lead to harmful effects such as lead poisoning and respiratory diseases. The main problem with pollution, however, is that of annoyance. Individuals may become distressed or anxious in the presence of pollutants, at levels well below those injurious to health. Most remedial environmental treatments (e.g. noise barriers and fences) are intended to overcome problems of annoyance. A useful conceptual model of the effects of traffic noise on households is due to Brown and Law (Brown, 1980). This model may be generalised to include other forms of pollution, as in Figure 10.1. The combination of physical and human factors leads to levels of annoyance with the pollutant. Annoyance depends on the behaviour and attitudes of the household, who may be made more sensitive to the pollution by other households, community values, and the relationship of the household to the rest of the community. Some pollution may be tolerated, but at some (threshold) level the household may begin to object to it.

The conceptual model of Figure 10.1 indicates the complexity of the analysis and appraisal of environmental impacts. The pollution is generated from a large number of individual sources (e.g. vehicles), which are interacting with one another. The total pollution load at a site is the result of all these individual generations. There is a physical measure of the pollution load (e.g. the concentration in parts per million of an air pollutant) at any point. The effect of the pollution load is, however, related to individual and community interpretation of that physical load (e.g. a measure of annoyance).

Environmental Impacts 215

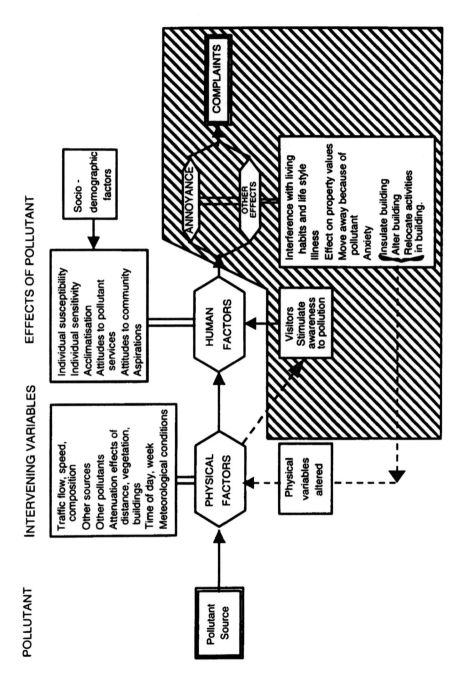

Figure 10.1 Conceptual model of effects of a pollutant (Brown, 1980)

Surveys of environmental effects may then be needed to determine firstly the emissions of individual vehicles in a traffic stream, the combined pollution load due to a number of traffic streams, and the level of annoyance in a community resulting from the pollutant load. Clear definition of objectives is essential, as discussed in Chapters 2 and 6, for only then can a particular survey and analysis methodology be selected.

10.3 Estimating the environmental impacts of road traffic

One procedural difficulty that has dogged planners and engineers has been how to assess the relative environmental effects, merits and disadvantages of alternative transport infrastructure proposals at the planning stage. Although survey methods are used for assessing levels of pollution for existing conditions, these methods cannot be applied to proposed developments, and so alternatives are required. This issue was previously addressed by Wigan (1976), who provided the following methodology for predicting the environmental impacts of road traffic:

1. collate data on a link-by-link basis on road type, width, number of households, amount of activity by category of land use, etc;
2. obtain traffic flow data, including traffic composition and travel time, speed and delay;
3. develop a database that can provide the required link-by-link data to apply models of fuel consumption, emissions and pollutant dispersion;
4. apply a framework that defines the conditions under which the consumption, emissions and dispersion models can be applied;
5. generate indices of pollutant loads and environmental impacts (e.g. number of households subjected to a given noise level over a specified time interval);
6. prepare tabular and graphical representations of this information as histograms, pollution load maps, etc, and
7. indicate levels of individual and community annoyance under different pollution loadings.

Given the logical, 'common-sense' nature of this methodology, it may come as something of a surprise to realise that it has seldom, if ever been fully applied in practice! All too often planners and engineers have considered only the generation of pollution at its sources, not where that pollution will end up and who will be affected by it. An implementation of Wigan's methodology is possible through the construction of a combined model system, comprising a traffic network model (capable of producing a number of alternative travel patterns in response to differing transport policies), a family of emissions and fuel consumption models, a pollution dispersion model, and a land use impact model. This implementation relies on the availability of GIS software as the means for spatial integration of the set of databases outlined in the methodology.

The basic scheme of the system is given in Figure 10.2. A traffic network model is used to produce (by simulation or forecasting) the levels of traffic flow and travel conditions on a study area network, under the given traffic management scheme. Models of vehicle fuel consumption and emissions under the modelled traffic conditions are then used to estimate the traffic system fuel usage and the levels and spatial distribution of pollution generation. This information, coupled with data on the meteorological conditions, may then be used as input to a pollution dispersion model, which estimates the spread of the pollution over the study area, so providing the modelled levels and spatial distribution of the pollution. The land use impact model superimposes the pollution levels on the land uses and populations in the study area to determine the likely sites and extent of environmental problems resulting from the traffic system.

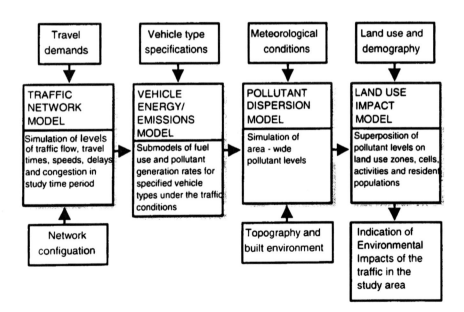

Figure 10.2   Integrated modelling system for assessing environmental impacts of road traffic

10.4 Survey methods for fuel consumption and emissions

On the basis of the previous discussion, surveys of fuel consumption and emissions may be divided into three basic categories: (a) individual vehicle-based surveys; (b) area-wide (system-wide) surveys; and (c) attitudinal surveys of households or individuals.

The methodologies for these categories are quite separate, and the results from one category may be difficult to transpose from one to the other. Indeed, the objectives of any studies that require one or other category of surveys will probably be quite different. Computer-based models offer one possible avenue for the integration of results from the categories, such as the Biggs-Akcelik family of models (Section 5.6). These models form an important component in the integrated model for environmental impact assessment presented in Section 10.3.

10.4.1 Individual vehicle surveys

At the level of the individual vehicle, it is possible to observe vehicle performance, fuel consumption, and emissions in detail. These data may be observed either on-road or in the laboratory (using *dynamometer tests*).

A dynamometer is a test bed on which a vehicle is fixed, with the driving wheels in contact with a system of flywheels, which simulate the inertia of the vehicle, and a power absorption unit which can be adjusted to reproduce different road conditions as speed is varied. The vehicle's acceleration, deceleration and speed performance can then be recorded directly. Measuring instruments are attached to the vehicle's engine, carburettor or fuel injection system and exhaust to measure 'instantaneous' fuel consumption and exhaust emissions respectively. The data are recorded directly by computer and time profiles extracted.

Legislative requirements and standards for vehicle design and performance have led to the development of standard speed-time profiles ('driving cycles') in many countries. These cycles are supposed to represent a 'typical' vehicle trip segment in an urban area. Their main use is as a standard for comparing the performance of different vehicles. The dynamometer can simulate the behaviour of a vehicle over a specified time-distance trajectory, and thus provides detailed data on gaseous emissions, particulate emissions and fuel consumption. The advantage of the dynamometer is that 'traffic conditions' may be strictly controlled and hence experiments replicated. Its disadvantage is that the correlation between dynamometer test results and actual performance in traffic is not properly known.

A major limitation of the dynamometer/driving cycle test is whether or not the results of laboratory tests represent observed field data. The alternative experimental method is then to use an instrumented vehicle, driven in a traffic stream, to record these *on-road data*. The use of instrumented vehicles was discussed in Chapter 8.

Extrapolation of individual vehicle fuel consumption and emissions data to more aggregated levels in a traffic system is difficult. The characteristics of the individual test vehicle need to be related to all of the driver/vehicle combinations in a traffic stream to permit this aggregation. Johnston, Trayford and van der Touw (1982) provided a methodology for experimental design which may eventually make the extrapolation possible. There are procedures which make use of the existing state of technology. For example, use of the 'chase car' technique in which a survey vehicle pursues a vehicle selected at random along the survey route (see also Chapter 8). The speed-time profile is logged, and then this driving cycle is used on the dynamometer to test the same model vehicle, yielding laboratory estimates of fuel consumption and emissions.

10.4.2 Surveys of system-wide consumption and emissions

Pollution is manifest at the area level, and the effects of the pollution may become pronounced at sites well away from points of emission. Thus there is a need to study emissions and fuel consumption across a study area.

If data are required on fuel consumption over periods in excess of a single trip, survey methods similar to those used to study vehicle usage may be employed. They fall into two groups: (a) questionnaire/interview surveys of a sample of drivers by a study team; and (b) diary surveys completed by a sample of drivers.

Questionnaires and interviews can provide data on people's attitudes, as well as their fuel purchases and travel behaviour over a (short) survey period, while diaries may be used to gather longitudinal data on travel distances, trip destinations and fuel purchases.

The introductory section of this chapter defined the primary local pollutants, CO, HC, NO_x and particulate lead, in urban areas. The derived pollutant, photochemical smog, results from the chemical reaction of hydrocarbons and nitrogen oxides in the presence of sunlight. Once emitted to the atmosphere, pollutants are dispersed by the wind. Wind speed, wind direction and atmospheric stability determine the level and extent of dispersion and dilution of the pollutant. In general terms, the concentration of a primary pollutant decreases as it moves away from its source. This is not the case with the derived pollutants. Peak concentrations of these pollutants may be found some distance from the emission source of their ingredients, and may occur some time after the generation of those emissions. For example, peak concentrations of photochemical smog may occur in mid-afternoon, resulting from exhaust emission generated earlier in the day. The peak concentration may be found in an area some kilometres away from the generation sites.

A variety of models may be used to assess air quality and pollution levels, see Taylor and Anderson (1988). The most widely used model is the Gaussian plume dispersion model. This model assumes that the pollutant plume from a single point source may be represented by a normal distribution of pollutant concentration about a centre line from the source, and drawn in the prevailing wind direction, as shown in Figure 10.3. Thus the spread (or dilution) of the plume, as a function of distance downwind of the source is given by an exponential function of downwind distance. For the Gaussian model the concentration of a gas $\rho(x, y, z)$ (g/m^3) at a point (x, y, z), due to a point source of that gas at (x_0, y_0, z_0) emitting pollutant at the rate H(g/s), is given by:

$$\rho(x, y, z) = \frac{H}{2\pi S_y S_z u_w} \exp\left(\frac{(y-y_0)^2}{2S_y^2}\right) \exp\left(\frac{(z-z_0)^2}{2S_z^2}\right) \exp\left(\frac{(z+z_0)^2}{2S_z^2}\right) \quad (10.1)$$

where z_0 (metres) is the effective stack height (i.e. centre line of the plume, S_y and S_z are the standard deviations (in metres) of the plume concentration distribution in the horizontal and vertical directions respectively, and u_w (m/s) is the mean wind speed. The wind direction is the X-axis. This model may be extended to include fine particulates (e.g. for lead emissions) and line sources (such as a road), as described in Taylor and Anderson (1988). For a given pollutant, S_y and S_z depend on meteorological conditions.

The advantages of the Gaussian model are:
- the relative simplicity of the modelling equation;
- general support for the plume spread function from field investigations;
- its ability to handle two- and three-dimensional source configurations;
- its ability to include multiple sources, and
- the relative cheapness and ease of implementation in computer models.

Its disadvantages are:
- the implicit assumption of steady state meteorological conditions for some minimum time period (perhaps ten minutes to one hour);

- an inability to account for chemical reactions between pollutant plumes, and
- a restriction to flat or gently rolling terrain.

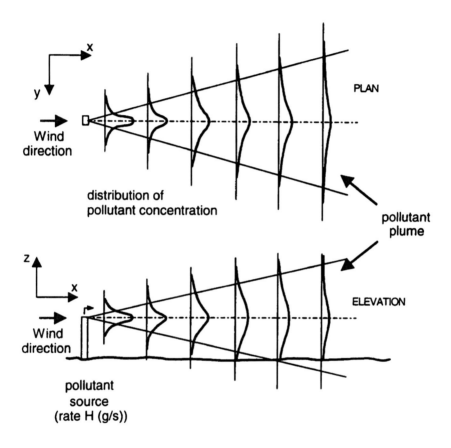

Figure 10.3 Gaussian plume model of pollutant dispersion

A pollutant dispersion model may be used to estimate area-wide levels of air pollution and air quality, given the generation sites of a pollutant. Air quality may be seen as the inverse measure of air pollution. Safe and 'annoying' levels of various pollutants have been set down in many places, often based on World Health Organisation recommendations. Local requirements should be checked with the relevant environmental protection organisation.

Many large cities now have networks of permanent observation stations for measuring air pollution levels. These systems have permanent recording stations across an urban area, which can provide periodic (e.g. hourly) air quality data to a central computer for monitoring and analysis. The area-wide systems record absolute levels of air pollution from all sources, not just road traffic. These permanent stations provide continuous base data. It is quite unlikely that the effects of traffic management schemes, for example, would be isolated (or even reflected) in their readings. Mobile testing stations have been used to provide more detailed local data, and some portable units for use in conjunction with automatic traffic counters are available. These data are still only point recordings, and this is all that seems possible with existing instruments. However, satellite remote-sensing systems can provide area-wide measurements of a variety of air pollutants.

10.5 Traffic noise

Noise may be defined as unwanted sound. Many international studies have concluded that noise is the main source of nuisance in the home, for most people (OECD 1986). Road traffic is the main source of noise in terms of numbers of people affected. Besides being a distraction or a nuisance, noise at high levels can also be injurious.

Noise should be seen as a stochastic variable for the noise level at a given site varies continuously. Variations in noise level are most usefully described in statistical terms on the basis of a probability distribution. Under heavy traffic conditions with low levels of background noise, the normal distribution may be used to provide a description of the distribution of noise (Hothershall and Salter 1977). Figure 10.4(a) gives a plot of sound levels over time, from a site close to a road. Considerable variability is apparent. The frequency distribution of noise levels may be extracted by dividing the range of noise levels into equal increments, and determining the percentage of total sample time for which the noise level was within each increment. Figure 10.4(b) shows the histogram corresponding to the time series data of Figure 10.4(a). These data may then be shown as a cumulative distribution. Figure 10.4(c) indicates the percentage of time that noise levels exceeded a given level. Several measures may then be taken:

- the L_{50} level is the sound level exceeded 50 per cent of the time, i.e. the median level. For the data in Figure 10.4, the L_{50} level is 63.9 dB(A);
- the L_{10} level (70 dB(A) in Figure 10.4), which is the sound level exceeded ten per cent of the time, and

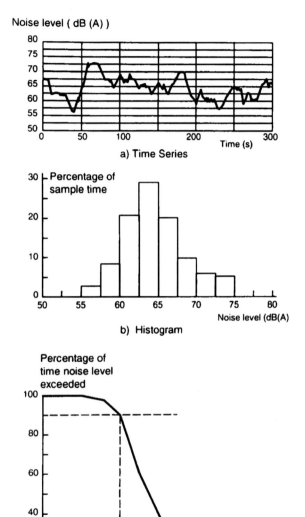

Figure 10.4 Observed noise levels (a) time series (b) histogram (c) cumulative distribution

- the L_{90} level (59.7 dB(A) in Figure 10.4), which is the sound level exceeded 90 per cent of the time.

These measurements are the basic measures of noise, taken directly from the observed data. As the response to noise varies widely amongst individuals, further measures have been sought to provide noise indices for the assessment of noise pollution problems (see Figure 10.1). Hothershall and Salter (1977) defined two conditions to be met by a noise index:
1. it must correlate strongly with expressed annoyance (determined by social surveys in a community), and
2. a comprehensive and definitive set of rules must be available for predicting the index.

The first rule is an obvious requirement for an empirical analysis of a complex phenomenon, such as noise pollution. The second rule enables planning decisions to be made, taking the effects of noise into account.

The L_{10} noise level, assessed over the 18 hour period 0600 to 2400 hours, is widely accepted as a useful index of noise pollution (OECD, 1986; UKDoT, 1986). This level is written as $L_{10}(18h)$. Sometimes a one-hour L_{10} level ($L_{10}(1h)$) is used. Although some authorities prefer consideration of a 24 hour L_{10} level, this correlates strongly with the $L_{10}(18h)$ value (Hothershall and Salter, 1977). Given the widespread acceptance of $L_{10}(18h)$, this measure is the most suitable value for present studies. The noise level 68 dB(A) for $L_{10}(18h)$ has become an acknowledged environmental criterion (OECD, 1986). The equivalent energy level (L_{eq}) has been proposed as more complete measure of environmental noise pollution: L_{eq} is the energy mean of the noise sample and is calculated from the expression,

$$L_{eq} = 10 \log_{10} \left[\sum_{i=1}^{n} P_i \, 10^{L_i/10} \right] \quad (10.2)$$

where P_i is the probability of the noise level lying in the ith measurement interval, and L_i is the mid-point of that interval. Given that noise levels are often well described by a normal distribution then, assuming that the standard deviation of the noise distribution is S, the L_{eq} dB(A) value is given by:

$$L_{eq} = L_{50} + 0.1152 S^2 \quad (10.3)$$

$$L_{eq} = L_{50} + \frac{1}{56}(L_{10} - L_{90})^2$$

Environmental Impacts 225

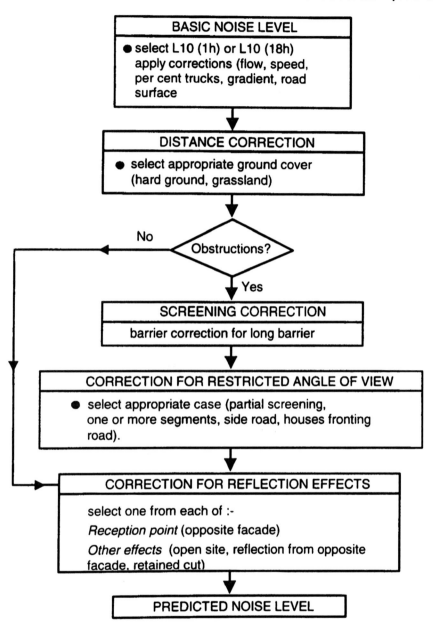

Figure 10.5 Flow chart for the CORTN procedure (UKDoT, 1986)

Considerable research has been undertaken to relate noise levels to traffic and environmental factors. Consequently, there are several useful models for predicting traffic noise levels. One of the most widely used of these models is the CORTN (Calculation of Road Traffic Noise) procedure, developed by the UK Department of the Environment (UKDoT 1986). The CORTN procedure is a multi-stage empirical model, as illustrated in Figure 10.5.

A basic noise parameter [either $L_{10}(18h)$ or $L_{10}(1h)$] is selected, and a value estimated at the kerbside, using the following mathematical relationships:

$$L_{10}^0(18h) = 28.1 + 10 \log_{10} Q$$
$$L_{10}^0(1h) = 41.2 + 10 \log_{10} q$$
(10.4)

where Q (veh) is the 18 hour traffic count for the period 06:00 to 24:00 hours, and q (veh) is a one hour traffic count; $L^0{}_{10}(nh)$ is the basic value of the noise parameter.

A series of corrections ($C_1, C_2, C_3, ..., C_n$) may then be applied for factors such as speed and percentage of heavy vehicles (traffic effects); road gradient and surface (road effects); distance from the road and nature of the ground between the traffic noise source and the observation point, and screening, facades and reflection effects (environmental effects). Details of the more commonly used correction factors and their calculation are as follows:

1. *speed (v)* and *percentage of heavy vehicles (p)*. The basic relationships of equation (10.4) assume a speed of 75 km/h and no heavy vehicles. For other values of these parameters, the correction is given by

$$C_1 = 33 \log_{10}(v + 40 + \frac{500}{v}) + 10 \log_{10}(1 + \frac{5p}{v}) - 68.8 \quad (10.5)$$

2. *road gradient*. Note that on a dual carriageway road, this correction is applied only to traffic climbing the grade. For percentage gradient G, the correction is

$$C_2 = 0.3G \quad (10.6)$$

3. *road surface*. The following correction factor is necessary for coarse-texture road surfaces,

$$C_3 = 4 - 0.03p \qquad (10.7)$$

4. *distance* and *nature of the ground* between the traffic noise source and the observation point. Two types of surface are distinguished: hard ground and grassland. When more than 50 per cent of the surface is non-absorbent (e.g. concrete, bitumen, or water), then the correction is

$$C_4 = -10 \log_{10}\left(\frac{d_s}{13.5}\right) \qquad (10.8)$$

where d_s (metres) is the line of sight distance from the effective source position to the observer. When more than 50 per cent of the surface is absorbent (e.g. grass), then

$$C_4 = -10 \log_{10}\left(\frac{d_s}{13.5}\right) + 5.2 \log_{10}\left(\frac{3h}{d+3.5}\right) \qquad (10.9)$$

if $1 < h \leq (d + 3.5)/3$; or

$$C_4 = -10 \log_{10}\left(\frac{d_s}{13.5}\right) \qquad (10.10)$$

if $h > (d + 3.5)/3$, where h (metres) is the height of the observation point above ground, and d (metres) is the distance along the ground from source to observer.

The full range of corrections is described in UKDoT (1986). Once all of the required corrections have been made, the final estimate of L_{10} (predicted noise level at a given site, generated by traffic on a nearby road) may be made using

$$L_{10} = L_{10}^0 + \sum_{i=1}^{n} C_i \qquad (10.11)$$

An aggregate noise level at a given location, based on traffic on several roads, may be obtained by logarithmic addition of the individual noise measures (Taylor and Anderson 1988). The following procedure is used.

Given two noise levels L_1 and L_2 dB(A) from two sources, where $L_1 > L_2$, then the total combined noise level is given by:

$$L = L_1 + 10 \log_{10} \left[1 + 10^{(L_2 - L_1)/10} \right] \quad (10.12)$$

The above CORTN model is for open country. When buildings and topographical features may affect noise levels, further corrections are possible. They are outlined in UKDoT (1986), and are included in Figure 10.5.

Another well-known noise prediction method is the US National Highway Cooperative Research Program (NCHRP) model (Gordon et al, 1971). This method is similar in principle to the CORTN procedure. Hothershall and Salter (1977) describe the NCHRP model and compare it to CORTN.

An area-wide noise model which can integrate noise pollution levels from individual links in a network to produce an overall noise level at a specified receptor point is described in Woolley (1998). This model, known as NETNOISE, uses CORTN to estimate the noise emissions from each link and then calculates the resulting total noise levels across a study area.

10.6 Surveys of traffic noise levels

Although measurements of noise emissions from individual vehicles are commonly performed by traffic and environmental regulatory bodies, they do not generally fall within the sphere of interest of the traffic engineer. Area-wide noise levels, however, are of considerable importance. Two aspects are of concern: the actual level of traffic noise and the perceived annoyance of traffic noise.

10.6.1 Actual noise levels

Actual noise levels in an area may be recorded by observations at a series of points in the area, in the same way that mobile air pollution readings are taken. Distributions of noise levels vary over time, so statistical analyses of noise distributions are made. Specific measures, such as the L_{10} weighted decibel (dB(A)) scale, are commonly used for traffic noise data. The basic configuration of a sound level meter for recording traffic noise is as follows:
- a microphone to transform sound-pressure changes into electrical impulses;

- an amplifier to raise the microphone output to a useful level;
- a calibrated attenuator to adjust the amplifier signal to an appropriate level for readout;
- weighted networks to enhance sounds in certain frequency ranges;
- a recording device;
- a meter for visual display of noise readings, and
- a calibration sound source.

In view of the random nature of noise and the intervention of many environmental factors, noise surveys must be conducted with great care, and standard procedures adopted. Many highway authorities will have their own procedures. The following description represents a typical traffic noise survey procedure. Measurement positions are fixed at locations at or within the boundaries of a selected site, as close as practicable to the place and time of annoyance. In residential area studies, measurements are usually taken at one metre from the house or building facade fronting the roadway. Some authorities also recommend that a measurement position should be at least five metres from the nearside running lane for traffic. Observations are sought when the road surface is dry, unless the effect of wet roads is to be investigated explicitly. Limits on maximum wind speed may be imposed, often with wind speed measurements taken at a prescribed height (e.g. wind speed not to exceed 7.2 km/h (2 m/s) at a height of 1.2 m). The microphone should be located so that the wind direction gives a (vector) component towards the measurement position exceeding the component parallel to the road. Meteorological conditions should be recorded as part of the survey data. The variety of factors and environmental conditions that could affect the data collected in a noise survey are clear reasons why standardised procedures should be used, for without these data from one survey could not readily be compared with that from another.

10.6.2 Community reactions

The degree of annoyance with pollution is usually ascertained from attitudinal surveys of residents to such; however social surveys based on questionnaires fall outside the scope of this book. The reader is referred to Ampt, Richardson and Meyburg (1995) for an introduction on the design and execution of such surveys. Brown (1980) developed a simple, seven-point semantic scale which succinctly measured all the noise annoyance effects reported by respondents: (1) not at all; (2) very little; (3) a small amount; (4) a fair amount; (5) quite a bit; (6) a lot, and (7) a great deal. When averaged over a group of households, this scale related in an acceptable fashion to observed noise levels.

Another method widely used to indicate the effects of traffic noise in residential areas is the comparison of real estate values between neighbouring subdivisions and streets subject to different levels of noise. If all other factors and features of residential property in different, neighbouring streets are the same, then the difference in noise pollution may be reflected in the difference in property values. Research by the OECD indicated that property values fell by 0.5 per cent for each decibel increase in noise level (OECD, 1986). Some care may be needed, however, to distil the effect of traffic noise from other traffic and socio-economic influences. If there are differences in the characteristics, an implicit price approval approach can be used. This method relates house price to the characteristics describing the house, using linear regression (see Chapter 17). The coefficients associated with the variable representing noise level in the results of the regression analysis will provide an indication of its impact on house prices and residential preference.

10.7 Environmental sensitivity

Environmental impacts from road traffic on surrounding land uses may be complex and may involve many factors. One multi-criteria method for assessing environmental impacts in urban areas in the SIMESEPT Knowledge-Based Expert System developed by Klungboonkrong and Taylor (1999). SIMESEPT considers a number of individual factors (e.g. noise level, pedestrian risk and difficulty of access to abutting land uses) to determine a composite index (CESI) which can be used to highlight environmental problems on specific links in a network and to provide a rank order of the most serious of these problems. The package is integrated with a GIS, both to hold the spatial database for the study area and to display output results. Figure 10.6 provides a sample map output from SIMESEPT.

10.8 Role of the traffic analyst

The systems planning process requires traffic engineers and planners to investigate the impacts on other systems of land use developments and modifications to traffic systems. The environment and energy consumption are prominent as areas in which traffic has significant impacts. Traffic surveys need to account for these impacts.

Environmental Impacts 231

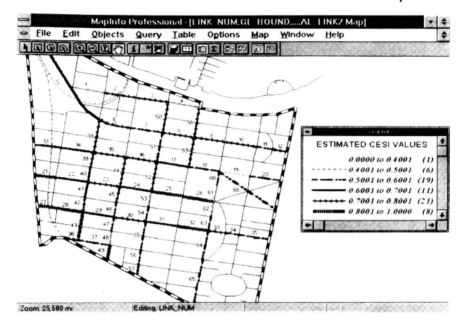

Figure 10.6   **SIMESEPT output indicating links in a network with potential environmental problems**

The survey methods described in this chapter are indicative of the procedures used, in what is a relatively new addition to the traffic surveying repertoire. The present methodology is not definitive nor necessarily completely satisfactory. What is apparent is that surveys of energy use and environmental impacts need to be undertaken through a multidisciplinary approach, to which traffic planners and engineers have much to contribute.

Part D

TRAFFIC STUDIES

11 Intersection studies

In most road networks, especially in urban areas, traffic performance at intersections dictates the performance of the rest of the network. Thus studies of intersection performance are of great importance in traffic analysis. Aspects of performance requiring attention include delays, queuing and capacity, with the related factors of gap acceptance and saturation flow rates. The performance of a junction is dependent not only on total traffic flows, but also on the pattern of traffic movement through the junction.

Junctions between roads, where traffic streams intersect and consequently interact, are the most common type of intersection encountered in traffic engineering, and much of the discussion in this chapter focuses on studies of flows, capacities, delays and queuing at junctions. However, a broader consideration of intersections in terms of disruptions to traffic flow, and hence capacity, delay and queuing, leads us to also consider bottlenecks, pedestrian crossings and railway level crossings amongst other external impediments to traffic flow. The main questions to be addressed in intersection studies concern the amount of traffic that can pass through a flow constriction over a given time period and the degree of difficulty (measured in terms of delays and queuing) accompanying that passage.

11.1 Turning movement flows

One of the particular features of traffic behaviour at a road junction is that road users approaching the junction have a choice about how they leave that junction. They may be able to proceed straight through, or turn to left or right, to follow the chosen paths for their journeys. Some movements through the junction may be prohibited if other traffic movements would be seriously inhibited by them, or if restrictions are sought on traffic movement into one of the departure roads as part of an area-wide traffic management scheme. The pattern of flows at a junction is important information for traffic engineers, especially for capacity analysis and junction design purposes. The pattern is called the turning movement flows matrix for the junction, and it can be regarded as a small origin-destination matrix. A schematic representation of a turning movement flows matrix is shown in Table 11.1 and in graphic representation in Figure 11.1. The rows of this table represent the (in-bound) approach road flow directions to the junction, the columns represent the (out-bound) departure road flow directions. Elements a cell of the table (e.g. q_{ijk}) represent the volume of type k vehicles from approach i turning into departure j. U-turns are seldom considered in turning movement studies, the expectation would be that q_{iik} would be zero only at a roundabout are u-turns made with ease if they are permitted at all. I_{ik} and O_{jk} are the total entry and exit flows on the approaches. Figure 11.1 indicates the relationships between these flows for a simple cross-road junction. Turning movement matrices are often disaggregated by vehicle type or road user class because such details are important in intersection capacity analysis and traffic signal timing calculations (see Section 4.3). Separate matrices are often determined for different times of day, such as morning and evening peak periods because the tidal nature of the traffic flows at such times will have a large bearing on the design requirements for the junction. Factors such as the ratio of traffic volumes between the intersecting roads will largely determine the appropriate type of traffic control for the junction, and how to deal with turns made in the face of opposing traffic.

There are some curious twists in the traffic engineering terminology used in various places, which we have tried to avoid if at all possible. It is worth noting at this stage that the approach roads to a junction are known as the *arms* of the junction in the UK, and as the *legs* of the junction in North America and Australasia.

One obvious way to assemble a turning movement matrix is to observe it directly, by counting the numbers of vehicles by vehicle type making each of the possible movements at the junction. These turning movement studies were discussed in Chapter 8.

The direct observation of turning movements has, in practice, to be performed manually and is therefore an expensive operation which cannot be undertaken routinely. The use of automatic counters is possible in theory, but is rarely practical, although this situation may change with the further development of automatic video analysis software.

An advanced traffic control system such as SCATS or SCOOT can be used for direct observation of turning movements, but these counts are often incomplete for planning and design purposes because some turns are made through separate slip lanes that may not be equipped with vehicle detectors and thus will not be seen by the traffic control system.

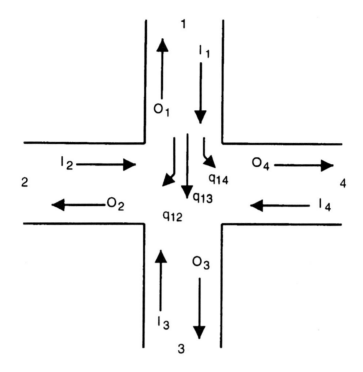

Figure 11.1 Turning movement flows at a junction (assuming left hand side driving)

Table 11.1 Schematic turning movement flows matrix

In-bound direction	Out-bound direction				Totals
	1	2	3	4	
1	q_{11k}	q_{12k}	q_{13k}	q_{14k}	I_{1k}
2	q_{21k}	q_{22k}	q_{23k}	q_{24k}	I_{2k}
3	q_{31k}	q_{32k}	q_{33k}	q_{34k}	I_{3k}
4	q_{41k}	q_{42k}	q_{43k}	q_{44k}	I_{4k}
Totals	O_{1k}	O_{2k}	O_{3k}	O_{4k}	

An alternative is to estimate the turning movements from directional link counts on the approach road (i.e. the I_{ik} and O_{jk} in Table 11.1). The mathematical procedure is based on the method for estimating origin-destination matrices from network link counts, as introduced in Section 5.2.1. The junction turning movement estimation problem is smaller in dimension and simpler than the general O-D matrix estimation problem, so that a simplified procedure is feasible. Rather than the multi-proportional problem of equation (5.7), a bi-proportional problem (i.e. involving two factors only) results, as discussed below. The most commonly used procedure for the problem is the Hauer-Kruithoff method (Hauer, Pagitsas and Shin, 1981), which is similar to the Furness technique (see Section 12.4.3). This method uses the principle of minimum information (maximum entropy) which was applied to the general O-D matrix estimation problem and to the derivation of the gravity trip distribution models (Section 5.2.2).

In the following discussion road user type (i.e. the subscript k in the $\{q_{ijk}\}$ of Table 11.1) will be ignored for purposes of clarity. The assumption is thus that entry and exit flows I_i and O_j are available for the junction. Given an initial estimate p_{ij} of the proportion of the flow from entry i going to exit j, Hauer, Pagitsas and Shin (1981) showed that the turning flow q_{ij} is give by

$$T_{ij} = p_{ij} a_i b_j \qquad (11.1)$$

where a_i and b_j are multipliers to be determined from the entry flows I_i and the exit flows O_j. These multipliers have to be found by iteration. The following algorithm may be applied to solve the problem.

Step 0 Set the iteration counter n to zero and calculate initial values a_i^0 such that

$$a_i^0 = \frac{I_i}{\sqrt{\sum_i I_i}} \qquad (11.2)$$

noting that the denominator in equation (11.2) is the square root of the total flow observed at the junction.

Step 1 Increment the iteration counter by one, i.e.
$n \rightarrow N = 1$

Step 2 Calculate values for the b_j^n, using

$$b_j^n = \frac{O_j}{\sum_i p_{ij} a_i^{n-1}}$$

where p_{ij} are the *a priori* turning proportions, and $0 \leq p_{ij} \leq 1$.

Step 3 Calculate new values for a_i^n, such that

$$a_i^n = \frac{I_i}{\sum_j p_{ij} b_j}$$

Step 4 Check for convergence, by testing if the new values for a_i^n are compatible with the old values. Given a maximum acceptable relative error ε, then if

$$\frac{|a_i^n - a_i^{n-1}|}{a_i^{n-1}} < \varepsilon$$

for all i then the procedure has converged and final values of the q_{ij} may be computed (step 5). If the procedure has not converged and the maximum number of iterations n_{max} has not been reached, go to step 1.

Step 5 The final values for a_i and b_j are available. Calculate the estimated turning flows using equation (11.1).

A typical value for ε would be 0.1 per cent and a typical value for n_{max} would be 20. Values of the *a priori* turning proportions (p_{ij}) are best determined for local conditions, but in the absence of any such data the values given by Hauer, Pagitsas and Shin (1981) for Toronto may be used. These values are given in Table 11.2. Note that the left and right turn proportions in that table are for driving on the right hand side of the road (as in Canada). Some care is needed with the choice of the *a priori* turning proportion values, for multiple solutions may exist to the bi-proportional problem and poor selection of the *a priori* values may lead to converge to an inferior solution.

Table 11.2 Turning proportions for Toronto (from Hauer, Pagitsas and Shin (1981))

To	Collector Road		Arterial Road	
From	Left Turn	Right Turn	Left Turn	Right Turn
Collector Road	0.18	0.20	0.35	0.17
Arterial Road	0.04	0.05	0.12	0.12

Note: left and right turns in this table are for Canadian traffic conditions, i.e. driving on the right hand side of the road.

The Hauer-Kruithoff procedure requires a balance of the total inbound and out-bound flows at the junction, i.e. that $\Sigma_i I_i = \Sigma_j O_j$ thus ensuring a conservation of flow at the junction. If the observed data do not balance, as could well be the case when the flows are collected independently by automatic means, then adjustments need to be made. In such circumstances the first action is to check that the entry and exit flows have been input correctly. If the data values are correct but an imbalance of the total entry and exit flows exists, then the recommended adjustments to the I_i and O_j are

$$I_i^* = I_i(1+\rho)$$
$$O_j^* = O_j(1-\rho)$$

where I_i^* and O_j^* are the adjusted flows and ρ is given by

$$\rho = \frac{\sum_j O_j - \sum_i I_i}{\sum_j O_j + \sum_i I_i}$$

The other constraint on the use of the Hauer-Kruithoff method is that the complete set of in-bound and out-bound flows needs to be known. In situations where only incomplete knowledge of these flows is available (i.e. there are missing values) then a solution can be found using the more general Van Zuylen-Willumsen method for estimating an origin-destination matrix from observed link counts (Section 5.2.1). What should be borne in mind here is a basic tenet of the mathematical theory of information: any estimation method using less-complete information will lead to solutions that are less accurate than those based on more-complete information, such as full knowledge of the in-bound and out-bound flows. This may also be extended to the case where some of the turning movement flows are available. Should one or more turning movement flows be known, then these can be included in the solution for the remaining unknown flows and this will lead to more accurate estimation of the unknown flows. The method is to subtract the known flows from the relevant in-bound and out-bound link totals, and solving the problem for the *estimated* turning movement flows matrix with the known turning movements set to always be zero in that matrix. Once the *estimated* matrix is known, it can be added to the known flows to produce the *final* matrix of turning movement flows.

Knowledge of turning movement flows at a junction is a basic information requirement for traffic engineers when examining a road junction. However, more information is required for the assessment of the traffic performance of the junction. In most instances delays and their distribution between the approaches will provide the primary information about performance, but queue lengths, the number of stops, fuel consumption and pollutant emissions may also be important. Queuing will become a major factor in network analysis, and the propensity for queues to grow and block upstream junctions may be an important factor in the performance of congested networks.

11.2 Delays and queuing

Delay is an important component of travel time, reflecting the additional amount of travel time imposed by the level of travel demand. In general, we can say that

$$T = T_0 + D \qquad (11.3)$$

where T is total travel time, T_0 is a 'free' travel time and D is delay. Free travel time is the absolute minimum time required to cover a given section of route.

11.2.1 Queuing

Before discussing the measurement of queue lengths, it is necessary to define just what a traffic queue is. A vehicle is in a queue when it is controlled in its actions by the vehicle in front of it, or has been stopped by a component of the traffic system. Queues can, therefore, occur in traffic moving along the open road as well as at constrictions in the traffic system. Bunching of vehicles, is by the above definition, a queue. This section, therefore, concentrates on queues forming in the proximity of a junction or a constriction in the road.

To illustrate the various measurements associated with a queue of stopped vehicles, consider a signalised intersection with vehicles arriving and departing uniformly. Figure 11.2 presents a trajectory diagram for such a case.

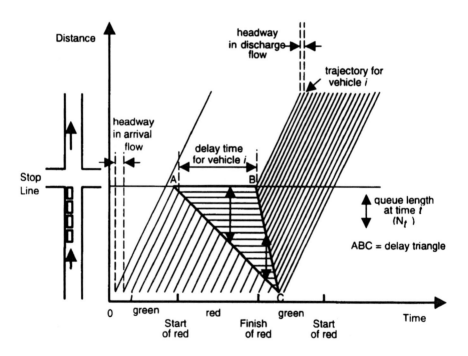

Figure 11.2 **Trajectory diagram at a signalised junction**

Vehicles arriving during the red phase are halted and are not able to proceed during the red time. The vehicles stopped at the intersection can begin to depart when the green phase starts. The vehicles stopped at the intersection when the lights turn green is the *maximum stationary queue*. During the green time, vehicles leave the intersection at a faster rate than they arrive. Hence the queue decreases in total size, but since it takes time for the leading vehicles to start moving, the later vehicles remain stationary for some time after the traffic lights turn green. The point in time when the last stationary vehicle in the queue moves determines the *maximum back of queue*. This of interest since it represents the end of the queue as perceived by the driver.

Another queue that is of interest is termed the *overflow queue*. This is the number of vehicles that are still present in the queue at the end of the green period.

The above discussion relates to traffic signals. In this case the event that determines the critical time of queue measurement is a change in phase. Queues forming at uncontrolled intersections or at other constrictions in the traffic system have the same characteristics as those present at a signalised intersection. The main difference lies in the critical time of queue measurement. In these cases it is likely to be the departure of a vehicle from the head of the queue that determines the critical queue lengths, as will happen under gap acceptance. In mathematical terms the general introductory queuing problem may be represented by a storage-output equation of the form

$$I(t) = \frac{dN}{dt} + O(t) \qquad (11.4)$$

where $I(t)$ is the arrival flow rate, $N(t)$ is the queue length, and $O(t)$ is the departure flow rate, at time t. Queues form if input flows exceed output flows. Output flows are limited by the discharge capacity of the approach leg to the intersection. At a signalised intersection, for example,

$O(t) = 0$ during the red time
and $O(t) \leq s$ during the green time
where s is the saturation flow. For a fixed time, signalised junction operation with a cycle time c and green time g on one approach, we can say

$$O(t) = 0 \text{ for } (n-1)c \leq t < nc - g$$
$$\text{and} \quad O(t) \leq s \text{ for } nc - g \leq t < nc \qquad (11.5)$$

where $n = 1, 2, 3, ...$ is the cycle number. The inequality in relation (11.5) comes from the rule that

$$O(t) = I(t) \text{ if } L(t) = 0 \text{ and } I(t) < s.$$

Queue length at time t is found by integrating equation (11.4) with respect to time, yielding:

$$N(t) = N_0 + \int_0^t I(u)du - \int_0^t O(u)du \qquad (11.6)$$

where N_0 is the initial queue length. Delays can also be found. The total delay time $D(t)$ (i.e. delays summed over all vehicles entering the system up to time t) is given by the area between the cumulative arrivals and departures curves. That is,

$$D(t) = \int_0^t N(u)du \qquad (11.7)$$

As discussed in Chapter 9, the measurement of the queue lengths involves an observer recording the number of stationary vehicles at a particular point in time. This can be done by physically counting the vehicles, or by placing marks along the road length to indicate the number of vehicles expected to be in a queue of a given physical length. Alternatively, video cameras can be used to record the queue lengths for later analysis.

11.2.2 Traffic delay

The second measure of performance of a traffic system considered in this section is traffic delay. There are many definitions of delay. Stopped delay is the delay experienced by vehicles that have actually stopped. This can also be referred to as queuing delay. It is one component of overall delay. Congestion delay can include both the delay due to queuing and that resulting from a vehicle having to slow down because of interactions with other vehicles, and is measured by the difference between journey time and the desired travel time, i.e. using equation (11.3).

Delays represent time that is non-productive, and when converted to monetary values, represent a large proportion of the cost to the community of inadequate transport facilities. Reductions in delay are thus part of the economic return that can be expected if a route is improved - therefore permitting priorities for transport link improvements to be set. At the same time some congestion delay is an inevitable consequence of traffic demand. Thus traffic engineers may seek to reduce delays, but can never eradicate them.

At this point we need to consider some of the definitions commonly used in delay surveys. These definitions extend the lists given in Chapter 9, for speeds and travel times. Then we can turn our attention to discuss delay in a general system and delay at signalised junctions. Some of the definitions adopted in delay surveys include:

- *delay section*. The section of road where all or most of the delay takes place. This section is defined by means of an upstream and downstream marker. The upstream marker should be placed where the vehicles have not started to slow down. This point may be difficult to find under some conditions. The downstream marker is placed at the front end of the queue. In intersection studies this is usually taken as the stop line, even though some delay is incurred while accelerating across that line;
- *desired travel time*. The minimum time for vehicles to traverse the delay section;
- *desired speed*. The length of delay section divided by desired travel time;
- *stopped delay*. The time the vehicle is stationary, due to intersection or other related activity. Stopped delay is the same as stopped time;
- *joining time*. That portion of the travel time during which the vehicle enters the delay section and comes to a stop;
- *motion time*. That portion of the travel time which occurs between two periods of stop time;
- *time in queue delay*. Time from when the vehicle first stops to when it exits the queuing section. In intersections the exit time is measured when the vehicle crosses the stop line;
- *unnecessary stopped time delay*. At intersections, this is that portion of the stopped delay which occurs when there is no vehicle entering an approach on the cross approaches of the intersection;
- *delay ratio*. Delay time divided by the actual travel time, and
- *system delay*. Stopped delay is simple to define and measure. In contrast, system or congestion delay is less precise. It is the delay caused by the constriction or slowing down effect of overloaded intersections, inadequate carriageway widths, parking and parked cars, crowded pavements and other factors.

246 *Understanding Traffic Systems*

The prime concerns when determining system delay are the determination of the travel time through the delay section and the desired travel time.

11.3 Intersection delay studies

The preceding discussion outlined the principal methods adopted for measuring delay in any part of a traffic network. Intersections are a particular network element for which delay information is often sought. Attention to particular methods for observations of intersection delays is therefore warranted. Intersection delays are usually taken separately for one or more of the approach roads to the intersection, although an overall intersection delay can then be found by summing the weighted delays on each approach, where the weights reflect the volume of traffic on the approaches. The methods used for observing delays at isolated intersections are similar in concept to those discussed above. They are often based on the study of the travel time over a length between two points, one upstream and one downstream of the junction. Recording of registration numbers, vantage point video recording and vehicle detectors can all be used to provide the basic information for analysis in the manner discussed above. However, there are a number of methods have been developed primarily for intersection studies, as outlined by Teply (1989). The methods relate to: (a) stopped delay; and (b) overall delay.

11.3.1 Stopped delay

Stop-time delay can be collected by manual recording or with a delay meter. The latter accumulates the number of vehicle seconds of stopped time delay. The manual method is similar to the input-output procedure discussed above. It is based on the calculation of stopped delay (D_s) from the formula

$$D_s = \sum_{i=1}^{n} [T_{Si} - T_{Ei}] \qquad (11.8)$$

where n is the total number of vehicles stopped, T_{Si} is the time when vehicle i stopped, and T_{Ei} is the time when vehicle i started. Because continuous functions can generally be represented by piecewise linear sections, the mean value of each section represents the best estimate for the time interval considered. The approach, therefore, separately records stopped vehicles in small time intervals, as well as all previously stopped departed vehicles. A matrix is drawn up as shown in Table 11.3, and the number of vehicles remaining in each time slice is calculated. The stopped delay D_s is thus determined by multiplying the sum of the remaining vehicles in each period by the duration of the time intervals (Δt), as shown by equation (11.9):

$$D_s = \Delta t \sum_{j=1}^{n} R_j \qquad (11.9)$$

where R_j is the number of vehicles remaining in the queue at the end of the jth time interval. For the data in Table 11.3 this yields $D_s = 10(6 + 10 + 12 + 8 + 2 + 0) = 380$ (seconds). Typically, intervals of ten to fifteen seconds are used. At traffic signals the time interval must not be a direct proportion of the cycle time, otherwise all observations will be made at the same time points in each cycle and misleading results will emerge.

Table 11.3   Vehicle numbers over time at a sample intersection

Vehicles	Time intervals in ten second slices of cycle time					
	0-10	11-20	21-30	31-40	41-50	51-60
Stopped	6	4	3	2	2	0
Leaving	0	0	1	6	8	2
Remaining in period (R_j)	6	10	12	8	2	0

11.3.2 Overall delay

Most of the techniques that have been discussed in the previous section can be used to determine overall delays at intersections. The QDELAY technique developed by Richardson and Graham (1982) utilises a point sample technique, and the observation of queue formation and dissipation to calculate overall delay on an intersection approach. The technique is based on the observation of the queues as depicted in Figure 11.2.

More specifically, the procedure entails the recording of queue lengths at different points in time during the signal cycle. Figure 11.3 shows the QDELAY form used to record data in the field. At the start of the green

248 *Understanding Traffic Systems*

phase, the time is recorded in column A and the number of vehicles stopped in the queue is recorded in column B. At this point it is necessary to make a mental note of the last vehicle in the queue at the start of the green time. If this vehicle crosses the stop line before the red phase starts, the time at which it crosses the stop line is recorded in column C. After the vehicle crosses the stop line, the number of vehicles crossing the stop line before the start of the red phase is recorded in column F. The time at which the signal changes back to red is then recorded in column D. Column E is left blank in this situation.

University of South Australia							Transport Systems Centre			
QDELAY Survey of Queuing and Delays at Traffic Signals										
Intersection:							Date: __/__/__			
Approach: _____			No of Lanes: ____				Observer: _____			
Start of Green				Last Vehicle			Start of Red			
Time			Queue Length	Time			Flow After	Time			Queue Length
HH	MM	SS		HH	MM	SS		HH	MM	SS	
A			B	C			F	D			E

Figure 11.3 Survey form for QDELAY

If, however, the end-of-queue vehicle does not cross the stop line before the light changes back to red, the time at which the light changes to red is recorded in column D. The number of vehicles in the queue at this point of time (including the last vehicle) is recorded in column E. Columns C and F are in this case left blank. This process is repeated for every cycle of the survey period.

These data can then be analysed to provide information on the signal settings; total flow; averages and standard deviations of the approach delay and stop delay per vehicle; average numbers and standard deviations of effective stops and complete stops; and average and standard deviation of maximum stationary queue length per cycle. It cannot, however, distinguish between delays to vehicles making particular manoeuvres if the lanes are carrying mixed turning and through vehicles.

11.4 Saturation flow studies

Saturation flows play an important part in intersection design, especially for traffic signals design. Saturation flow occurs when there is a discharge of vehicles from a standing queue, as happens at the start of a green period in a traffic signal cycle, and it is regarded as the maximum possible instantaneous flow rate for a traffic element such as a traffic lane. The use of saturation flow rates in the capacity analysis of signalised junctions was described in Section 4.3, where a number of methods for calculation of saturation flows based on knowledge of intersection geometry and traffic volume and composition were outlined. However, it is often necessary to make direct observations of saturation flows for existing situations, as local factors peculiar to a particular intersection can greatly affect saturation flows. Indeed there are cases where junctions seen to be operating satisfactorily could not do so if their saturation flows were those suggested by the standard results! In addition, the methods for calculating saturation flows are all based on an assumed standard value, e.g. '2080 pcu/h/lane', and this value had to be determined by direct observation and data analysis at some stage. What is also known is that saturation flows have been changing over time, perhaps under the twin influences of improved vehicle technology and increasing pressure from traffic congestion. Thus there are needs for survey methods that can provide measurements of saturation flows.

Two main methods are used for measuring saturation flows. These are the *headway ratio method* and the *regression method*. Both methods require detailed observation of a traffic stream, including the headway between successive vehicles and identification of the vehicle type and turning manoeuvre. Indeed, the measurement of saturation flows is well regarded as an exceptionally demanding task, particularly if the data are collected to a single observer. The difference between the two methods lies primarily in the analysis to which the data are subject.

11.4.1 Headway ratio method

This method was originally used in the UK in the 1960s and has been steadily refined since then. It requires the measurement of headways between successive vehicles in a traffic stream (e.g. a single lane) under saturated conditions (e.g. vehicles leaving a standing queue). The normal definition adopted for headway is the time spacing between the passage of the rear wheels of the previous vehicle over the stop line until the passage of the rear wheels of the vehicle of interest over the stop line. Accuracy of measurement to within 0.1 seconds or better is sought. If the average headway between a succession of passenger cars under saturated conditions is h_i then the saturation flow in pcu/h is given by:

$$s = 1/\overline{h_i}$$

If the traffic stream consists of a mixture of vehicle types then pcu equivalents can be determined for each type by considering the mean headways for the types. This requires the additional recording of a vehicle type code with each observed headway. Conventional practice is to ignore the headways of the first few (four, sometimes five) vehicles leaving the queue, as some time lag may be expected in building up to saturation flow rate as the queue starts to move. Observations should only be made of queues that appear to be operating at saturation. Remember that saturation flow rate is the maximum possible instantaneous flow rate for the queue. Any observations taken of lower flow rates will yield misleading results. The success of the method is dependent on having a well trained observer and a good observation position. Cuddon (1993) identified five rules for site selection:

(1) a clear view must be available of the traffic signals for the lane(s) being measured, so that changes in the signal aspect on display can be seen immediately;

(2) the stop line for each lane being studied is visible and the tyres of the vehicles crossing the stop line can be seen clearly with only small parallax errors;

Intersection Studies 251

(3) the vehicles at the back of the queue must be clearly visible and identified;

(4) sources of possible interference to the traffic are in view (especially the downstream conditions), so that the observer can determine if flow away from the queue is being impeded, and

(5) observers must be in positions where they cannot obstruct the pedestrian or traffic flow nor should they be in conspicuous locations that may case drivers to modify their behaviour.

In situations where no location satisfying all of the above rules can be found, a trade-off may be necessary. A clear view of traffic crossing the stop line is essential, because this provides the basic headway data used to calculate saturation flows. Perhaps the best locations will tend to be found close to the stop line. A choice of site next to the signal controller box (if any) at the junction would be a good starting point. A typical observer location point is shown in Figure 11.4. In this position the observer could collect data for both the eastern and southern approaches, provided that the signal stages for these lanes do not overlap. In addition, observers should be completely familiar with the signal phasing arrangements before starting to record data.

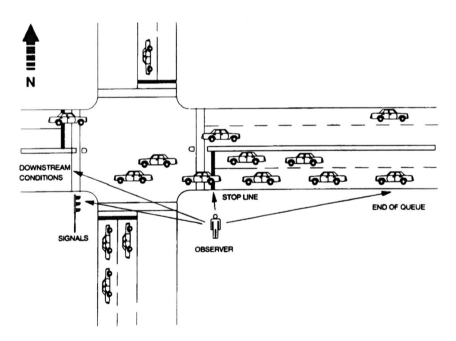

Figure 11.4 Observer location in saturation flow studies (Cuddon, 1993)

Observation and data recording is often done using the traditional pencil, paper and stopwatch but increasing use is being made of event recorders, video and portable computers. The success of *manual methods* dependents on the availability of a well designed data recording sheet. The sheet illustrated in Figure 11.5 is based on a design by Adams and Hummer (1993). Each surveyor requires a stopwatch and pencil in addition to the field sheet. The observation method is as follows:

1. at the start of the green time, identify the last vehicle in the queue;
2. start the stop watch when the rear tyres of the fourth vehicle in the queue cross the stop line;
3. record the times at which the rear tyres of the seventh, eighth, ninth and tenth vehicles in the queue cross the stop line, but not including any vehicles joining the back of the queue after the start of the green;
4. circle the position in the queue of any heavy commercial vehicles (trucks, lorries or buses). This includes the first six vehicles in the queue, even though their crossing times are not recorded explicitly (see Figure 11.5), and
5. stop the watch after the passage of the tenth vehicle in the queue or when the back-of-queue vehicle at the start of green crosses the stop line.

The method waits for the first four vehicles in the queue to pass, ignores queues of six or less, and does not record data for more than ten vehicles in a queue. Saturation flow is calculated by converting the recorded times into vehicles per hour. Simple vehicle equivalents can be determined by comparing records containing heavy commercial vehicles with those only containing passenger cars.

The manual method is difficult and error prone, particularly if the site conditions are less than ideal or the traffic flow is mixed. These difficulties have stimulated interest in the alternative methods.

The use of *event recorders* attached to vehicle detectors laid in the lane of interest provides one way to obtain accurate vehicle headway information, but the limitation of this method may be in identifying vehicle types for the headways. An observer might still be needed for this operation.

Video recording of the flow away from the queue can provide good data although the data extraction process is time consuming and tedious. An advantage of video recording is that it can provide a complete history of the traffic movements under observation. Any apparent ambiguities or discrepancies later discovered in the headway observations may then be checked by examining the video record.

Intersection Studies 253

	University of South Australia			Transport Systems Centre	
	Saturation Flow Survey				
	Intersection: _____			UBD ref: _____	
	Approach: _____				
	Date: __/__/__			Time: __:__ until __:__	
	Observer: _____				

Cycle No	Time between 4th Vehicle and ...				Commercial vehicle positions
	7th vehicle	8th vehicle	9th vehicle	10th vehicle	
1					1 2 3 4 5 6 7 8 9 10
2					1 2 3 4 5 6 7 8 9 10
3					1 2 3 4 5 6 7 8 9 10
4					1 2 3 4 5 6 7 8 9 10
5					1 2 3 4 5 6 7 8 9 10
6					1 2 3 4 5 6 7 8 9 10
7					1 2 3 4 5 6 7 8 9 10
8					1 2 3 4 5 6 7 8 9 10
9					1 2 3 4 5 6 7 8 9 10
10					1 2 3 4 5 6 7 8 9 10
11					1 2 3 4 5 6 7 8 9 10
12					1 2 3 4 5 6 7 8 9 10
13					1 2 3 4 5 6 7 8 9 10
14					1 2 3 4 5 6 7 8 9 10
15					1 2 3 4 5 6 7 8 9 10
16					1 2 3 4 5 6 7 8 9 10
17					1 2 3 4 5 6 7 8 9 10
18					1 2 3 4 5 6 7 8 9 10
19					1 2 3 4 5 6 7 8 9 10
20					1 2 3 4 5 6 7 8 9 10
21					1 2 3 4 5 6 7 8 9 10
22					1 2 3 4 5 6 7 8 9 10
23					1 2 3 4 5 6 7 8 9 10
24					1 2 3 4 5 6 7 8 9 10
25					1 2 3 4 5 6 7 8 9 10
26					1 2 3 4 5 6 7 8 9 10
27					1 2 3 4 5 6 7 8 9 10
28					1 2 3 4 5 6 7 8 9 10
29					1 2 3 4 5 6 7 8 9 10
30					1 2 3 4 5 6 7 8 9 10

Figure 11.5 Saturation flow survey form (source: Adams and Hummer, 1993)

Problems with the traditional manual methods, with event recorders and with video have led a number of analysts to develop survey procedures based on *portable computers*. This approach offers the possibility of an accurate time base, collection of more detailed data than was traditionally possible (e.g. further disaggregation by vehicle type), instant data checking and data analysis on site, and electronic storage of detailed data ready for further analysis if necessary.

An example of this software is Cuddon's (1993) SATFLOW program. SATFLOW requires only one observer working with a portable PC. The observer initialises the program and then collects event time and vehicle type information by using a restricted set of keys on the computer keyboard. Data are recorded by pressing a key each time the rear-most axle of a vehicle crosses the stop line. Separate keys are used for different vehicle types and turns, and for signal timings, the end of saturation and pedestrian movements opposing the flow. The following data are logged:
- the start of the green time for the movement;
- the vehicles crossing the stop line throughout the green and amber periods;
- the end of saturation, and
- the end of the green period.

At the start of the green time the observer must glance to the rear of the queue to identify the last vehicle to arrive at the end of the queue. If vehicles continue to arrive at the tail then the observer must update this observation whilst recording the passage of vehicles across the stop line. Given that a gap of at least one second can be anticipated between vehicles at the stop line the observer will have opportunities to identify the last *stopping* vehicle. A stopping vehicle is one that is stopped by the queue or one which would have stopped had it not slowed down to avoid stopping. Any vehicle whose progress is governed by the queue is thus a stopping vehicle. Identification of the last stopping vehicle is a demanding task. It is also essential, for saturation is deemed to have finished once the last stopping vehicle crosses the stop line. A certain level of skill and expertise is thus necessary for saturation flow observers, who should thus be trained and assessed before undertaking survey work.

SATFLOW allows collection of vehicle data after the end of saturation, if this is desired. The end of green must also be recorded; data for the movement are then saved. This offers some protection against catastrophic data loss should any mishap occur thereafter. The observer also has the opportunity to log comments about each cycle and to reject a given green period as necessary.

The output from the program includes both saturation flow rates and vehicle (pcu) equivalents.

11.4.2 Regression methods

An alternative method which may require less intensive data collection is the application of multiple linear regression modelling to a set of observed headways. The general method was described by Kimber, McDonald and Hounsell (1985) and exists in two forms: *synchronous* and *asynchronous*.

In the synchronous method, n_i vehicle departures of class i vehicles are recorded over a total time period Δt, where Δt starts and finishes with the departure of a vehicle. If the mean headway between type i vehicles is η_i then the model

$$\Delta t = \sum_i \eta_i n_i + \varepsilon$$

where ε is an error term may be fitted by multiple linear regression. If the estimated values of the η_i are a_i and vehicle class number one is for passenger cars then the basic saturation flow rate in pcu per unit time is given by $1/a_1$ and the vehicle equivalent for class i vehicles is given by a_i/a_1.

The synchronous method is limited by the requirement for the start and end of the time period Δt to coincide with the departures of vehicles. In cases where this cannot be assured in data collection an alternative procedure, the asynchronous method, may be used. In this method the time period can begin and end at any instant, independent of vehicle departures. The number of passenger cars n_1 is regressed on Δt and on the numbers of other vehicles n_i, $i > 1$, to obtain estimates of b_0 and b_i for the parameters β_0 and β_i in the model

$$n_1 = \beta_0 \Delta t - \sum_{i \neq 1} \beta_i n + \varepsilon$$

Again ε is an error term. The values of b_i are then the estimated pcu equivalents for vehicle types i. The base saturation flow rate in pcu per unit time is given by b_0.

11.5 Gap acceptance studies

The theory of critical gaps as a model of driver behaviour at unsignalised intersections was described in Section 4.1. Measurement of critical gaps is accomplished in somewhat similar fashion to that of saturation flows, for it is based on observation of the headways between vehicles in a traffic stream. When a minor stream unit is attempting to complete a manoeuvre, the headways in the major stream are measured. In additional, the behaviour of the minor stream unit is noted. Assessment of the distribution of critical gaps is complicated, for only the following information can be observed for minor stream unit i: (1) the size of the largest rejected gap (x_i), and (2) the size of the accepted gap (y_i). The critical gap cannot be measured directly, rather it is assumed to lie between x_i and y_i, implying that $x_i \leq y_i$. (For observations where $x_i > y_i$, the usual assumption is that the subject was inattentive and x_i is just less than y_i.) The method of maximum likelihood can be used to determine the distribution parameters for critical gaps, as described by Troutbeck (1993) and in Section 17.3 of this book. Given that the cumulative distribution function for the critical gaps is $F(t)$, then $F(y_i)$ $F(x_i)$ is the probability that the critical gap is between x_i and y_i. The likelihood function for the set of n observations is the product of these probabilities, i.e.

$$\Theta = \prod_{i=1}^{n} \left[F(y_i) - F(x_i) \right]$$

and the logarithm of this function is

$$L = \log_e \Theta = \sum_{i=1}^{n} \log_e \left(F(y_i) - F(x_i) \right) \qquad (11.10)$$

Maximum likelihood estimates are then found by assuming a functional form for the cumulative distribution function $F(u)$, and determining the relevant parameters for that distribution by equating to zero the derivatives of the log-likelihood function L in equation (11.10) with respect to those parameters. The log-normal distribution has found wide acceptance as a suitable distribution for critical gaps. It can usually be fitted to observed data, has positive values only, and is skewed to the right (thus agreeing with the trend of observed data sets).

This process is involved and requires a significant amount of data. In certain circumstances a simpler procedure may be used in its stead, although this will yield less accurate results. This method is to plot the number of road users who accept a gap of size t against gap size t. A step function should be drawn on the plot, as shown in Figure 11.6. The slope of the line of best fit drawn through the mean gap sizes for each step is $1/t_f$ (i.e. the reciprocal of the move-up time), and the intercept on the t-axis is equal to $t_a - \frac{1}{2}t_f$, where t_a is the critical gap. The method can only be applied as along as there is a continuous queue in the minor stream.

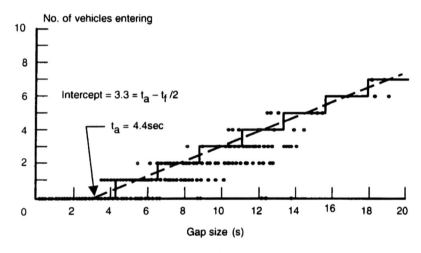

Figure 11.6 Estimation of critical gap and move-up time from observed accepted gaps for minor stream traffic leaving a standing queue

12 Origin-destination and route choice studies

Origin-destination (O-D) matrices describe the basic pattern of demand across a network and so provide an essential input to analyses of the way in which traffic might adjust to traffic management measures or the provision of new capacity. The nature of O-D data was described in Section 5.2. For some purposes it will be necessary to know the complete (door-to-door) O-D matrix, and possibly other information such as the purposes of the journeys or the frequency with which they are made, whereas for others it may be sufficient to know just the pattern of demand within a sub-area in the network. In such situations a cordon can be drawn around the area of interest with origins and destinations defined at the cordon crossing points.

An understanding of route choice is, of course, an essential input to the efficient design of networks and of a wide range of traffic management measures from traffic control to directional signing. Designers will typically make use of assignment models which purport to reflect route choice behaviour but, to this point in time, these models have rarely attempted to reflect the effect of such aspects as directional signing. Against this background it is particularly important to provide the designer with insights into the route choice process and to provide the modeller with information on which to base more behaviourally oriented models.

12.1 Issues in O-D and route choice studies

Two broad issues confront the traffic analyst when considering the collection of O-D and route choice data. These are (1) the problems and difficulties associated with the data collection and analysis, and (2) the nature of the information to be gathered (especially the specific information needs of a given study).

12.1.1 Problems and difficulties

Unfortunately it turns out to be very expensive to collect data on the underlying O-D matrix and on route choice. It is very easy to spend quite considerable sums and still not have data which will pass statistical scrutiny, or indeed which may be quite erroneous. The expense is a consequence of the need to obtain quite detailed data on large numbers of vehicles - often employing large teams of surveyors in the process. The statistical problems are due to the relative complexity of the data, the practical and economic constraints which stand in the way of obtaining sufficiently large sample sizes, and the errors inherent in some data collection procedures (unless great care and organisation is undertaken).

12.1.2 A categorisation of data sources

Table 12.1 indicates a range of data sources and indicates their potential use in O-D surveys and route choice studies. A distinction has been made between door-to-door O-D and cordoned O-D matrices.

Table 12.1 Data sources for origin destination and route choice studies

	Purpose		
	Door-to-door O-D matrix	Cordoned O-D matrix	Route Choice Studies
Interviews and questionnaires	✔	(✔)	✔
Registration plate matching		✔	✔
Vantage-point observation and video		✔	(✔)
Tag surveys		✔	(✔)
Headlight surveys		✔	
Traffic counts		✔	
Vehicle tracking	✔	✔	✔
Car following			✔
Simulator studies			✔

12.2 Methods of obtaining O-D data

Table 12.1 indicates that there are many techniques available for collecting O-D data, and that different techniques may be used to collect different types of information. The analyst has to determine the most suitable and efficient technique to provide the information required for the particular study.

12.2.1 Interviews and questionnaires

Travellers can be asked to record the origin and destination of one or more trips via interviews or questionnaires administered at the traveller's home or trip destination or at the roadside. The length of time available during *household interviews* allows for details for all trips during a specified time period to be recorded along with a wide range of contextual information such as the traveller's, age, sex, socio-economic status and car ownership. *Trip end interviews* might be conducted at the traveller's workplace or other attractors such as shopping centres (see Section 13.4). They are normally constrained to be shorter than household interviews and so typically collect details on fewer trips and provide less contextual information. Household and trip end interviews can provide data on a sample of trips within a study area but the sample is restricted by the high costs involved in such interviews (each household interview will typically cost an amount equivalent to one or two days employment of a surveyor). A useful and much cheaper variant on the trip end interview is the *trip end mailback survey* whereby prepaid postcards are handed out asking for the location of places visited before and after the place where the card was issued. This can be a particularly effective way of collecting information about O-D matrices associated with special events such as festivals and sporting fixtures but may result in a biased sample.

A problem with all of these surveys is that they cannot provide details of through traffic which neither starts not ends its journey within the study area. *Roadside interviews* and *roadside mailback questionnaires*, on the other hand, can pick up such traffic. A roadside interview will typically involve stopping a sample of traffic at some convenient location, such as a layby or wide stretch of road where an interview bay can be coned off, and asking for details of each driver's origin, destination, journey purpose and perhaps frequency. Figure 12.1 shows a typical interview form. Note that it contains only five questions per driver and so the interview can be completed in one or two minutes. The problem is that, even if the interviews are kept relatively brief, such surveys may disrupt the traffic flow and so be unpopular with drivers and with highway authorities.

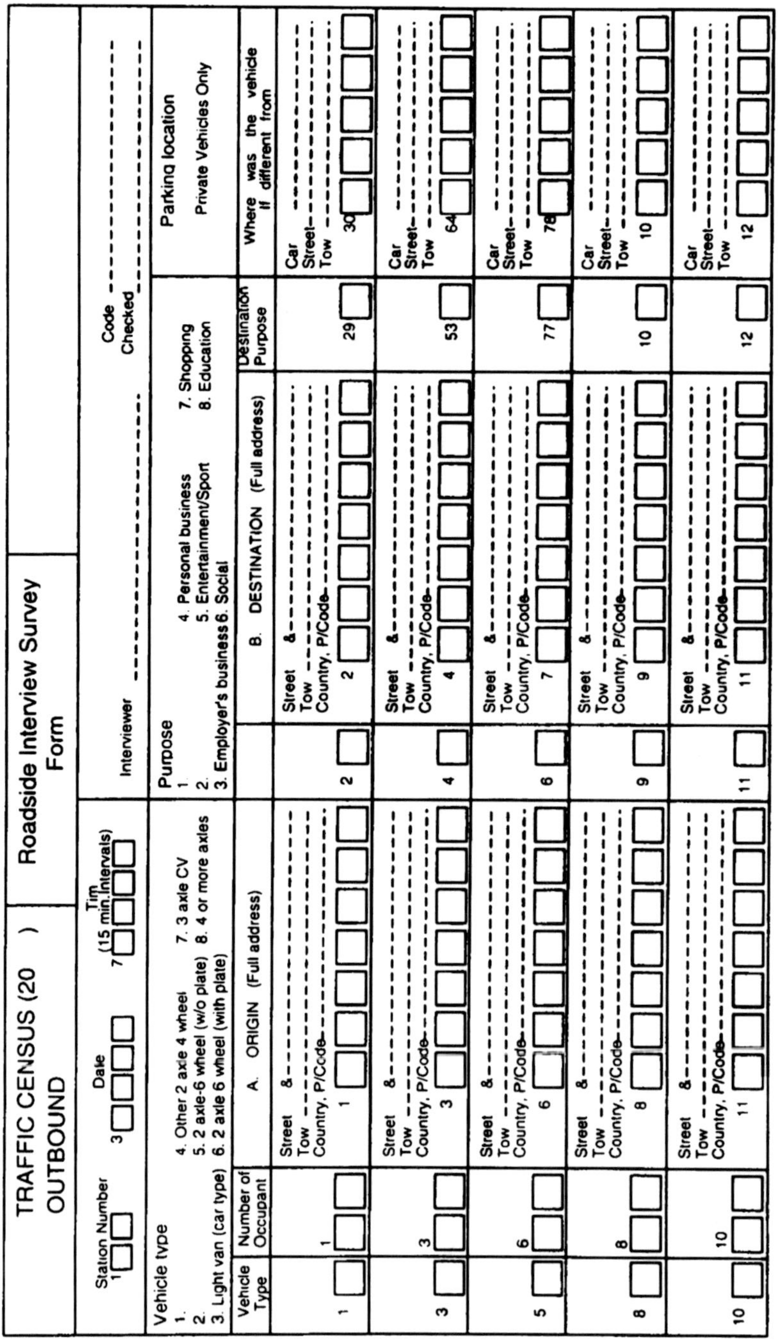

Figure 12.1 Typical roadside interview form

Origin-destination and Route Choice Studies

A roadside mailback survey involves handing out brief questionnaires, usually printed on a prepaid postcard, to traffic at stoplines or in 'natural' queues and therefore causes less disruption to the traffic. The problem with this type of mailback survey is, however, that the response rate is likely to be low (typically around 30-40 per cent) and may give a distorted picture of the underlying matrix (Bonsall and McKimm, 1993). Nevertheless they are becoming more widely used because of the impracticability of conducting roadside interviews on increasingly busy networks.

One problem with roadside interviews and mailback surveys is that convenient locations at which to conduct the survey may not coincide with the locations required to ensure that all relevant traffic streams are sampled. Some success has been reported with an alternative method of distributing mailback questionnaires; the *registered owner mailback survey*. This involves recording the registration plates of vehicles passing target locations and then sending a questionnaire to the registered owners of those vehicles. Privacy legislation or bureaucratic difficulties may in practice prevent the use of this method but, even where it is possible, it is fairly expensive and produces low response rates with potential inherent bias. It is not therefore generally recommended if the only data sought are the traveller's origin and destination.

12.2.2 Registration plate matching

If a cordon is drawn around the study area with screenlines separating the area in subzones and the registration numbers of all vehicles crossing the cordon or screenlines are recorded and then matched, it is possible to create a cordon O-D matrix. The selection of cordon points will depend on the purpose of the survey and great care should be taken over their precise location (see Figure 12.2). The methods of recording will be essentially the same as those described in the context of travel time surveys (see Section 9.4.1) except that less accuracy is required in the recording of time. [It is however, still useful to record the time period within which the recording was made because this can assist in ruling out infeasible matches during the matching process.] The matching software will differ from that used in travel time surveys in that it will be able to deal with several sites simultaneously.

The method is very susceptible to error and so the very highest standards of data collection must be maintained. If a vehicle which passes two observation points has its registration plate incorrectly recorded at one of the observation points it will not be correctly matched and, instead of one through trip, the matrix will contain two trips - one inbound and one outbound. This possibility leads to the conclusion that, in order to reduce observer error, attempts should be made to lighten the workload by reducing the sampling rate and by recording only part of the registration plate. However, if the sampling rate is too low some cells in the matrix will not be observed and, if the records of each plate are too brief, the risk of spurious matching, and hence over-representation of the dominant flow, will be increased. Notwithstanding this last problem, experience suggests that it is best to record partial plates, but that great care needs to be taken in data collection, editing and analysis (see Section 12.4).

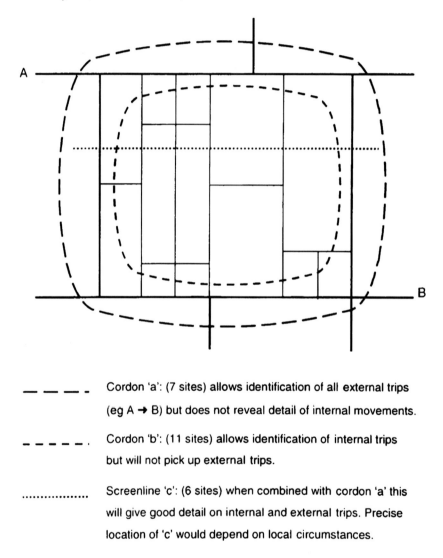

— — — — Cordon 'a': (7 sites) allows identification of all external trips (eg A → B) but does not reveal detail of internal movements.

- - - - - Cordon 'b': (11 sites) allows identification of internal trips but will not pick up external trips.

.... Screenline 'c': (6 sites) when combined with cordon 'a' this will give good detail on internal and external trips. Precise location of 'c' would depend on local circumstances.

Figure 12.2 **Setting a cordon line for a study area**

Origin-destination and Route Choice Studies

Provided that the correct survey procedures are adhered to at all observation points the resulting matrix will be a reasonable sample of the true pattern and it can be factored up to the observed control totals (directional flows of vehicles at each observation point) by means of an iterative procedure such as the Furness technique (see below). However, if this leads to significantly different factors for different cells there must obviously be the suspicion that the data are flawed.

Registration plate matching is probably the most commonly used method for the direct observation of O-D data in the field, but the inherent possibilities of data errors must be recognised. See Section 12.4 for a description of methods for analysis of registration plate data.

12.2.3 Use of vantage point observers or video

When the study area is very small, such as when the task is to determine the pattern of flows through a complex intersection, it may be possible to find a vantage point from which the traffic movements can be recorded. A human observer will be cheaper than video recording and will have a wider effective field of view than a video at the same location. However, a human will only be able to deal with a small, possibly biased, sample of movements within a give time period whereas a video can be reanalysed to track all vehicles. Another potential advantage of video is that, using a telescopic mast, tethered balloon or model aircraft, a better vantage point may be achievable.

12.2.4 Tag surveys

This method traditionally involves 'tagging' parked vehicles with small *stickers*, with different colour stickers being used at different locations. By logging the subsequent appearance of tagged vehicles elsewhere in the network it is possible to reconstruct the pattern of routes away from each tag station. The method is particularly appropriate as a means of gathering data on the dispersal of vehicles from different car parks in a CBD, shopping centre, transport terminal or leisure complex. It is important that the stickers remain affixed for long enough but can be easily removed by the vehicle owner.

See Section 12.2.7 for a discussion of *electronic tagging*.

12.2.5 Headlight surveys

This technique can yield information on the patterns of dispersion of vehicles from special events and on O-D patterns more generally. It involves putting up a sign asking drivers to turn on their headlights and to keep them lit until they finish their journey or until told otherwise. An observer just downstream of the sign then counts the proportion of vehicles which have complied with the instruction. Further observers at various downstream locations elsewhere in the network then count the number of vehicles passing with their headlights on. Multiplication of these numbers by the inverse of the compliance rate then gives an estimate of the flow of vehicles from the original site to each of the downstream sites. The method depends on the assumptions that, having turned on their lights, drivers do not subsequently extinguish them and that the pattern is not distorted by other factors that might influence drivers to turn on their headlights (such as tunnels, or rainstorms). Also, if a significant proportion of vehicles have their headlights on anyway (some vehicles have them permanently on while in motion) an extra observer, upstream of the signs, would need to calculate this proportion so that it can be subtracted from the estimates.

By moving the instruction signs to each of a series of input sites on successive days an estimate can be made of the complete O-D matrix. The estimate can be improved, for example to correct for variation in flow or in the O-D matrix between the different days, by using exit flow counts on each day as a basis of factoring to a common base. The Furness procedure, described in Section 12.4, could be used to do this.

12.2.6 Traffic counts

Traffic counts are much cheaper to undertake than O-D surveys and since, as indicated in Section 5.2.1, it is possible to use maximum likelihood methods to estimate matrices from traffic counts, this is becoming a very popular method of deriving matrices. The methods, several of which are available via proprietary software, generally require a prior matrix (supposedly obtained in some halcyon days when full O-D surveys were affordable), together with network structure, costs and flows on key links. The more links for which flows are available the more reliable will be the result.

The method has been a great boon to traffic analysts but it must not be forgotten that it only provides an *estimate* of the true matrix and theoretically requires all the traffic counts to have been taken during the same time period.

12.2.7 Vehicle tracking

An increasing proportion of the vehicle fleet is now equipped with devices which can determine the vehicle's location and display it within the vehicle or, via a radio link, at a remote monitoring station. This equipment, which

Origin-destination and Route Choice Studies 267

employs triangulation from GPS satellites or terrestrial beacons, is used to support on-board navigation/guidance or fleet management/security functions. If access can be gained to centrally held records, and if movement of the equipped fleet are thought to be of interest to traffic analysts, this can be a potential source of O-D and route choice data (see Van Aerde, Mackinnon and Hellinga, 1991).

The recent development of the self-locating mobile phones such as the Personal Handiphone System (PHS) appears to offer an attractive alternative to GPS in this role because it provides a relatively cheap device which can locate itself with respect to radio beacons and regularly report its location to a base station. PHS does not have quite the same precision as GPS but it does not suffer from shadowing in urban areas (Asakura et al, 1999).

It is possible to envisage a future scenario in which real-time data from vehicles requesting guidance for a particular O-D journey could be fed to a traffic management control centre and used as an input to an on-line model of the traffic system and hence to influence real-time signal control. However, the use of such data, and of data from fleets equipped with GPS or PHS, is likely to cause great concern for personal privacy and so may not become an important source of O-D data except in the context of special studies where the drivers' permission has been granted.

12.3 Methods of obtaining route choice data

As well as O-D data on where (and perhaps when) vehicles enter and leave a study area (i.e. cross the cordon line), information on the routes taken through the area may also be required. For example, an understanding of route choice is essential to the design and evaluation of traffic management and control schemes that may result in route deviations (e.g. if turns are to be banned at some junctions, or if route guidance information is to be given to drivers for (say) the avoidance of bottlenecks or congested junctions).

12.3.1 Observation of flows

Some information on route choices can be deduced from data on the flows on key links in a network. The data are enhanced if disaggregated by vehicle type or, better still, by using the 'tag' or 'headlight' techniques described above to distinguish vehicles which have passed, or originated at a specific point in the network. The deduction of routes from traffic counts is achieved in a manner analogous to the derivation of O-D matrices from traffic counts (see Section 5.2.1) and is similarly dependent on input assumptions in this case assumptions about the O-D matrices. Given the day-to-day variability in O-D matrices the route choice deductions cannot be more than approximate.

12.3.2 Observation of individual vehicles

Clearly, if the number of observation points for a registration plate survey is increased to include, not only the entry and exit points for a network, but also key links within it, it becomes possible to track the path of individual vehicles through a network and so observe their route choices. Similarly it may sometimes be possible to observe the local area route choice of individual vehicles from a vantage point, or to use records from electronic tagging or probe vehicles as described in Section 12.2.7 above.

These sources would indicate what routes are chosen but not *why* they were chosen. Subsequent analysis might be able to deduce that the drivers were selecting shortest time or distance routes or were following signposts etc. but there would clearly be room for misinterpretation particularly if the driver's ultimate origin and destination were not known.

12.3.3 Interviews and questionnaires

Interviews or questionnaires clearly have the potential to record actual routes taken together with drivers' statements as to their reasons for choosing a particular route. Where the route choice question is simple (e.g. 'Did you use bridge A or bridge B to cross the river?') it may be included in a roadside interview, but otherwise it will need to be administered in a longer interview, e.g. a household interview, or via a self completion questionnaire. Since some drivers have difficulty describing their routes verbally and others find it difficult to retrace their routes on a map, it is wise to offer more than one medium for recording the information.

Drivers can have problems of recall, particularly if they are asked to report on several journeys some of which may have been undertaken some time previously or if they are asked to report a journey, even a recent one, on which they got lost. Where subjects can be identified in advance of making their journeys (e.g. by recruiting a panel) they can be asked to record journeys as soon as they complete them or can even be provided with voice-actuated tape recorders onto which to record the routes while travelling. The use of electronic tags or probe vehicles offers the prospect of more reliable data with less effort on the subjects' part. For instance, the installation of a self-locating mobile phone or a GPS receiver and dedicated data logger in a vehicle can be used to record the locations of that vehicle over a period of days, and full route information can thus be obtained. Asakura et al (1999) combined data on drivers' route choices derived from self-locating mobile phones with data from questionnaires seeking their reasons for selecting particular routes. An interesting outcome from this study was that the mobile phone data revealed more journeys than the drivers had recalled in their questionnaires.

12.3.4 Car following studies

Drivers' route choices can be logged by an observer in a car following the target vehicle wherever it goes within a defined area. This is a fairly expensive procedure, particularly if the sampling is done with statistical rigour. Low productivity is inevitable because, if it is to be unbiased, the sample must be based on selection and successful tracking of one vehicle per unit flow. If any vehicles are 'lost' (as may quite easily happen in heavy traffic with traffic lights), the flow unit ought to be increased to avoid biasing the sample to the less busy periods, strict adherence to the rules could result in only a dozen or so observations per survey day. Nevertheless, this technique can be quite effective for studying complex movements within a relatively small area. It is therefore a useful method of studying parking search behaviour - particularly if, once the target cars are parked, their drivers are offered a questionnaire containing questions about their search strategy, final destination and other relevant information.

The method relies on the pursuing driver being able to keep up with the target vehicle, without influencing the target vehicle's speed-time profile. It is therefore common practice to employ taxis as the pursuit vehicles.

12.3.5 The use of route choice simulators

A number of research studies have sought their route choice data from computer-based route choice simulators rather than from the real world (see, for example, Mahmassani and Stephan, 1988; Bonsall and Parry, 1991; Koutsopoulos et al, 1994; Adler and McNally, 1994 and Bonsall et al, 1994). These simulators incorporate a representation of a network though which volunteers are asked to 'drive' – selecting their route as they think fit. The software provides a journey-like experience via, for example, an appropriate sequence of driver's-eye views of the dashboard instruments and through-the-windscreen views. The speed with which the simulation proceeds can be made proportional to the time which each manoeuvre would take to complete in the prevailing traffic conditions – thus making the 'drivers' aware of the consequences of choosing a slow or tortuous route. Figure 12.3 shows the interface from one of the most widely used simulators.

270 *Understanding Traffic Systems*

Figure 12.3   The VLADIMIR simulator (Bonsall et al, 1997)

Route choice simulators offer the analyst an opportunity to explore, within a sound experimental design, route choice responses in a range of situations which are not easily observed in the real world. They have been particularly popular as a means of exploring drivers' reactions to route guidance advice, other in-vehicle traffic information and roadside variable message signs. The indications are that a well designed simulator can engender very realistic responses (Bonsall et al, 1997).

12.4 Analysis of registration plate data for O-D surveys

The registration plate technique is widely used in traffic survey practice, which demands the simultaneous collection of information from a number of sites, by a number of observers. Its success depends on the consistency of the data collected at the sites, and the methods chosen for data analysis and interpretation. Errors can arise from two separate sources: *errors in data recording*; and *spurious matchings* due to combinational effects in the analyses of partial registration plates (e.g. the last three characters of a six character plate).

12.4.1 Errors in data recording

Several errors are commonly made during the collection of registration plate data. They are, in decreasing order of frequency:
- the plate is missed completely;
- one or more characters are misrecorded, or two characters are transposed (typically the middle two characters in a group of four);
- the direction of movement is recorded incorrectly (where the observer is recording more than one direction or movement);
- the type of vehicle is misrecorded (where vehicle classification data is sought), and
- the time at which the vehicle passes the site is wrongly recorded.

The occurrence of any of these errors will normally preclude a successful match with another observation, (although misrecording of time is only crucial if it puts the plate outside the time window designated as feasible for a match). Such problems can be reduced, albeit at the cost of increased risk of spurious matches (see below), by adopting wide tolerances during the matching process. But it is by far preferable to make every effort to minimise errors in the first place.

The *estimation of observer error rate* is a first step when selecting suitable survey personnel. The following procedure may be used to test the accuracy rate of observer O. Three observers (i, O and j) are stationed at separate sites along a road section, with O located between i and j. Observations by i and j are compared, and in each case that a match occurs, the observations by O are scanned to see if a correct recording was made - on the assumption that all vehicles travelling from i to j must have passed O (this is not hard to organise). The accuracy rate A_O of observer O can then be defined as the proportion of correct recordings C_O made by O, as in equation (12.1),

$$A_O = \frac{C_O}{Q_{ij}} \qquad (12.1)$$

where Q_{ij} is the number of passing vehicles which fall within the nominal sample (e.g. those with registration plates with the specified sampling digit(s), not all of the traffic).

Start-up and shut-down errors are important considerations in registration plate matching. As every trip in the study area will involve a finite travel time, the first vehicles leaving the cordon at the start of the survey will be interpreted as local trips, even if they are in fact through trips. As they entered the study area before the start of the survey, there are no records of their entries. A similar problem occurs at the end of the survey. Without an adjustment for these errors, the results will be biased against through traffic, especially if the survey area is large or the survey duration is short. The solution is to discard those observations leaving the cordon at the start of the survey, or entering it at the end of the survey, which are deemed too close to the end of the survey period. An appropriate cut-off for early leavers and late entries should be chosen by estimating the time that a vehicle *should* take to pass through the study area. The best way to do this is in fact to use manual matching of registration plates at the start of the survey to determine the time at which the first 'wave' of vehicles, entering the cordon at the start, passed each station. Observations prior to this wave can be discarded. A similar technique may be applied at the end of the survey.

Confused characters are another source of error in data recording. Depending on the recording technique, there will be pairs of characters that can easily be confused - for example, 'S' and '5', '2' and 'Z', 'I' and '1', 'O' and '0'. One possibility is to replace one of each of these pairs of characters with its partner (e.g. code all characters resembling 'S' as '5'. For audio cassette tape recordings, 'B' and 'D', 'M' and 'N', and other similar sounding characters may be similarly treated. This method will allow matchings that might otherwise be missed because of recording or transcription errors, but will increase the risk of spurious matches.

12.4.2 Spurious matchings

Errors due to spurious matchings may occur in the analysis of partial registration plates. For instance, three different registration plates 'ABC123', 'BAC123', and 'CAB123' could be observed. Assuming that these are correct observations, then they represent three different vehicles. If analysis were performed on the basis of the last four characters only, the partial registration plates would then be 'C123', 'C123' and 'B123'. The first two observations would now match. This is termed a *spurious match*. If only the last three characters were used, the observations could all be interpreted as the same vehicle.

The problem of spurious matchings has been subjected to statistical analysis (see, for example, Hauer (1979), Shewey (1983), and Maher (1985)). Spurious matches may be a significant source of error, particularly when only a small number of characters are used from each registration plate. In many circumstances only partial registration plate data are recorded, so some account needs to be taken of possible errors. Hauer (1979) used an approximate, numerical method to try to estimate the expected number of spurious matches on the basis of probability theory. Shewey (1983) indicated that the use of supplementary information, such as consideration of observed journey times, could provide logical constraints which would reduce the probability of spurious matchings. Maher (1985) used maximum likelihood and weighted least squares techniques (see Chapter 17) to estimate the occurrence of spurious matching.

Practical experience suggests that, provided that at least half of the characters on each registration plate are observed, errors in observing and recording are likely to outweigh errors due to spurious matching.

12.4.3 Analysis of results from registration plate surveys

The compounding of errors in O-D registration plate surveys means that specific consideration of errors needs to be built in to the analysis procedure. Figure 12.4 provides a recommended procedure for analysis which includes all of the major issues. Where there is a 'perfect' match between two observations (i.e. all characters match, the vehicle type matches and the direction and timing of the observation indicate a feasible movement) then deduction of flow is straightforward. An initial pass through the data should identify all such matches and remove the matched observations from the lists of plates at each location.

The matching criteria should be progressively relaxed to pick up further 'matches' which can be assumed to have failed due to an observer error (e.g. misrecording of similar looking characters). This process will obviously increase the number of apparent matches but must be used with care so as to avoid generating spurious matches. If there is no reason to believe, for example, that the direction of movement may have been misrecorded then it would be wrong to relax the matching criteria to assume that it was. A sophisticated matching algorithm would be able to make allowance for the fact that different degrees of misrecording may be apparent at different sites.

At the end of the matching process some unmatched observations will inevitably remain. If the observations were made at a leak-proof cordon and if appropriate allowance has been made for the time taken by vehicles to move from one site to another, these unmatched observations can be assumed to represent one end of trips which started or ended within the cordon (i.e. 'local' trips) and may be categorised as such. Note however that a common cause of apparently one-ended trips is simply a failure to record it at the 'other' end. Some adjustment for this can be made allowing for each observer's recording abilities.

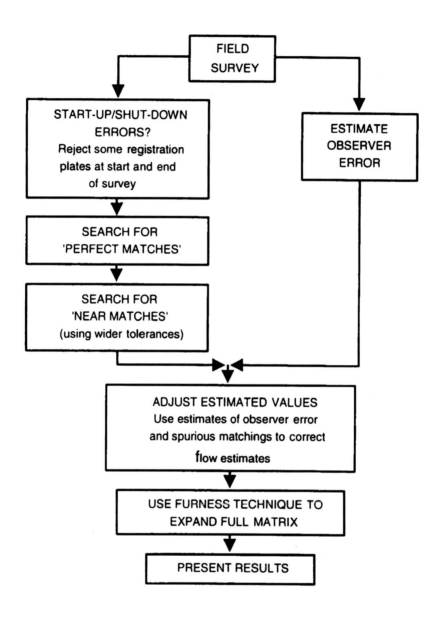

Figure 12.4 General procedure for correction of errors in registration plate surveys

Origin-destination and Route Choice Studies

Using each observer's known ability (their accurate recording rate (A) as defined in equation (12.1)), the accuracy of raw matching data can be enhanced. For example, the matched flow (m_{ij}) between any pair of sites i and j can be factored up by dividing by the product of the two observers' accuracies to produce a revised flow R_{ij} using equation (12.2).

$$R_{ij} = \frac{m_{ij}}{A_i A_j} \qquad (12.2)$$

Note that, if one is dealing with matches between more than two survey points (i, j and k) the denominator will be the product of more than two terms (e.g. $A_i \times A_j \times A_k$).

Note also that the definition of the observers' accuracies (e.g. A_i and A_j) should be defined to correspond to the tolerances used in the matching process (e.g. if the matching process allows B and P to be interchangeable, then so too should the calculation of observer accuracies).

The proportion of unmatched records (U_i) at a given site that are probably due to local trips (L_i) rather than failure to record the 'other end' can be approximated by equations 12.3 and 12.4 as follows:

for trips inbound $\qquad L_i = U_i \sum_j (R_{ij} - M_{ij}) \qquad (12.3)$

for trips outbound $\qquad L_j = U_j \sum_i (R_{ij} - M_{ij}) \qquad (12.4)$

Once these adjustments have been made the next step is to gross up the sample matrix to represent the total flow within a given time period.

The process described above can produce estimates of flows but these are not constrained to fit control totals of vehicles observed at the survey sites. In particular the flows relate to targeted vehicles (e.g. those of a particular colour or having registrations ending with particular characters) rather than to the total vehicle population. The recommended procedure for expanding the R_{ij}, L_i and L_j flows to match control totals is the Furness technique (which is similar to the Hauer-Kruithof method described in Section 11.1). A worked example of the Furness technique is provided as Table 12.2.

12.4.4 Computer analysis of registration plate O-D data

Computer programs for the analysis of O-D data have obvious applications, and there are many PC programs available which provide easy analysis on the analyst's desk top. However, each program will have its own internal logic and, given the complex nature of the matching decisions in the face of error-prone data, it is possible that two different programs will generate quite different results. Until recently few programs accounted for observer errors. Usually they search for exact matchings of registration plates, with no editing facilities once the data are entered into the program. Spurious matches are rarely considered. Thus a simple, straightforward matching procedure will involve two compensating errors. The effect of data recording errors is, however, likely to far outweigh errors from spurious matching. A detailed discussion of these issues, including case study analyses of the impacts of different error correction procedures on the resultant O-D matrices, is given in Greenwood and Taylor (1990).

12.5 Presentation of results

The main result from an O-D survey is usually a matrix showing the trips from each origin to every destination defined in the study. This matrix is the origin-destination (or trip-interchange) matrix. This matrix will be the primary output of an O-D survey. A useful extension to the conventional two-dimensional O-D table is the three-dimensional table in which the third axis defines some classifying factor, such as:
- internal stations, thus indicating the routes taken through the area;
- time, so that the changes in trip movements over the survey period can be assessed, or
- vehicle type, thus indicating the different patterns of movement for (say) private and commercial vehicles.

The three-dimensional table may be pictured as a book of O-D matrices, with each page containing a separate matrix, split from the others in terms of the third variable.

A convenient pictorial representation of the O-D matrix is the desire line diagram superimposed on a study area map. The desire line diagram shows trip movements between pairs of zones as straight lines (of variable thickness to denote amount of flow) joining the origin and destination. An example is included as Figure 12.5.

Route choice information can also be displayed pictorially. Observed paths between specified zone pairs are plotted as separately coloured 'tapes' following as closely as possible the appropriate links in a base map. The tapes may have a thickness to indicate the number of trips using that path. This type of presentation indicates the continuity, divergence and recombination of paths in a way that is not apparent from a standard desire line diagram or from a simple link-usage plot where the flow associated with all O-Ds are combined.

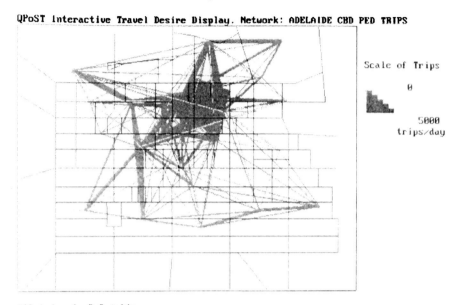

Figure 12.5   Example of desire line diagram

Table 12.2   Worked example of Furness Technique

The Problem

Assume that a matrix has been estimated (e.g. from registration plate matching) as follows:

		To		
		x	y	z
	x	10	20	40
From	y	30	20	30
	z	40	20	20

and that control totals at sites x, y and z show:
  inbound $x = 140$ $y = 80$ and $z = 120$
  outbound $x = 160$ $y = 90$ and $z = 90$

The Method

Since the outbound flows sum to the same total as the inbound flows, we can begin.
- Calculate A_i factors for each 'origin'

$$A_i = \frac{\text{control total inbound at } i}{\text{sum of elements relating to origin i in current best estimate matrix}}$$

Thus $A_x = 140/(10+20+40) = 140/70 = 2.0$

$A_y = 80/(30+20+30) = 80/80 = 1.0$

$A_z = 120/(40+20+20) = 120/80 = 1.5$

- Multiply matrix elements by appropriate A_i factors to produce new matrix:

   20   40   80
   30   20   30
   60   30   30

- Calculate B_j factors for each 'destination'

$$B_j = \frac{\text{control total outbound at } j}{\text{sum of elements relating to destination j in current best estimate matrix}}$$

Thus $B_x = 160/(20+30+60) = 160/110 = 1.45$

$B_1 = 90/(40+20+30) = 90/90 = 1.0$

$B_2 = 90/(80+30+30) = 90/140 = 0.64$

- Multiply matrix elements by appropriate B_j factors to produce new matrix.:

29	40	51.2
43.5	20	19.2
87	30	19.2

- Calculate new A_i's

$A_1 = 140/(29+40+51.2) = 1.16$

$A_2 = 80/(43.5+20+19.2) = 0.96$

$A_3 = 120/(87+30+19.2) = 0.88$

- Multiply matrix elements by A_i's to produce new matrix:

33.6	46.4	59.4
41.8	19.2	18.4
76.6	26.4	16.9

- Calculate new B_j's

$B_1 = 160/(33.6+46.4+59.4) = 1.05$

$B_2 = 90/(46.4+19.2+26.4) = 0.98$

$B_3 = 90/(59.4+18.4+16.9) = 0.95$

- Since all these factors are within the range 0.9 to 1.1 we can regard this as a converged solution and can adopt the most recently produced matrix (with cells rounded if appropriate to nearest whole numbers). Otherwise method continues until convergence is reached.

13 Traffic generation and parking studies

The interface between transport networks and land use is a key aspect of any transport system. An important facet of the planning and design of new facilities, such as shopping centres, sports and recreation centres, office blocks, or multi-unit residential development, is the provision of adequate access. For road traffic this involves access to and from car parking at the development. Besides the immediate connections of the development to the road system, traffic planners and engineers are also concerned with the impacts of the new development on the traffic system in the surrounding area. Thus a fundamental part of traffic analysis is the consideration of the amount of traffic generated by, and the parking required at, a development. This provides the additional load to be placed on the traffic system, and indicates the need for the upgrading of parts of the road network or the provision of new traffic facilities. This chapter considers traffic generation and parking surveys together since they are generally carried out in conjunction with one another.

13.1 An introduction to traffic generation and parking

Since drivers start and end their trips at parked vehicles, traffic generation and parking are closely related.

13.1.1 Traffic generation

Traffic generation is the measured level of vehicular activity associated with a given site, development or land use. As such, it provides the connection between the land use system and the traffic system. In the context of transport planning, the general travel demand modelling system often contains a specific trip generation model. Traffic generation is one manifestation of trip generation. Some of the concepts of trip generation are necessary for a clear understanding of the processes in traffic generation: in particular, a given area or site may be seen as a sink or source of trips. The number of trips originating at a site or zone is known as the trip production, while the number of trips finishing at a site or zone is known as the trip attraction. The total trips connected with the site, the sum of the production and the attraction, is the number of trip ends. In transport planning, the levels of trip production and attraction are important in their own right and are kept separate. In traffic generation analysis, the analyst is more concerned with the total level of movement associated with a given land use or site. The number of trip ends thus becomes the principal measurement.

Brindle and Barnard (1985) described traffic generation as a measure of vehicle (occasionally person) movement associated with a given site or land use, incorporating the number of one-directional vehicle movements arriving at or leaving a study area per unit time. Thus traffic generation may be seen as the number of trip ends at a site, per unit time.

The inclusion of the time dimension into the traffic generation measure is very important, as it was in the definition of traffic flow rate. Time provides the means of assessing the travel demand generated, against the capacity of the system to cater for that demand. Concepts similar to the 'highest hourly volume' (see Section 8.1) may also be applied to traffic generation. In assessing the impact of a new development on the surrounding area, the traffic analyst will need to estimate the peak rate of traffic generation of the development, and the period of time for which the peak rate will persist.

Additional traffic generated at the site of the activity will be present over a specified time period, corresponding to (say) retail trading hours at a shopping centre, or the duration of a sporting event at a stadium. This traffic may be observed in an area surrounding the activity site; the time of observation and the impact of the traffic will depend on the proximity of the observation point to the activity site, and on the level of activity which takes place. One or more peak periods may be expected. A useful conceptual model for exploring the phenomena of traffic impacts may be gained by considering the ideas of the 'trafficgraph' for the traffic generating activity.

Apelbaum and Richardson (1978) used the 'unit hydrograph' theory of hydrology to develop the idea of a 'trafficgraph' to describe the traffic generating impact of an isolated event (their specific event was the arrival of an aircraft at an airport terminal) on the traffic activity on a road section. Figure 13.1 indicates the concept. The event triggers a lagged steady increase in traffic activity, which then builds up to a peak level, then drops away. The lag reflects the time for the influence of the event to become apparent at the observation site. In traffic analysis the lag may be either positive or negative. Figure 13.1 shows a positive lag. For the aircraft arrival considered by Apelbaum and Richardson, Figure 13.1 would indicate the departure of traffic from the airport after a flight arrival. If we considered the arrival of ground traffic to greet the flight, then an equivalent **trafficgraph** with a negative lag would result. The lag is the time difference between an event and the first occurrence at the traffic observation site of traffic activity associated with that event.

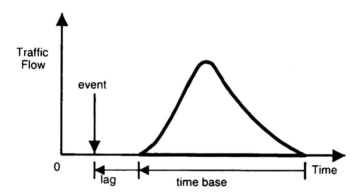

Figure 13.1   The trafficgraph for a traffic generating event

The following basic conditions underlie the trafficgraph concept:
- there may be a base flow of traffic, which is independent of the event under study;
- the trafficgraph for the event may be superimposed on the base flow;
- a given type of event at a nominated site will have a constant time base for the resulting trafficgraph at the observation point;
- the given event will have a constantly-shaped traffic flow profile (i.e. the trafficgraph) at the observation point;
- the peak height of the trafficgraph is directly proportional to the magnitude of the event. Thus two occurrences of the same event, with different magnitudes (such as the arrival of two different aircraft), will have the same shaped trafficgraphs, with the heights of their peaks in proportion to the magnitudes of the events, and
- the principle of superposition will apply to successive (or simultaneous) events. That is, the total traffic impact at the observation point will result from adding the trafficgraphs of the events, taking account of the times at which the events occur. The principal limitation on the simple

application of this principle is that level of traffic flow cannot exceed the traffic carrying capacity at the point.

The conceptual model of traffic generation is thus that the effect of a traffic generating event at a site is seen as a wave of increased traffic activity (flow) at a given point in the road system. When multiple events take place at one site or a collection of sites, the resulting effect at the observation point is made up of the sum of the effects of each separate event. Figure 13.2 indicates the combined effects. Figure 13.2(c) is of particular importance, as it shows the capacity restraint effect of the road system. Once the capacity is reached, the resulting traffic graph is stretched, as the flow cannot exceed capacity and the effect of the event is prolonged in the traffic system.

The trafficgraph model may be used to describe many events that lead to peaking in a traffic system. These include the arrival and departure of people at sports or recreation events, and at shopping centres; or a long sequence of similar events, that leads to the build-up of general peak period traffic. The model can also include the effects of congestion occurring when traffic demands exceed the traffic-carrying capacity of some part of the road system for a period of time. It is this combination of demands and effects that the analyst tries to include in a traffic generation analysis. The trafficgraph concept indicates that the principal parameters of interest are the magnitude of the traffic generating event, and the time duration of the event itself, and its traffic-inducing effect.

13.1.2 Parking

Traffic professionals also plan, design, operate, manage and monitor the parking system. To carry out this task adequately they require accurate information. Specific data requirements will, of course, depend on the precise nature of the investigation but an indication of the possible range of requirements is given in Table 13.1. Perhaps the most commonly collected data describe the supply of, and the demand for, parking space in each block of a study area. Data on parking behaviour and supply provides the basis for this information. Many different procedures are available for collecting the considerable variety of data types required to understand the parking process. Recent developments in microprocessor technology have substantially enhanced both the magnitude and quality of the data collected by reducing the cost of collection and increasing the achievable level of detail.

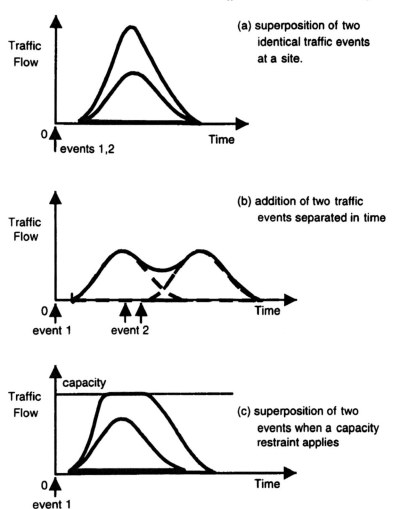

Figure 13.2 Combination of traffic events

Parking studies have long been an important part of traffic analysis but recent years have seen a subtle shift in the types of information sought. According to Bonsall (1991), the main factors influencing this development are:

Table 13.1   Information needs for parking analysis (Bonsall, 1991)

Inventory	• number, location, type, size and ownership of spaces
	• operating arrangements, physical restrictions, technology installed, current fee structure
	• condition of physical structure, facilities, signs etc
	• maximum throughput at entries, exits, ramps etc
	• queue storage space
Demand profile (per unit time)	• arrivals, departures and accumulation
	• duration of stay and turnover
	• types of user; journey purposes, frequency of visit, origins and destinations
	• estimates of future demand
Operational performance (per unit time)	• search times
	• search and fill patterns
	• speed of manoeuvres and transactions
	• congestion points
Costs and revenues (per unit time)	• fee income from tickets of the various values
	• income from contracts, leases etc
	• potential income from other sources (advertising, penalty charges etc)
	• operating costs; staff, infrastructure, rents etc
	• income lost through illegal parking acts
	• estimated price elasticity
Enforcement	• number and type of illegal acts
	• method of enforcement (clamping, fixed penalty fines, court actions etc)
	• effectiveness of enforcement (various measures including: number of tickets issued, number of vehicles clamped, percentage of fines collected, percentage of illegal acts identified, percentage of parking acts that are illegal)
	• costs of enforcement
Secondary impacts	• on the local street [including: location and timing of queues to enter, volume and timing of entry and exit flow, volume and timing of search traffic, obstruction of the highway by on-street parking and loading/unloading (legal or otherwise), on economic activity, evidence of impact on retail turnover and commercial activities]
	• on the local environment and community [including: visual intrusion, damage to verges, sidewalks etc, ease of parking by local residents]

- a greater need for efficient management and operation of parking facilities;
- a more urgent need to consider the influence of parking on road network performance;
- increased concern about the effect of parking provision on economic activity, and
- a general rising expectation of information availability.

Against this background of increasing demand for data there have been a number of important technological and methodological developments affecting its availability. Those of particular significance are:
- increased use of computers to control and monitor the operation of parking facilities;
- development of more effective automatic and semi automatic vehicle detectors and identifiers for use at parking facilities;
- increased use of hand held data loggers in parking enforcement work, and
- new technology alternatives to pencil and paper parking surveys.

13.2 Supply of parking and entrance facilities

An important part of any parking and trip generation study survey is an inventory of the parking facilities and the possibilities for new development in the area of concern. Such an inventory should detail the type of parking and its location. Supply studies seek information on: parking spaces, type of parking, method of operation of off-street facilities, parking restrictions, entrance conditions, and parking fees.

A first step in any traffic study is to check existing sources of data; even if they are not up to date they may provide a useful base on which to work. It would, however, be unwise to assume that previously collected data are necessarily correct; even if they were correct when originally collected they may not have been properly updated.

If existing inventory maps are not appropriate, the required data must be collected in the field. In compiling the record of the street facilities the data can be first entered, in the field, on prepared sketch plans. These sketch plans may be based on existing local maps or each street can be sketched onto graph paper to the correct scale, with respect to length. The use of a suitable key enables the exact location of parking and parking restrictions to be marked on the map. The location of off-street parking could be marked on the map but the layout of these facilities is usually detailed separately.

Normally the road network is coded in relation to road lengths with each block given an identity number. This coding provides a basis for recording and analysing the data. A sample inventory map is shown in Figure 13.3.

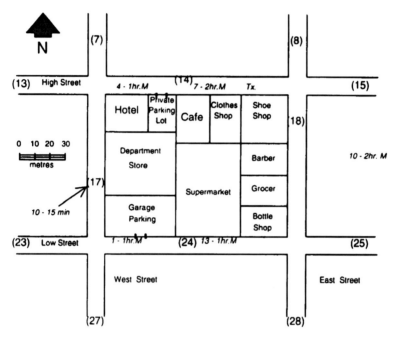

Figure 13.3 Simple inventory map showing available parking facilities

The inventory survey is traditionally carried out by a two-person team, equipped with a base map, a supply of kerb and off-street inventory survey forms, a clip board, pencils, a 20 metre tape and a measuring wheel. Increasing use is being made of portable computers, of video and of audio tape to assist in inventory recording. The use of portable computers linked to GIS files and perhaps to GPS is a particularly promising development. The errors which can beset the use of keyboards to record details of moving traffic are avoided in inventory surveys because the surveyor can control the rate of data entry. When using audio tape it is advisable to use a voice-actuated system in order to avoid generating vast lengths of blank tape.

The inventory survey may involve inspection of signs and markings; if so it should be carried out by a trained person familiar with the appropriate standards for traffic signs and markings. It should include a record of deviations from the standard and of poorly located, confusing, conflicting or misleading signs and kerb markings. Data on signs placed without authorisation should be included in the inventory. The data collected in the field should be recorded on a substantial map. In turn this map can compared with the existing facilities in the field to ensure its accuracy.

13.3 Traffic generation and parking demand surveys

It is important to distinguish between revealed demand at the present time and latent demand when considering traffic generation and parking demands. *Revealed demand* is the observed use of the facility. *Latent demand* is a measure of the total desired use of the facility. The full extent of latent demand will only be revealed when supply exceeds demand. Most parking demand and traffic generation data collected in the field are revealed demand data, whereas future planning data reflect the latent demand. The required demand data would include:
- temporal distribution of traffic and parking demand;
- spatial distribution of traffic and parking demand;
- spatial distribution of traffic and parking demand generators;
- total number of parkings in study area over study period;
- parking duration;
- trip purpose and destination, and
- trip origin.

The techniques for collecting traffic generation and parking demand data may be split into three broad types: existing data held by local or regional authorities or by parking operators; observations of traffic movements at a site, and interview or questionnaire surveys of people involved in a given activity in a given study area.

13.3.1 Existing information

Before any study is undertaken it is always advisable to determine what data are available. Data on parking in a given area may have been collected by various authorities, agencies and car park operators. Each may have collected the data for their own purposes and may have used different definitions and standards. Some attempts have been made to reconcile the available data; perhaps the best known of these are the handbooks on trip generation and parking usage, published by the Institute of Transportation Engineers (ITE 1997).

Reliance on published databases is controversial. On one hand, the databases may be regarded as a definitive source of information. Others dismiss them as being unsuitable for use outside the specific regions in which the data were collected. The use of data, collected in another place, at another time and for other purposes, always presents the possibility of misleading results, but there will be no alternative if local data are unavailable and the collection of new data is infeasible. The main advantage of databases, such as the ITE database, is that they indicate the known ranges of traffic generation parameters. This knowledge provides invaluable back-up to the collection, analysis and interpretation of local data. Comparisons of local and external data sets need to be made with some caution, particularly when concerned with peaking effects. Peak traffic generation periods for many activities differ widely between regions, cities, states and countries and may change over time - for example in response to changed regulations on shop trading hours.

Information collected at the relevant site but for other purposes can be used to estimate traffic generation levels. For instance, a successful technique in surveying off-street parking facilities is the use of cancelled parking dockets, where these are retained by the parking operator. These dockets usually show the exact time of entry and exit and hence, parking duration and traffic generation. The advantages of this data source are that no field surveys are necessary, a complete sample is obtained, and that the data can be collected over an extended period of time and over a number of different parking locations. The disadvantage of the method is that it only gives information on parking times and will not provide a complete sample if some users (e.g. parkers who hold a permanent parking place) do not use the returnable tickets.

Many off-street car parks will monitor the entry and exit of vehicles (detected via induction loops and/or barrier movements) and will record all financial transactions. Arrival by special types of users (e.g. season ticket holders and contract parkers) can be logged separately. Some systems involve human attendants at the entry and exit while others are entirely automatic relying for the most part on tickets (having been punched, bar coded or magnetically encoded with time of entry) being machine read on exit. In either case the central computer can not only look after real-time functions such as access control and information display but can also provide valuable off-line data, such as entry flow, exit flow, accumulation, fee paid (disaggregated by user type), number of tickets sold and duration of stay.

There are current trends for on-street meters and ticket dispensers either to be networked to a central computer or for a record of their transactions to be stored on the unit until extracted through a portable enquiry terminal. It must, of course, be recognised that data from parking meters and pay-and-display ticket dispensers relate to transactions rather than to actual usage and may produce biased estimates of both the number of parkers and the average duration of stay. Recent evidence, however, suggests that estimates of this bias can be made and correction factors can be applied.

As more and more of the on-street and off-street parking stock comes under the control of networked computers with facilities for automatic transaction logging, the resulting databases will become a major source of information for parking demand analyses. For the foreseeable future, however, there is still an important role for the more traditional survey techniques, which we now consider.

13.3.2 Interview surveys

If the parking demand or trip generation is to cover a large geographic area, and it is expected that changes in parking supply could cause substantial change in the total number of parkers or their spatial distribution, data collected from an interview technique may be required. Four techniques commonly used are *parking person interview*, *reply paid questionnaire*, *home interview* surveys and *site specific interview* surveys.

In the *parking person interview survey* an interviewer is assigned to a pre-determined number of parking spaces. The interviewer records each parking incident in the area and attempts to interview the people parking. Questions asked in the interview may relate to: trip purpose, final destination of trip, origin of trip, places visited, duration of parking, alternate parking locations considered, and frequency of parking in the study area. Other details that can be collected by observation are: the vehicle registration number, vehicle type (car, taxi, truck, lorry etc), nature of parking (legal, kerbside, off-street, garage etc), and time of arrival or departure.

The information obtained can be recorded on an appropriate survey form (e.g. Figure 13.4) and then transferred to computer for further analysis. Increasingly, however, the responses are being coded directly in to handheld computers. The latter approach reduces the number of times the data must be manipulated, the chance for errors in coding, and the labour costs. The personal interview can be used to obtain data on people's attitude to various parking policies (e.g. changes in parking fees, parking restrictions etc.). Care is need to keep the length of the interview to tolerable limits.

292 *Understanding Traffic Systems*

Vehicle Type	Time In	Time Out	Trip Origin	Trip Destination	Trip (Circle one)	Parking Type	No.
					1.Shop 2.Work 3. Friend 4.___		
					1.Shop 2.Work 3. Friend 4.___		
					1.Shop 2.Work 3..Friend 4.___		
					1.Shop 2.Work 3.Friend 4.___		
					1.Shop 2.Work 3.Friend 4.___		
					1.Shop 3.Friend 4.___		
					1.Shop 3.Friend 4.___		

Facility Number:_____   MONASH PARKING STUDY   Interviewer's Name:_____   Date:_____

Supervisor: _____

Figure 13.4   Example of a parking interview form

Interviews of on-street parkers can be carried out on parker arrival or departure. The departure interview has a number of advantages. Firstly, places visited can be reported more accurately since the parker has already visited them. Secondly accurate duration of stay information can be obtained. Thirdly the parker is less likely to be in a hurry and therefore more likely to complete the interview. The major disadvantage is that the interviewer has less time to catch parkers before they leave.

The personal interview technique used on the road system can often be expensive since the interviewers are limited in the number of parking spaces they can handle. The number depends on the length of the interview, arrival and departure rate of parking vehicles, and the physical dimension of the area. The size of the area that can be covered by an interviewer can be determined by a preliminary pilot survey, but in typical central city parking conditions is not likely to exceed 200 metres.

The non-response to the interview could introduce bias into the population, for example the characteristics and parking behaviour of people who are too busy to respond may be quite different from the rest of the population. Some attempt should always be made to determine basic characteristics (age, sex, occupation etc.) of those not completing the questionnaire because this may enable corrections to be made to the results of the responding sample.

Sampling of parking vehicles can be carried out but care must be used not to introduce bias into the analysis. Bias may occur if the interviewer selects only (say) approachable white collar workers or females. Sampling should be carried out by selecting the driver of perhaps every third parking vehicle.

In the case of off-street facilities interviews can be carried out when the vehicle is entering or exiting the facility. Such facilities may have a large proportion of long term parkers and may be subject to peaks. Greater numbers of interviewers may be required over this period. Interviewing people as they leave has the advantage of avoiding vehicles having to queue onto the adjoining roads. If interviews are carried out upon entry, provision for queuing vehicles should be made.

When detailed information is not required, *reply paid questionnaires* may be used. An example is shown in Figure 13.5. These questionnaires can be inserted under the windscreen wipers on the parked vehicles. Personnel costs for this method are lower than for a personal interview since one person can cover a larger number of parked vehicles. Information on the parking location and arrival time of the vehicle can be obtained by marking or pre-coding the questionnaire. If pre-coding is used the person distributing must note the time of distribution and location. This can be recorded in a logbook or on a portable computer. The computer has the advantage of quicker access and information transfer time.

The above surveys do not measure the latent demand for parking, rather they measure the revealed demand. Many people wishing to visit an area may be turned away by the lack of parking facilities. Indications of latent demand could, however, be obtained by a *home interview survey*. The large cost associated with such an approach usually results in questions on parking being grouped with other questions on a large transport questionnaire. This approach has been shown to be a useful substitute for those mentioned previously (Schulman and Stout, 1970). Sampling of the population may however be a problem if the study is concentrating on improvements in a particular part of the urban area. Home interviews can be carried out in several ways, as described by Richardson, Ampt and Meyburg (1995). The home interview survey addresses the entire population of the urban area and is the only approach that can be used to determine latent demand in multi-use parking lots. However, some parking lots are only used by people working at specific locations. In such a case the total population of possible users can be defined. This population can then be used as a basis for determining the latent demand for the particular site, using a *site-specific interview survey*. This approach will not, however, indicate the latent parking demand of visitors to the site.

PARKING SURVEY

The rapid growth of population in Knox has reached a point where parking facilities are inadequate. Your answers to the following questions will aid in determining the need for parking facilities.

1. Where did your trip start before parking here?

   _____
   No.   Street   Suburb

2. Where did you go after parking?

   _____
   No.   Street   Suburb

3. What time did you park today? _____ am/pm

   What time did you move your car from this parking space? _____ am/pm

4. What did you do while your car was parked here (e.g. work, shop, visit friend, etc)?

   _____

Figure 13.5   Typical reply paid questionnaire form

13.3.3 Observational surveys

Given the nature and manifestations of travel demand, traffic generation and parking information derived from personal interview data should, in theory, be superior to information deduced from observations of vehicle movements. Unfortunately, however, interview-type surveys are far more expensive and difficult to design and administer than the observational surveys. The design of questionnaires and their administration to a survey population is beyond the scope of this book - but see Richardson, Ampt and Meyburg (1995).

Most surveys of traffic generation are conducted as observational surveys, particularly when the land use activity can be ascribed to a well-defined area or a site, such as a shopping centre or fast food restaurant. The information that can be collected by the different types of observational parking surveys is shown in Table 13.2.

Cordon surveys can be used to collect both parking demand and trip generation data. The study area is surrounded by a closed cordon and observation stations are established on all cross roads entering and exiting the cordon in the designated period. Specification of the boundaries (in space and time) of the survey is crucial; in a central city study a useful spatial definition might be the area devoted principally to business establishments together with the adjacent streets, and the temporal boundaries would usefully be wide enough to pick up early and late arriving workers.

Table 13.2   **Data items obtained from different observational parking survey methods**

Data item	Input-output survey (cordon survey)	Patrol survey Traditional	Video
Duration of parking	✔	✔	✔
Total number of parkers	✔	✔	✔
Arrival rate	✔	✔	✔
Departure rate	✔	✔	✔
Composition of population	✔	✔	✔
Parking accumulation	✔	✔	✔
Parkers' spatial distribution		✔	✔
Vehicle entrance point	✔		✔
Vehicle exit point	✔		✔
Vehicle speeds			✔
Gap acceptance times			✔
Conflict points			✔
Parking time			✔
Unparking time			✔

In its basic form the cordon survey simply involves counting. A separate count is made of vehicles entering and leaving the defined area in each specified subperiod (e.g. each hour) within the overall study period. The algebraic summation of entering and leaving traffic gives the accumulation of vehicles in the area. This accumulation represents the sum of vehicles parked and on the move inside the study area. After removal of the moving vehicles a measure of the required parking facilities is obtained which can act as a controlling framework. Counting can be carried out either manually or by automatic counters (see Sections 8.3 and 8.4 for a description of the relevant technologies). Manual methods are more expensive but may be required in special surveys or in order to check, or make corrections to, the automatic counters. By recording the times of entry and of exit of each vehicle it is possible to estimate the mean time spent inside the cordon (see Sections 9.4.1 and 12.2.2 on input-output surveys) and thus to estimate parking durations.

The cordon count method can be used to survey individual parking lots, city zones or short stretches of street. Where the method is used to survey an individual parking lot the observation points will be at the entrance(s) and exit(s), and the survey may be termed a 'sentry survey'. Where the method is used to study an individual stretch of road the turnover will determine the feasible maximum length of road to be monitored by each surveyor   in practice it is rarely desirable to ask one surveyor to monitor more than 100 metres of kerb. The effort required to conduct cordon counts is, of course, proportional to the number of observation points and to the complexity of turning movements at each one.

Even with this more intensive type of survey, details are only recorded as a vehicle enters or leaves the cordon. This information gives no indication of how time is spent in the cordon. It is, therefore, assumed that the time spent inside the cordon is predominantly spent parked. Unfortunately, this may not always be the case. Some factors that are not taken into account are:
- *search patterns*, differing driver behaviour creates large variations in the time spent searching. Aspects such as speed and preferred space location may also influence the total time spent inside the cordon;
- *parking ability*, some drivers are more adept at parking than others. Less confident parkers may forgo the use of a vacant space if it is too difficult to drive into, so increasing their travel time;
- *congestion*, if there are insufficient exits, not enough space or obstructions due to unparking vehicles, congestion may arise, increasing the time spent searching for a vacant space, and time taken to exit from the parking lot, and

- *moving spaces*, occasionally vehicles do not park in the same spaces for their entire stay within the cordon. Each movement from one space to another increases the travel time component of the time spent inside the cordon.

These factors combine to cause cordon surveys to produce a net overestimation of the parking duration.

A popular type of observation study is a *patrol survey*. In its simplest form it involves an observer walking, or being driven, along a predetermined route at fixed time intervals noting the presence of each parked vehicle using either a tally counter, pencil and paper or handheld computer. This is known as an accumulation survey and is a cheap means of determining the parking accumulation profiles in different parts of the study area.

Much more information can be obtained if, instead of merely recording the presence of each vehicle, the surveyor makes a note of its registration number. The time the vehicle is noted or the start time of the patrol of particular sections can be recorded and used as a basis for calculating the parking duration as well as the accumulation, because the number of times a vehicle is observed in the same parking place multiplied by the observation interval, gives an indication of its parking duration. The information that can be obtained using this approach includes: total number of parkers, arrival rate, departure rate, parking accumulation, parking duration, and spatial distribution of parkers within the study area.

The study area must be divided into sections, where each section is small enough for the surveyor to cover its length and return to the start in the time allotted for a tour. If it is possible to divide the area so that different sections complete a circuit, the time spent in returning to the start can be eliminated, and the surveyor used more efficiently. A conservative estimate of the time taken to walk between two adjoining 90 degree parking spaces and record the first three digits of a licence plate is five seconds.

Patrolling by car enables longer sections to be considered in a given time interval but both a driver and a surveyor are required. For making simple counts one surveyor is required and the vehicle can drive at the speed of the surrounding traffic. The surveyor can use an appropriate mounted tally counter or calculator to record the vehicle and parking types of interest. If registration numbers are to be recorded it may be necessary to have two observers, one calling out the registration numbers for the other to record. It is possible to reduce personnel numbers required for the survey by the use of tape recorders. When recording vehicle numbers, vehicle speeds of about 10 km/h would be appropriate for normal close right angle parking.

The first tour is the most arduous and should start before the survey proper gets under way. This tour sets the base data for the study. On the first tour the observer should record every vehicle parked at the kerb, whether parked legally or illegally. Double parked vehicles should also be recorded. If registration numbers are being recorded, it may be sufficient to only record part of the registration plate. Vehicle type can be recorded by using different codes for the different types. The subsequent tours are carried out at regular intervals, using a similar approach to the first.

Figure 13.6 shows how survey effort may be reduced on subsequent visits if, instead of writing down each registration number on each visit, a tick is placed in the appropriate column when the same vehicle occupies a space on successive tours. The success of this method is dependent on the surveyors keeping track of where they are on the survey form not an easy task if when there are long stretches of empty parking spaces or a high turnover along particular sections of the tour. Initial labelling of individual spaces (as shown in Figure 13.6) may be a solution where there are distinct parking bays but many survey managers prefer to avoid the problem by reverting to the method of recording the registrations of all vehicles on each tour. If this method is being used then there are some advantages in recording the data on audio tape or directly into a portable computer rather than on paper form. The use of audio tape brings considerable savings in surveyor time (thus reducing costs or allowing a more frequent tour cycle) but is associated with high transcription costs. Some years ago Dew and Bonsall (1991) reported accuracies of up to 99 per cent but only in the most favourable conditions. More recent technologies achieve high accuracy in typical field conditions. The use of portable computers in the field does, of course, bring very considerable savings in data transcription effort.

To avoid potential biases, the patrolling observer should only record details of vehicles which were present, and stationary, in a parking space when the observer reached that space; parking events that have just finished or are just about to occur should not be included. In addition, to avoid influencing drivers to modify their normal behaviour, the patrolling observer should record registration plates as inconspicuously as possible.

A patrol survey can yield very useful information about parking durations, parking discipline and illegal parking in the study area as a whole, and within identified subsections of it. The data can be further enriched if details are kept of the behaviour of particular types of vehicle (e.g. taxis, commercial vans, or cars bearing parking season tickets). An advantage of patrol surveys over cordon or sentry surveys is that they can exclude cars circulating within the cordon area.

Traffic Generation and Parking Studies 299

Figure 13.6 **Typical form for patrol survey**

There is considerable interest in the use of *video surveys* for parking surveys. Its potential advantages include the ease with which an accurate and complete record can be made and the fact that it is unobtrusive and therefore unlikely to affect driver behaviour. Section 8.3 contains a discussion of some of the issues involved in recording and transcribing video data.

A comparison of cordon, patrol and video techniques by Young and Thompson (1990) indicated that the cordon method provides the largest sample size, the traditional patrol survey provides the lowest sample and the video patrol produces a sample between the two. The low sample received from the traditional patrol survey results from the fact that parkers can enter the parking lot and leave during a tour and be missed. Figure 13.7 shows the number of short term parkers missed. With a large proportion of these durations being less than five minutes, it was possible that two or more vehicles could utilise a space and not be recorded on either of the two consecutive patrols.

Methods of correcting patrol survey data were suggested by Cleveland (1963) and Richardson (1974). Cleveland assumed that the distribution of parking duration was exponential. He suggested that the 63rd percentile duration could be used to estimate the mean duration. The ratio of the patrol duration and the mean duration is then used to determine the percentage missing. Bonsall (1991) describes a procedure for correcting for small durations when uneven breaks are not used. The smaller sample size obtained from the video patrol results from the difficulty in matching cars by location and colour and problems in seeing vehicles parked a distance from the camera.

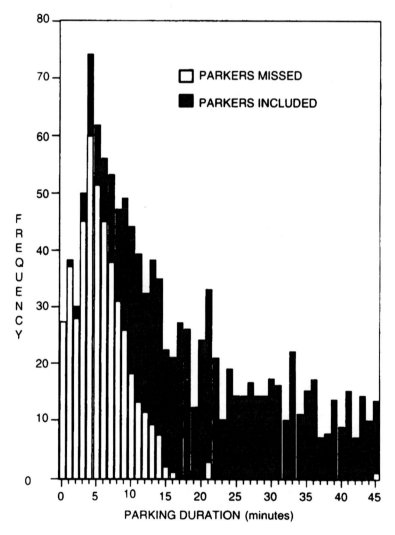

Figure 13.7 Vehicles missed by patrol survey (Young and Thompson, 1990)

Comparison of the video and the input-output surveys (Figure 13.8) shows only random differences between the two distributions. The major differences in the distributions derived from the two methods are that the video does not collect as many high value durations and it measures the time parked rather than the time in the parking lot. The parking durations determined from the video and the travel times in and out of the parking lot should agree with those determined from the input-output survey.

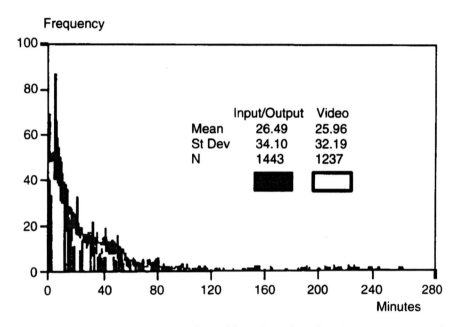

Figure 13.8 Comparison of parking duration for input-output and video patrol methods

Even though the cordon method provides the largest sample it still underestimates the total parking population. Errors in the version of the cordon survey which involves recording vehicle registration numbers result from:
- *reading errors*, due to hurried observations, obscured registration plates or character inversion errors;
- *missed readings*, due to overloading of observer at peak periods and the need to look at the survey form (rather than the traffic) to record data;
- *wrong choice* between two identical entries, and

- *misreading* of handwriting on survey forms.

The comparison of the three survey methods also requires consideration of the cost, information obtained, flexibility of the systems and their accuracy. In terms of the total cost the cordon survey is generally more expensive than the traditional patrol survey. The highest cost survey is the video patrol survey. The video is the least flexible of the three methods, for example it can only consider a small self contained parking lot. Similarly the cordon method is best suited to self contained, uncongested parking lots with relatively low travel times to and from the parking space. The presence of through vehicles can influence the accuracy of the method. The traditional patrol can be used in any situation, if sufficient survey personnel are available. The video patrol, however, gives the most accurate estimate of parked times and provided the opportunity to collect data on more aspects of the parking system than the input-output method or the traditional patrol survey. It provides a permanent record which can be used for comparative studies. The traditional patrol is the least accurate. The accuracy of the method can be improved by reducing the patrol cycle time but missed observations are always possible. Corrections should always be applied to data from the traditional patrol survey.

13.4 Summarising the requirements for parking surveys

The general principles for traffic survey design and management are important in traffic generation and parking studies. The following particular requirements must be met:
- *exact definitions must be provided* for the traffic variables to be measured. For example, in a shopping centre survey, is it all people arriving at the centre, or all vehicles arriving, or only those vehicles that park there that should be recorded?
- *does the measured generation includes all vehicles, or just private vehicles*? For example, should service and goods vehicles be included in (say) a shopping centre or residential area survey? Are data required on taxi parking at taxi ranks?
- *area-wide surveys need to account for 'through' traffic movements separately*. Through traffic would be defined as that part of the traffic not stopping in the survey area;
- the *possibility of internal trip movements*, i.e. trips not crossing the external cordon, must be considered in area-wide studies. The larger the survey area, the more likely are such trip movements;

- *precise definitions of traffic generation rates are needed.* For example, the measurement of trip ends (the sum of trip production and attraction) is preferred to a measurement of trips *per se*;
- surveys seldom collect data over extended time periods, so *there is currently little information on the variations in traffic generation* over days of the week and weeks of the year. The collection of long term data is recommended, and
- as traffic generation is often related to measures of activity, such as floor area, *care is needed to ensure that consistent measurements and definitions are used.* For example, there are several alternatives for area; gross area, net floor space, leasable floor space, etc.

13.5 Models of traffic generation and parking

The theories of trip generation and parking choice are less well developed than those applicable to some other areas in traffic planning, despite the significant advances in understanding travel behaviour phenomena – see for example, Jones et al (1983) and Ettema and Timmermans (1996). Conventional traffic generation analysis, however, considers only incremental or marginal effects. It assumes that the specific land use activity site represents a small proportion of the overall level of that activity. This is a reasonable assumption when, for example, assessing the impacts of a new service station or a small shopping centre. It is unreasonable when new large regional centres are being considered, since these may attract significant amounts of traffic from existing centres. Where such effects are likely it is more appropriate to go beyond traffic generation models and to explore a wider range of possible responses via a strategic transport planning model of the type described in Section 5.1.

For the practising traffic analyst, the use of empirical data and relationships represents the current state-of-the-art in traffic generation and parking demand analysis. Two basic approaches are available. The first is the use of mean trip generation rate or mean parking requirement, and the second is the use of regression relationships. Mean trip and parking rates are commonly used when considering dispersed traffic generation, such as that of households across a residential area, whereas regression relationships are often used to predict the traffic generated by or parking at a single large development, such as a shopping centre, where some physical measure of that centre (such as its gross floor area) is available.

13.5.1 Trip generation and parking rates

A trip generation rate is an average number of (vehicle or person) trip movements per unit of activity, per unit time. The parking rate is the number of parking spaces required per unit of activity. The unit of activity could be a household, square metre of retail floor space, hospital bed or some other easily discernible unit. The unit of time is typically either the hour or the day. Daily trip rates give a total measure of generation, while an hourly rate indicates a peak load. The approach is only viable if the activity and its unit can be defined explicitly, as according to Brindle and Barnard (1985):

> movement occurs in most cases as a result of some sort of human activity occurring at a particular location (buying, drinking, meeting, sleeping, etc.). The problem is then to measure the amount of such activity at a particular site. Activity, however is an elusive concept and even when obvious measures exist (for instance the number of workers occupying an office building) these tend neither to be predictable nor stable.

A consequence is that trip generation and parking rates tend to be expressed in units of physical land use, rather than activity. For instance the floor area of an office building might be used, rather than the number of workers employed. The floor area can be assessed readily from the plans for the development - the number of workers cannot. Even if some indication of employment level is presented with the plans, this may well change in the future: floor space is more permanent. The difficulty with this approach is that the land use unit *per se* does not generate trips; it is only a proxy for the activity. The office block generates no traffic if it remains unoccupied. The following typical measures of land use units may be employed:
- gross floor area (e.g. shops, offices);
- net retail area, or customer space (e.g. bars, restaurants);
- number of seats (e.g. concert hall, theatres, cinemas);
- number of beds (e.g. hospitals, hotels, motels);
- enrolments (e.g. schools, colleges, universities), and
- number of dwelling units (e.g. flats, apartments, houses).

In addition, the number of households represents an important socio-demographic unit, which can be estimated for a given area with a fair degree of reliability. This unit is commonly used for estimating the traffic generation of a residential area. The simplicity of the trip rate approach makes it very attractive, and it is often used in traffic planning (ITE 1997).

In a well-defined residential area, in which the demographic characteristics of the residents are known, it is possible to use category analysis models for trip generation. These models present a cross-tabulation of trip rates, in terms of one or more socio-economic and demographic

characteristics of a household. This is an effective way of expressing the differences in travel generating activities between different households. Category analysis models may be constructed as follows. Given a set of survey data on household trip generation:
1. divide the households into the cells of a matrix, based on the level of each independent variable (i.e. form an n-dimensional histogram where there are n independent variables);
2. calculate the mean number of trips per household per unit time for each cell in the matrix;
3. plot the results and construct smooth curves to obtain consistent results (alternatively, fit a polynomial surface to the data, if desired), and
4. construct a matrix using the adjusted values.

The advantage of the category analysis approach is that it accounts for the variation in traffic generation between different types of households in an area, and thus may be used to estimate future traffic generation based, for example, on ageing of a population. This can be most useful in the analysis of developing residential areas. The disadvantages are that data on the distributions of household characteristics are needed, and that the trip rates for each cell are assumed constant over time. Given the usual availability of comprehensive demographic data at the municipal level, or finer, from the national census or from population prediction models, the first of these disadvantages is less important. The second is more difficult to reconcile in the general absence of time series data.

For most applications, the use of simple mean trip rates is the usual procedure in residential area traffic generation analysis. The analyst is advised, however, to use a 'sensitivity analysis' approach by considering a range of trip rates, rather than a single value.

13.5.2 Regression models

The regression approach is to develop a simple linear relationship between the total trip generation or parking demand (G) of an activity, and a set of independent variables where X_i, i = 1, 2, ..., n. The relationship is thus

$$G = b_0 + \sum_{i=1}^{n} b_i X_i \tag{13.1}$$

where the b_i, i = 0, 1, 2, ..., n, are constants. The methods described in Chapter 17 are used to develop the model. The major application of this type of model is in the impact of isolated developments, such as shopping centres.

An example would be the following relationships given by Hallam (1980), for retail centres in New South Wales,

$$GF = 184 + 0.0658\, FA \tag{13.2}$$

$$GS = -95 + 0.0503\, AD + 0.2517\, AM \tag{13.3}$$

where GF and GS (vehicle trip ends/h) were the peak numbers of vehicle trips per hour at a shopping centre on Fridays and Saturdays, respectively. The independent variables were FA (gross floor area in m^2), AD (department store floor area in m^2), and AM (supermarket plus small shops floor area in m^2). These relationships were two from a large number of models developed by Hallam for different site land uses in urban areas of New South Wales. Relationships of this type allow the inclusion of a number of independent variables, and may be ascribed a certain level of statistical fit through the well-established means of regression analysis.

There are a number of possible problems with this. For example, the constant coefficient b_0 in equation (13.1) provides a means of exploring the statistical fit, but care is needed that this coefficient does not exert an undue influence on the particular model. In ideal circumstances b_0 should be zero: a non-zero value indicates the possibility of variables not in the equation having an influence on the generation, or a non-linear underlying relationship. This is expected, for the models are simple, and other problems may result if too many variables are included or more complicated model types are fitted. Care is needed in interpreting the estimated values of the coefficients. For example, equation (13.3) contains a negative value of b_0. Does this really mean that the land use can ever generate 'negative traffic'? Similarly, the signs of the coefficients b_i may need to be checked on logical as well as statistical grounds.

A major concern with the application of regression models lies in their range of applicability. There are significant problems in extrapolating any regression model outside the ranges of the observed data. Consider the hypothetical example given by Figure 13.9, which indicates surveyed levels of traffic generation for a set of shopping centres with floor areas between 10,000 m^2 and 20,000 m^2. The figure has been drawn to accommodate centres with floor areas up to 40,000 m^2. For purposes of illustration it indicates the observed data, a 'true' underlying (non-linear) relationship, and a regression line fitted to the observed data. How useful would the regression result be if applied to a centre of floor area 2,000 m^2, or to one of 40,000 m^2?

Despite these traps, regression models provide useful descriptions of traffic generation and hence of the parking requirements, when they are applied under the conditions for which they were developed. It is the responsibility of the analyst to ensure that a particular model is correctly applied.

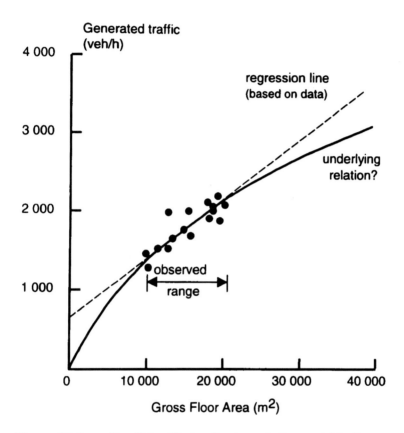

Figure 13.9   Possible effects of extrapolation outside the range of the observed data

14 Road safety studies

Road accidents constitute an important and emotive issue which no decision maker can afford to be seen to ignore. Improvements in vehicle design, in the delivery of medical care and in traffic engineering have reduced the accident rate (per unit distance driven) and, in some countries, even their absolute number and severity, but road accidents are still a major cause of death and disablement, particularly for young people. In addition to causing human grief and suffering they represent a significant cost to the economy through lost working time, traffic delays, use of emergency services and repairs to vehicles and property. Reduction in the occurrence of accidents through traffic systems design measures can often make a major contribution to the economic benefits of a given project.

Most road accidents result from a combination of factors; poor driving and poor vehicle maintenance are usually among the contributory factors. Many road safety professionals advocate that the term 'road accident' should be replaced by 'road crash' in order to get away from the notion that they occur without any particular cause – we will adopt this new terminology throughout the remainder of this chapter.

The engineer can contribute to road safety through appropriate design of networks, layouts, surfaces, materials and control strategies. The aim will be to adopt designs which tend to reduce the number of crashes and to reduce the severity of those that do occur. Typical features will therefore include good sight lines and lighting to aid visibility, refuges and space for acceleration and deceleration at junctions on high speed roads, clear signing and road markings to separate potentially conflicting traffic streams, rough pavement textures to assist braking and provision for vulnerable road users (cyclists and pedestrians).

Traditional wisdom was that safety would be enhanced by promoting the smooth uninterrupted flow of segregated traffic streams but it is now recognised that, particularly where vulnerable road users are involved, it may be better to adopt traffic calming measures which, by judicious use of obstacles and interruptions to flow, cause drivers to reduce their speed and proceed with greater caution. There is often a potential conflict of interests between safety on the one hand and capacity or traffic throughput on the other. Many authorities are dealing with this by adopting safety audit procedures whereby all new designs are subject to scrutiny by safety specialists before being approved.

Clearly it would be unreasonable to hope that, however good the road design and maintenance, crashes can ever be eliminated. Even with better driver training, more effective enforcement of regulations, and more rigorous standards for vehicle design and maintenance, crashes will still occur – some of them may be genuine accidents due to adverse weather conditions or perhaps with no discernible cause. Traffic engineers have an important role in reducing the disruption and delay caused whenever and wherever they occur. Here is another conflict of interests; in 'normal' (non-incident) circumstances system efficiency may be maximised by fine tuning it to make full use of all available road capacity, but if this results in the elimination of reservoirs of spare capacity, the system may then be very susceptible to disruption and an incident at a key location may have severe consequences throughout the network. If such locations can be identified in advance, for example by simulation modelling, it may be possible to 'detune' parts of the network or to design flexible control strategies which can adapt to incidents in such a way as to reduce their impact and propagation through the network.

The design of measures to reduce the occurrence of crashes is clearly helped by analysing the occurrence of past crashes. This analysis will involve long term study of crash statistics to identify trends and correlations which point to the need for action in respect of a particular type of site, vehicle, driver or manoeuvre. For example, the statistics may reveal that drivers in a particular age group may be particularly at risk on certain types of road or driving particular types of vehicle. Statistically speaking, crashes are rare events and since each one is the result of a unique combination of factors at the site (the ambient traffic and weather conditions, the vehicles, the drivers, their state of mind, details of the manoeuvres they were making, and so on), it can take several years to build up sufficient data to have any confidence in the findings in respect of a particular site and, since some factors will be evolving over time, statistical confidence in the results for a particular site may in fact never be achieved. Statistical analysis is therefore supplemented by in-depth study of individual crashes at specific sites, with a view to establishing the need for particular remedial measures at that site. Prioritisation of sites for detailed investigation and action should where possible be based on objective criteria (such as the occurrence of an unusually high number of crashes) but will also need to recognise local political pressures. Many authorities now have crash database software which will automatically flag *blackspot sites* where the number of crashes exceeds a given threshold.

In-depth investigation of individual crashes uses detailed data collected at the time and may involve specialist software to reconstruct the incident.

14.1 Some definitions

Crashes are usually categorised according to their severity. A distinction is made between those involving *personal injury* (which may be *slight, serious* or *fatal*), and those that involve *damage only*. The precise definitions vary from country to country (e.g. as to whether, in order to qualify as a *fatal crash*, the fatality must have occurred at the scene of the crash or within 3, 30, 100 or 365 days).

Road safety studies may be concerned with absolute numbers of crashes or with *crash rates per unit exposure*. Commonly used measures of exposure are the population, the number of vehicles, the number of trips, the number of vehicle kilometres driven or the number of person kilometres travelled. The last of these are arguably the most revealing of true levels of risk but are clearly more demanding of data. Different measures will be appropriate in different circumstances. Thus the rate per head of population may be appropriate for international comparison and economic analyses while the rate per person kilometre travelled may be particularly revealing of the risk to vulnerable road users.

14.2 Statutory crash records

The prime source of crash data is usually the records kept by highway authorities or the police. It is, in most countries, a statutory duty of one of these bodies to collect data on the occurrence of crashes involving fatalities or personal injuries and to provide summary statistics relating to them.

The system is typically dependent on records made by the police when they attend the scene of a crash. The data are usually recorded on a special crash report form which seeks information about the circumstances surrounding the crash and of the injuries and damage caused. Some authorities also provide space on the form for the attending officer's assessment of the causes of the crash; this can be useful but is clearly subjective. Table 14.1 lists the main items typically included on a crash report form. A crash involving several vehicles and numerous casualties can, of course, generate several pages of information.

Table 14.1 Components of typical crash report form

Date and time of crash

Site details
- location (e.g. road number, distance from junction, map or grid reference)
- road layout/features (e.g. pedestrian crossing, type of junction, width of road)
- provision of lighting
- local speed limit
- road surface and condition

Ambient conditions
- weather conditions (noting, in particular, any precipitation, ice, fog or dust)
- visibility

Vehicles involved (for each vehicle)
- manoeuvres being undertaken (descriptive text e.g. 'turning right from side road X into main road Y', or 'road user movement' type)*
- type, age and condition of vehicle (noting in particular any obvious defects in lights or tyres)
- driver identity and characteristics (age, sex, driving licence details, evidence of drug/alcohol use)
- other occupants identity and characteristics

Damage/injuries sustained
- nature of damage or injury
- actions taken (eg hospitalisation)

Comments
- (e.g. reporting officer's attribution of cause, details of any prosecution pending)

Reporting officer's identity

* *movements of all vehicles involved may be recorded on an annotated diagram of the site.*

Given the volume of data to be recorded at the site, and the other priorities which the police may have at the scene, it is not surprising that the quality of this information sometimes leaves something to be desired. Interesting attempts to reduce the scale of the misrecording problem include studies in Dublin and Adelaide in which police officers code the information directly into a portable computer at the crash site and the computer immediately conducts logic checks on the data and adds a location code based on its own knowledge - via GPS - of where it is.

Even if the data entered by the police are current there is still another problem in that, typically, the system does not 'catch' all crashes. It is generally accepted that the system is not intended to cover crashes which result in damage-only but it is supposed to cover all crashes involving fatalities or personal injury. However research (e.g. James, 1991) has shown that the system under-represents crashes which involve only slight injury and those involving only one vehicle. The reason is obvious; people involved in such crashes may not see any need to alert the police and, under the terms of their insurance policies, may indeed have an incentive not to do so.

The data collected by the police are usually entered into a database and, as a minimum, basic tabulations of trends in different types of crash are produced by, or for, the highway authority. Where government functions are centralised, as in the UK, the data are usually forwarded to the responsible national body and they in turn will produce reports and summaries based on statistical analysis of the data (e.g. the annual report *Road Accidents Great Britain*). Provincial or state authorities usually provide similar information for their jurisdictions in federal systems of government.

14.3 Other sources of crash data

Other useful sources include hospital and/or ambulance authorities, insurance companies, towing services and special surveys. These sources can be used in special research studies to check or supplement the information in the police records relating to specific crashes or can be used to gauge the number of crashes which have been missed by the police system. Obtaining this data can be very time consuming and is understandably hampered by a concern for the privacy of individuals involved.

The special surveys referred to might be of various kinds. At their simplest they might involve the addition of extra questions onto the police report form for a specified period. For example if a special investigation was to be made of the effect of car-phones on crash occurrence, a question on whether or not vehicles involved in the crash were equipped with car phones, might be added to the police report form for a year.

Another form of special study might involve very detailed investigation of a sample of crashes. This might necessitate interviews with the people involved in order to gain greater insight into the factors contributing to the crash and detailed inspection of the site and the vehicles involved. Interviews with crash victims can, of course be traumatic if conducted insensitively or too soon after the event but can suffer from selective recall if delayed too long. A major task in such investigations is often to reconcile the accounts of the different parties involved (e.g. Tight et al, 1990).

At a quite different level, another kind of special study would be an attempt to estimate the number of damage only crashes at a specific site by regular collection and analysis of vehicle debris from the scene.

14.4 Site investigations

Statistical information derived from one or more of the sources referred to above can be very useful but, if remedial measures are contemplated at sites which have a particularly poor crash record, it is of course important to carry out more detailed investigations of the sites and of the crashes occurring there.

A detailed site investigation could include the collation of detailed inventory data including photographs of sight lines, measurements of the visibility of key signs, and the collection of traffic flow data (disaggregated by turning movement and, perhaps, class of vehicle), pedestrian movement data and spot-speed data. An experienced engineer, armed with this information and details of recent crashes at the site, would be well placed to identify dangerous features at the site and to recommend remedial measures.

14.5 Conflict studies

As has already been mentioned, a problem with crash data is that crashes are insufficiently frequent to allow statistically robust conclusions to be drawn about crashes at a particular site unless several years worth of data is available. This makes it particularly difficult to identify changes in risk following changes in the layout, signing, lighting or signalisation at a site. This makes it difficult to assess the effectiveness of specific remedial measures or to identify the safety consequences of policy actions undertaken for reasons unconnected with safety. By the time any change in the crash record becomes statistically significant other factors affecting the site may have altered and it may be impossible to attribute the change in the number or nature of crashes to any specific cause.

In order to overcome this problem a technique known as conflict analysis has been developed (e.g. Grayson and Hakkert, 1987). The idea is that, the number of 'near misses' recorded by trained observers at a site within a short time period (typically less than a day) is indicative of the level of underlying risk and will therefore be equivalent to the number of actual crashes that might be expected at the site over a period of years. Thus it is possible, using this technique, to assess the relative risk at a number of sites (e.g. in the context of an attempt to prioritise remedial measures at several candidate sites) or to assess changes in risk (by applying the technique before and after a specific treatment).

The validity of the technique is crucially dependent on proper training of the survey staff and strict adherence to the procedures laid down. The technique involves classification of manoeuvres on a scale from minor vehicle-to-vehicle interactions to ones involving severe evasive action. There are a number of variants on the technique.

Perkins and Harris (1968) were pioneers in the field, working in the General Motors Research Laboratory, they defined a conflict as any manoeuvre involving the application of brakes. Twenty objective criteria were then used to subdivide these conflicts into left turn, weaving, cross-traffic and rear end conflicts, with the nature of any evasive movements being noted. For a four-arm junction the Perkins-Harris technique involved three study periods of twelve hours with two observers in each period; the first had the observers on opposite arms of the junction, observing from 300 metres back from the intersection itself, the second repeated this procedure for the other two arms and the third had the observers noting conflicts within the intersection itself.

Subsequent work by Older and Spicer (1976) suggested that it was not necessary to record events involving mild or medium braking and that conflicts should be judged according to a subjective assessment of their severity or potential severity as judged by consensus between a pair of trained observers located so that they could get a good overview of the site. A major criterion in this classification was whether all drivers involved had adequate time for a steady and controlled manoeuvre. This technique required subjective assessment of the whole scene and therefore was more reliant on highly trained staff than was the Perkins-Harris technique. It subsequently evolved into the 'TRL technique' (TRL, 1987) which used internationally agreed definitions and which concentrated on four components of the scene; the timing, severity and complexity of the evasive action and how close the vehicles came to one another.

Hyden (1987) has proposed a simpler, more objective, technique (which has subsequently become known as 'the Swedish technique') based on estimation of 'time to impact' (TTI) assuming that no evasive action is taken. The seriousness of the conflict is indirectly proportional to the TTI. The justification of the method is based on evidence which indicates great similarity between the events and behaviours which precede conflicts and those which precede crashes.

Some authorities suggest that conflict studies can be undertaken using video and that this allows for more objective analysis of the scene with the aid of slow-motion playback. Research is currently under way to see whether computer analysis of the images involving trajectory identification and automatic calculation of TTI can be helpful in this process. Other authorities are sceptical of such developments pointing out that it is rarely possible to get an adequate view of everything that is happening at a site from one single video camera. Nevertheless, it is widely expected that developments in computer imaging using multiple cameras and in micro modelling are likely to revolutionise the conduct of conflict studies and thereby promote them from a 'clever idea' that may or may not be reliable, to become the mainstay of crash analysis work.

The conflict technique is currently very labour intensive and is never likely to be cheap, it can therefore only be applied at sites where a problem is thought to exist or where a change in underlying risk is suspected. It is not a practical method of identifying such sites in the first place. A potential method of doing this has been suggested based on the analysis of the deceleration profiles of instrumented vehicles; the idea is that if sufficient vehicles were equipped, perhaps as a by-product of a route guidance system involving probe vehicles to monitor current traffic conditions, then locations which exhibit a significant increase in deceleration severity could be targeted for more in-depth study. Research so far (Bonsall and Palmer, 1993) has suggested that there is a correlation between the distribution of severe decelerations and of underlying crash risk as manifest in recorded crashes over a period of years but that the technique is not a practical proposition until and unless a significant proportion of the vehicle fleet are equipped as probe vehicles.

14.6 Driver behaviour studies

Useful insight into the behaviour of drivers can be obtained by detailed monitoring of drivers' actions, intentions and physiological state whilst driving on the road or in a sophisticated driving simulator. Information can be gathered about their interaction with other drivers, with features in the road environment and with the controls and accessories in their own vehicle. Such studies are currently making a particularly valuable contribution to our understanding of the safety implications of new human-machine interfaces associated with in-vehicle guidance systems and have similarly contributed much to research into the design of vehicle control fascias and road signs and into the effect of drugs and alcohol on driver performance.

The necessary observations can be made by a trained observer supplemented where appropriate by questionnaires or objective measurements of such factors as use of vehicle controls, eye movements, nervousness (as measured by skin moisture levels) and brain activity. On-road observation is preferred where possible because of its greater realism but driving simulators have an important role when special conditions are required as part of an experimental design or when hazardous situations are to be reproduced. The driving simulators comprise a real or mock-up car placed in front of a screen on which a road view is projected. Traditionally this view was a cine or video view but much greater flexibility is now achieved using computer generated images. All simulators should make the road view move in response to the speed and trajectory of the simulated car, some add extra realism (albeit a considerable extra cost) by also tilting and panning the vehicle itself to induce a sense of the internal and gravitational forces which act on a moving car. The most sophisticated simulators are typically owned by aircraft and car manufacturers (e.g. Daimler Benz and the Hughes Aircraft Corporation) while cheaper ones have been developed, for example at the University of Leeds for use in traffic research.

14.7 Controlled crash tests

Much can be learned about the factors influencing the severity of crashes by conducting controlled crash tests. These tests involve driving, or catapulting, carefully instrumented vehicles into obstacles in order to see how the design of the vehicle or of the road furniture influences the outcome in terms of damage to property or injury to humans.

Humans are represented in such studies by carefully designed and instrumented dummies (the use of cadavers for this purpose now being very rare). The instrumentation allows the forces acting on the vehicle and dummy humans to be recorded and analysed in detail. Such studies have contributed much to our knowledge of the benefits of good design for the vehicle and the road furniture- leading, for example, to the widespread adoption of seatbelts, side-impact protection and collapsible steering columns within the vehicle and of effective crash barriers, plastic bollards and collapsible lighting poles at the roadside.

14.8 Analysis of crash data

Data from police reports on crashes have been being stored on databases for several years and many highway authorities now have an apparently rich source of data on which to conduct analyses. Simple analysis of trends in different types of crash and identification of sites with more than a given number of crashes per unit time has routine and straightforward. Many authorities now have software which will automatically alert them to 'blackspot' sites (which have more then the specified number of crashes) and which using GIS-like facilities will display maps showing the geographical distribution of crashes and allow details of crashes at specified locations to be called up and displayed or printed out.

These analyses are, of course, much more revealing of underlying risk database also contains time series data relating to exposure (e.g. vehicle or pedestrian flows at the site, land use and frontage details for each link, or population and car ownership in different parts of the area). Some authorities have now developed their databases to the extent that they have a link between their crash database, their traffic flow monitoring database and their traffic model (e.g. Mackenzie and McCallum, 1987).

Identification of significant change in the crash risk at individual sites is notoriously difficult because of the sparse nature of the crash data and the variability (often unrecorded) in exposure. This difficulty has spawned a number of famous controversies in the analysis of data of which the most interesting is perhaps the 'crash migration' debate which originated when it was observed that the crash rate at sites where crash remedial measures are installed often improves while that elsewhere deteriorates. One explanation of this is that drivers, deliberately or otherwise, might adjust their driving caution to suit the ambient conditions so as to keep their risk at an acceptable level and, if the most dangerous sites are improved, they will drive less cautiously and have their crashes at those sites which are now most dangerous hence the crashes 'migrate'. Sceptics point to the widespread phenomenon of regression to the mean whereby, due to random variation among rare events, sites which have a high crash rate in one time period will tend to have a lower crash rate in the next and vice versa. Since the treated sites will tend to have been those with the highest crash rates they are almost bound to show an improved crash record and untreated sites are almost bound to show a deterioration. Interested readers are advised to consult the evidence on these phenomena for themselves. Opposing points of view are expressed by Boyle and Wright (1984) and Maher (1987).

Part E

DATA ANALYSIS AND MODELLING

15 From data to information

The results of traffic analysis must be transmitted to the user or client in a form that is understandable and efficient. Chapter 2 described the recent developments in database systems, spreadsheets and statistical packages that assist in this process. These tools are of considerable assistance in developing a picture of what a data set is trying to tell us. The important step is the transformation of data into information. To do this requires appropriate examination of the data, for which methods of data presentation, often graphically based, are used.

The first task in any analysis is to become familiar with the data. It is useful to have answers to questions such as: are there any outliers? Is the distribution of the data symmetric? Do data values accumulate in the middle or at the end? Are any values repeated? Answers to these questions can often be obtained by having a look at the data. Data may be viewed through the astute use of tables (Section 15.1), but one of the most powerful methods is the construction of diagrams (Section 15.2). This forms the basis of Exploratory Data Analysis (EDA, see Section 15.3).

The correct analytical procedure with any data set is to undertake EDA and then move into statistical analysis. EDA can also be used throughout the statistical analysis process to enrich understanding and interpretation (see Chapter 17). Modern methods of interactive data analysis allow analysts the opportunity to have many different views of the data, quickly. Then follows the use of descriptive statistics (Section 15.4).

15.1 Tables

The tabular form of presentation is a common and useful first step in data analysis. The simplest table is just a simple listing of the data. The listing of vehicle speeds in Table 15.1 is an example of this form of presentation. These speeds are those from the data set presented in Appendix B.

Table 15.1   Vehicle speeds (km/h) in residential streets

50.2,	61.1,	56.0,	55.4,	54.7,	51.2,	61.1,	56.8,	58.7,	60.3,
51.5,	52.6,	49.1,	54.8,	58.4,	58.4,	51.2,	64.8,	56.3,	54.6,
58.7,	57.1,	59.5,	52.8,	54.1,	48.8,	50.8,	47.7,	41.1,	50.4,
49.9,	36.0,	51.8,	46.8,	48.8,	34.4,	48.6,	50.8,	40.0,	45.9,
31.6,	57.6,	44.0,	49.3,	42.4,	50.1,	52.4,	44.5,	47.2,	47.8,
59.5,	50.9,	46.6,	52.0,	52.0,	49.2,	61.6,	54.4,	49.6,	44.5,
51.0,	48.8,	47.2,	53.0,	52.0,	52.0,	45.6,	52.8,	57.9,	55.6,
50.4,	52.0,	54.8,	39.7						

The unsystematic manner in which the data are presented in Table 15.1 makes it difficult to gain insights into the character of the data. A better view may be gained by arranging the data in some manner. The ordering of the data from the lowest to the highest value is presented in Table 15.2. This table provides more insight. It shows the extremes of the data and provides an indication of where the data tend to group together.

Table 15.2   Table of speeds (km/h) sorted in increasing order

31.6,	34.4,	36.0	39.7	40.0	41.1,	42.4,	44.0,	44.5,	44.5,
45.6,	45.9,	46.6,	46.8,	47.2,	47.2	47.7,	47.8,	48.6,	48.8,
48.8,	48.8,	49.1,	49.2,	49.3,	49.6,	49.9,	50.1,	50.2,	50.4,
50.4,	50.8,	50.8,	50.9,	51.0,	51.2,	51.2,	51.5,	51.8,	52.0,
52.0,	52.0,	52.0,	52.0,	52.4,	52.6,	52.8,	52.8,	53.0,	54.1,
54.4,	54.6,	54.7,	54.8,	54.8,	55.4,	55.6,	56.0,	56.3,	56.8,
57.1,	57.6,	57.9,	58.4,	58.4,	58.7,	58.7,	59.5,	59.5,	60.3,
61.1,	61.1,	61.6	64.8						

These tables present information on one variable and are, therefore, referred to as one-dimensional tables. One-dimensional tables can be divided into the following components: the table number, the title or caption, the (column) headings, the stub (or left-hand column - a vertical listing of categories about which data are given in the other columns of the table) and, if necessary, the source. The table number, title/caption, column headings, stub and other columns of data make up the body of the table. Table 15.3 is an example of a simple one-dimensional table with the components marked.

Table 15.3   One-dimensional table

Table number, title/caption	**Table 2 Pedestrian fatalities**
(Column) headings	Year — Number
	1994 — 255
	1995 — 231
	1996 — 220
Stub	1997 — 215
	1998 — 195
	1999 — 195
Source	*Source*: National Bureau of Statistics

Two-way classification tables can be obtained by subdividing the stub. Two-way classifications are sometimes referred to as cross-tabulations. They are possibly the most common analytical method in the social sciences. The first step in the construction of cross-classification tables is to determine the primary emphasis. The data with the primary emphasis should be placed in the columns and that with the secondary emphasis places in the rows. The nature of cross-tabulations can best be illustrated by a typical example. Consider a study which hypothesises a relationship between the level of alcohol in a driver's bloodstream and the location of accidents. In a survey, information was collected from five locations, yielding the results presented in Table 15.4. It can be seen that most of the accidents occur at location D and that most accident victims have a blood alcohol level between 0.001 and 0.050 parts per litre. The distribution of accidents across sites also appears to vary.

Three-dimensional tables can be created by subdividing the stub, as in Table 15.5. This table provides detailed information on the types of passenger vehicles in the vehicle fleet and those scrapped.

The presentation of information in tables becomes more complicated and hence less clear as the analysis goes above three dimensions.

The preparation of most tables can be carried out without the aid of statistical packages. Statistical packages, word processors, spreadsheets and similar software packages can, however, decrease the time involved in carrying out this task. Useful procedures for inputting data and presenting it in tabular form are available in spreadsheet programs (see Section 2.4.2).

Table 15.4 Two-dimensional table: survey results for blood alcohol levels of drivers involved in accidents at five locations

Blood alcohol level (parts/litre)	Location					
	A	B	C	D	E	Total
No reading	13	5	8	21	43	90
0.001 to 0.050	18	10	36	56	29	149
above 0.050	16	16	35	51	10	128
Total	47	31	79	128	82	367

Table 15.5 Three-dimensional table: motor vehicles – summary of selected items

ITEM	1970	1975	1978
Passenger car factory sales			
Total	6667	7921	4258
4 door sedans	3247	3044	1481
2 door sedans	2794	1691	705
Others	626	3186	2072
Truck and bus factory sales			
Total	3230	4392	4173
Trucks	2594	3773	3635
Buses	636	619	538

15.2 Diagrams

Tables provide information on the distribution of the data. However, many data sets are not amenable to analysis using tables. The construction of diagrams is a relatively easy and quick task. The information gained from diagrams displaying the data in graphical form is an invaluable aid in analysis, made more useful by the availability of interactive computer graphics. There is, therefore, great interest in diagrams and the information they can provide. Velleman and Hoaglin (1981) and Chambers et al (1983) provide useful introductions to the rich literature on this topic and we have drawn heavily from them.

15.2.1 One-dimensional plots

A variety of one-dimensional plots may be used to provide simple yet informative displays of the nature of a data set. The plots include quantile plots, dot diagrams, histograms and probability distributions.

A good preliminary look at one-dimensional data can be obtained from a *quantile plot* (Figure 15.1). This plot displays the data from Table 15.1. The concept behind the plot is similar to percentiles. If 60 km/h is the 85th percentile speed, 85 per cent of the vehicle speeds are below 60 km/h. Similarly, the definition of the 0.85 quantile divides the data set into two parts, such that a fraction of 0.85 of the data lies below and 0.15 lies above the specified speed. To determine the quantiles, the raw data should be ordered from the smallest value to the largest and then plotted in this order. Many important features can be seen from the plot. The median is the 0.50 quantile, and provides a measure of the centre of the data set. The 0.75 quantile is known at the *upper quartile*, and the 0.25 quantile is known as the *lower quartile*. The difference between the upper and lower quartile, the *interquartile range*, is a measure of the spread of the middle half of the data. The local density of points is demonstrated by the slope of the curve. The flatter the slope, the greater the density of points.

Another method of presenting data diagrammatically is the *dot diagram* (Figure 15.2). This diagram is formed by drawing dots onto a line to represent the values of measured variables. It is a valuable device for displaying the distribution of a small set of data. The maxima, minima, centre of the data, density of points, symmetry and outliers are easily seen. Its main advantage is its compactness. This allows it to be used in the margins of other displays to add information.

326 *Understanding Traffic Systems*

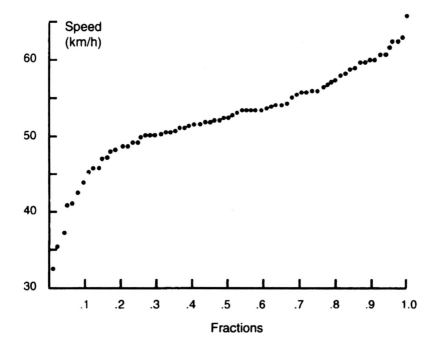

Figure 15.1   Quantile plot of vehicle speeds

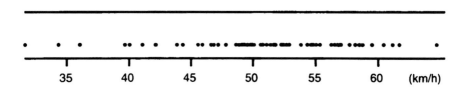

Figure 15.2   Dot diagram of the vehicle speed data

For large data sets it is possible to get overlapping dots. This may introduce difficulties in discerning the density of points. In such cases it is possible to stack repeated data points on top of one another (Figure 15.3). However, this does not overcome the problems associated with dots which partially overlap one another.

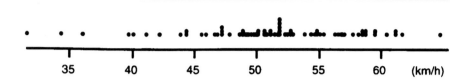

Figure 15.3 Stacked dot diagram of the vehicle speed data

One diagram that can overcome this partial overlap is the jitter plot (Figure 15.4). The jitter plot displays each point on a separate horizontal plane. The height of the plane is obtained by randomly allocating numbers, between one and the number of data points, to each data point. These numbers represent the height of the plane above the base plane. No units are marked on the planes. The dots are then plotted in terms of their variable value and their random index. The resultant plot is still very compact but presents a clear view of every data point.

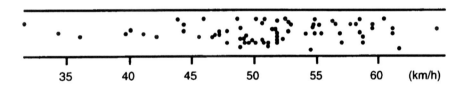

Figure 15.4 Jitter plot of the vehicle speed data

Another approach that is used to present one-dimensional data is to use frequency diagrams, or *histograms*. In these diagrams the data are plotted along the base axis in class intervals of appropriate size. Data that fall into each interval are grouped together, and a rectangle proportional to the number of observations in each interval is drawn along the base axis.

Consider the speed data in Table 15.1. The speeds can be grouped into particular ranges, say, every 2 km/h. Since the highest speed in 64.80 km/h and the lowest is 31.60 km/h, it would be necessary to have 18 class intervals of 2 km/h, starting from 30 km/h. The number of occurrences in each of these class intervals is called the frequency. Table 15.6 shows the frequency table for this situation. The corresponding histogram is shown in Figure 15.5.

Table 15.6   Frequency table for vehicle speed data (km/h)

Class interval	Frequency	Class interval	Frequency
30.0-31.9	1	48.0-49.9	9
32.0-33.9	0	50.0-51.9	12
34.0-35.9	1	52.0-53.9	10
36.0-37.9	1	54.0-55.9	8
38.0-39.9	1	56.0-57.9	6
40.0-41.9	2	58.0-59.9	6
42.0-43.9	1	60.0-61.9	4
44.0-45.9	5	62.0-63.9	0
46.0-47.9	6	64.0-65.9	1

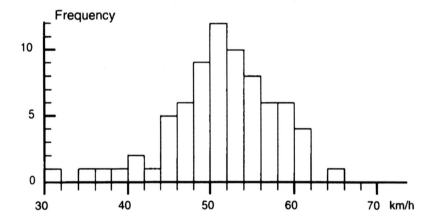

Figure 15.5   Histogram of vehicle speed data using 2 km/h class intervals

The choice of class intervals is important in determining the character of the data. Figure 15.6 shows a histogram that uses class intervals of one km/h for the vehicle speed data. Comparison with Figure 15.5 shows that the second plot is much less regular in shape.

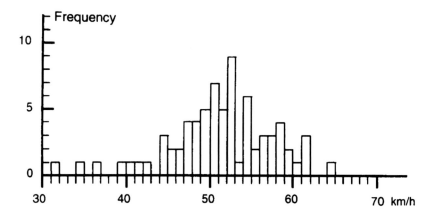

Figure 15.6 Histogram of vehicle speed data using 1 km/h class intervals

A number of important points apply to the use of class intervals:
- when a frequency table has too many class intervals, it may defeat the purpose of the simplification. Generally, it is convenient to have about 10 to 20 intervals. However, even with this number of class intervals, a histogram represents a simplified view of the data since information on individual speeds has been lost. To provide this extra dimension, it is possible to include a jitter plot along the base axis;
- the class intervals need not be equal. Here again there are no fixed rules, and one must use judgement in connection with what is to be presented. For example, when presenting a distribution of income, perhaps most of the values might be between $0 and $50,000 per annum. However, there are incomes in excess of $500,000. Presentation of these incomes, using the same class interval as for the bulk of the population, would result in a rather large base axis and a diminution of the detail on the bulk of the population. If the main point of the study were to look at the bulk of the population, then a category of 'above $50,000' would be advisable, and
- the midpoint of a class interval can be obtained by adding the top and bottom values of the interval and dividing by two. However, if information on the overall distribution of the data is available, more accurate interpolations can be made.

Continuous curve probability distributions may provide theoretical models for comparison with observed histograms. Section 3.3.2 described the use of continuous distributions in modelling such traffic variables as headways and speeds. The theoretical distribution curves of Chapter 3 can be used in the analysis of data samples as well. The total set of observations that may occur is the population. It is often convenient to think of the population as infinite, with the set of observations used in the analysis being termed a sample. For large populations the bumps in the frequency distributions may become very small (e.g. Figure 15.7). If the area of each vertical column (j) is given by the relative frequency (n_j/n, where n is the sample size), then the area under the entire relative frequency distribution is equal to one. This means that, given the class interval width is Δx, the height of each rectangle ($p(y)$) is $n_j/(n\Delta x)$.

The relative frequency distribution is also called the probability distribution, or *probability density function* (pdf). The probability of an event taking place is the area under the relative frequency curve associated with an event. For instance, the probability of obtaining a speed less than 62 km/h is the sum of the probabilities for each class interval below 62 km/h in Figure 15.7. The vertical ordinate is denoted as $p(y)$ (see above) and is called the probability density. Since the population is supposed to be very large, it is possible to consider the distribution as consisting of many small grouping intervals. When the grouping intervals are very small, the distribution can be taken as being continuous. In turn, the continuous curve associated with the distribution can be represented by a mathematical function, as in Section 3.3.2. This function is the probability density function. Just as the discontinuous function (relative frequency) has a total area of one, so does the probability density function. The probability of a particular event is also the area under the curve surrounded by the event boundaries.

An interesting question associated with continuous functions relates to the probability of obtaining a particular value. If the point corresponds to a precise value, then the answer is zero. However, what is usually meant is: what is the probability over a very small range? A comparison of the base histogram and the continuous function for the vehicle speed data set is given in Figure 15.8.

Another method of presenting frequency data is to look at the proportion of the sample greater (or less) than a particular value (Figure 15.9). Presentations in this form are termed *cumulative frequency distributions* and are similar in concept to the quantile plot discussed earlier. The main difference is that the quantile plot does not use grouped data. The cumulative frequency plot can use both grouped data and individual data values. The cumulative distribution can be used to provide, for instance, the number of vehicle travelling below the speed limit. Cumulative distributions of this type are also commonly used in gap acceptance problems.

From Data to Information 331

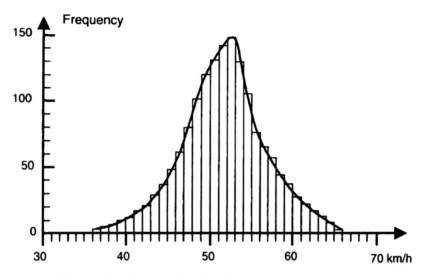

Figure 15.7 Continuous distribution

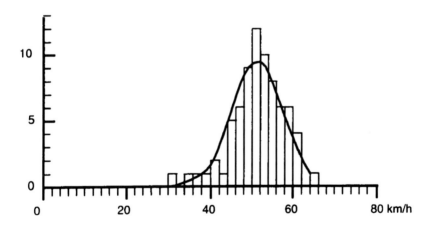

Figure 15.8 Comparison of histogram and continuous function

332 *Understanding Traffic Systems*

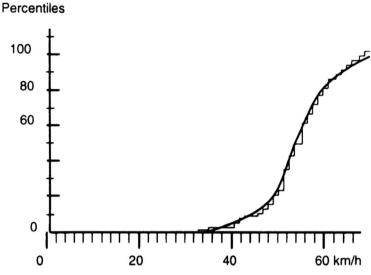

Figure 15.9   Cumulative frequency distributions

While the dot plots present information on every data point and the histogram summarises that data, sometimes it is useful to summarise the information even further. One method of doing this is to draw a *box plot* (Tukey, 1977). Figure 15.10 presents such a plot.

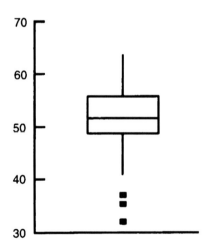

Figure 15.10   Box plot of the vehicle speed data

The upper and lower quartiles of the data are displayed at the top and bottom of a rectangle. The median (the 50th percentile) is marked inside the rectangle. Lines (termed *whiskers*) extend from the ends of the rectangle. The length of the dotted line is determined by first calculating the interquartile range (*IQR*). The upper extremity is the largest observation less that 1.5 times the *IQR*. The lower extremity is defined as the smallest observation greater than or equal to the lower quartile minus 1.5 times the *IQR*. If any data values fall outside the extremities of the whiskers, the individual values are plotted.

The box plot provides a quick impression of certain prominent features. The bulk of the data is seen from the size of the box. The lengths of the dotted lines show how much the tails of the distribution are stretched. The individual plotted points provide the opportunity to consider outliers. [*Outliers* are extreme values of a variable that are either much smaller or much larger than nearly all of the other observations.] The general use o box plots in data analysis is strongly recommended.

The *stem and leaf diagram* (Figure 15.11) is a hybrid between a table and a graph since it uses numerical values, but its profile is very much like a histogram. The diagram is constructed by firstly writing down to the left of a vertical line all possible leading digits in the range of data. Each individual data point is then represented by writing down the trailing digit in the appropriate row to the right of the line. Thus the first column in Figure 15.11 represents the speed 31.6 km/h, while the last column represents the values 47.2, 47.2, 47.7 and 47.8. The stem and leaf diagram is a compact way of presenting the data and at the same time providing a visual impression of the spread of the data.

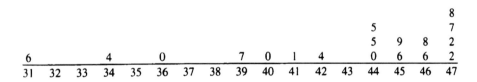

Figure 15.11 Stem and leaf diagram for the vehicle speed data

A *pie chart* is a convenient way of presenting information on the proportions of a particular data set. Figure 15.12 shows a pie chart of the proportion of fatal pedestrian accidents by blood alcohol content. It is possible to see at a glance that pedestrians with no alcohol reading make up a

334 *Understanding Traffic Systems*

large proportion of the total pedestrian fatalities. An approach similar to a pie chart is a bar chart. This chart represents the proportions in each groups by shaded areas along a line. Pie charts, bar charts and similar displays can readily be produced by spreadsheets, amongst other software packages.

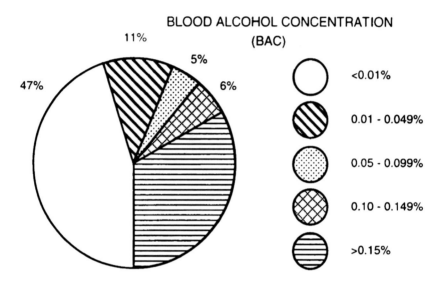

REF: H.S.R.I. (1998)

Figure 15.12   Pie chart of fatally injured pedestrians by blood alcohol content (BAC)

15.2.2 The comparison of one-dimensional data using diagrams

From time to time it may be necessary to compare the same variable over a number of situations. This can be carried out by plotting back to back stem and leaf diagrams, overlapping histograms or overlapping probability density functions (as introduced in Section 3.3.2). Such comparisons provide an idea of the differences between two data sets, before statistical tests are used to quantify those differences. The box plot can be used to extend this comparison to more than two data sets. This may be useful in comparing time series data, e.g. see Figure 15.13.

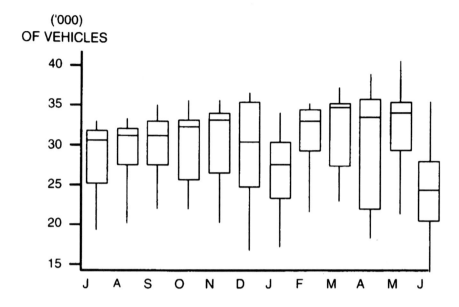

Figure 15.13 Box plot of monthly variations of arterial road flows

Further, the box plot can be extended to provide an indication of differences in the median values by introducing notches into the rectangle (Figure 15.14). The upper and lower points of these notches are determined using the formula

$$Median \pm 1.47\,(IQR)/\sqrt{N}$$

An informal interpretation of the box plot method is as follows: if the notches for the two boxes do not overlap, there is strong evidence that the medians are different; if there is overlap, the evidence points to a similarity in the medians. These conclusions are founded in the concepts of hypothesis testing – see Section 16.2.

15.2.3 Two-dimensional plots

Two-dimensional graphs present the relationship between two variables. The most common two-dimensional graphs use a grid system where the grids are

336 *Understanding Traffic Systems*

at right-angles. Other coordinate systems, such as polar, may be of use in some situations.

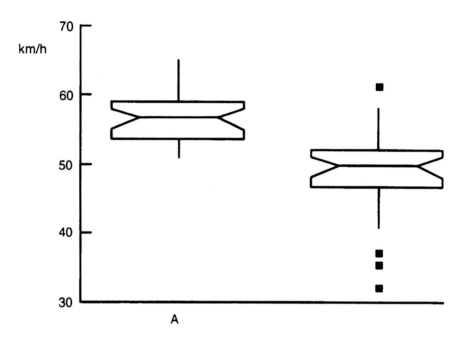

Figure 15.14   Comparison of median speed values using box plots

Graphs, if drawn properly, provide an immediate view of the data. Figure 15.15 is a scatterplot indicating the relationship between vehicle speed and the percentage of local traffic from the data set in Appendix B. There is little relationship between the two variables to be seen. The final relationship is usually, however, presented in the form of an equation, together with the degree of fit of the line (see Chapter 17). One should, however, be wary of an analysis which merely plots a line or gives an equation without providing some indication of the spread of the data. Lines and equations can give false impression of the reliability of a data set.

When preparing a graph, the following information should be provided:
- *origin*. The origin of the axis is the point where the variables plotted on each axis equal zero. If the origin is remote from the points of concern, it is possible to use a new set of axes parallel to the original one. If this is

done, it is necessary to draw the attention of the reader to the fact, otherwise erroneous conclusions can be drawn. Figure 15.16(a) shows what appears to be a considerable decline in the average speed of vehicles over a nine year period. A look at the actual axis scale in Figure 15.16(b) shows that this decline is only marginal. This problem can also occur with the construction of a histogram;

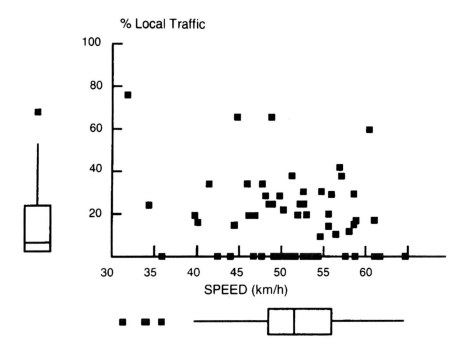

Figure 15.15 **Relationship between vehicle speed and per cent local traffic**

- *graduation of axes.* Axes must be graduated. A statement of scale at the corner of a sheet is no substitute for graduation of each axis. If graduations are too far apart, it will hamper visual interpolation. If they are too close, they may be difficult to read. The incorrect use of scales can also lead to an incorrect interpretation of the results;
- *labelling of the axes.* Axes should be labelled to indicate what physical quantities are represented. The statement of units is equally important. The lack of units would considerably reduce the value of the graph;

338 *Understanding Traffic Systems*

- *title*. The title is a necessary adjunct, which is all too easy to omit. The title should be as brief and to the point as possible, and
- *legend*. The legend, or key (as it is sometimes known), enables the user to specify the character of particular components of the graph.

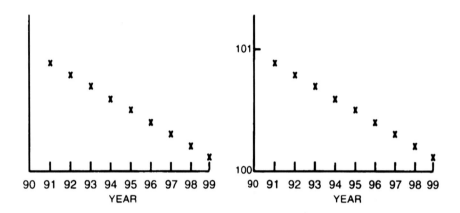

Figure 15.16 Change in average vehicle speed over time with (a) unlabelled and (b) labelled axes

Scatterplots can be enhanced in a number of ways to improve the understanding of the data, for example to:
- include jitter plots along the axes;
- group the data and introduce box diagrams (Figure 15.13); and
- smooth the data points using averaging (Chambers et al, 1983).

15.3 Multi-dimensional plots

Although scatterplots and related displays enable the study of paired comparison data, many serious data analysis problems involve the study of several variables. A number of techniques are available to study such data.

15.3.1 Interactive mapping in data analysis

Maps and diagrams have always been the fundamental communications medium for transport analysts and engineers. Further, transport and traffic data often have distributions over space, the most obvious example being origin-destination data. Means of representing and displaying such data on a map base of the study region are inevitably sought at some stage in an analysis, and this task is well handled by GIS software – as discussed in Section 2.4.

The use of interactive colour mapping on video screen or hardcopy provides an important tool for data analysis, by supplementing the more conventional methods of analysing spatial data. For the purposes of this discussion, a map is defined as a two-dimensional plot showing the shape of a region, and indicating the subareas (zones) into which the area is divided for analysis purposes. The plot shows the value of the mapped variable for each zone, coded into a preset range. Besides the plot, the map contains a title to define its contents, and a key block to relate the values of the displayed variable to the specified ranges. Figure 15.17 provides an example map. GIS software packages provide a full range of spatial data fusion and analysis capabilities, as discussed in Chapter 2.

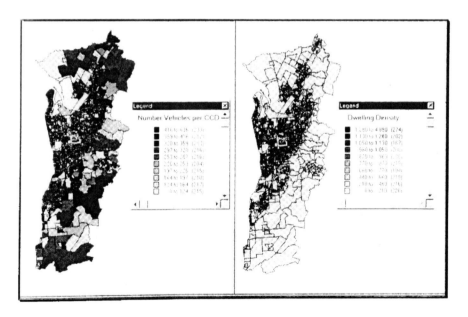

Figure 15.17 GIS maps showing vehicle ownership and dwelling density by zone in an urban area

15.3.2 Three-dimensional data

Several methods are available for the viewing of three-dimensional data. The first is to view the data through a series of paired comparisons (Figure 15.18). This method allows relationships between each of the variables to be viewed and provides an indication of the character of the data in three dimensions. For instance, Figure 15.18 indicates that the displayed data set only presents information on two densities of intersections. This does not allow the general shape of the relationship to be determined. More data are required for a proper study of the effects of intersection density.

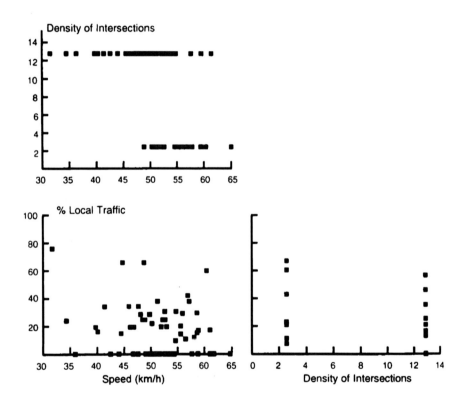

Figure 15.18 Comparison of three variables using pair-wise comparisons

Symbolic and coloured plots can be used to provide an indication of the three-dimensional character of the data. Interactive colour graphics make this approach very useful. Another method is to partition one of the axes (Figure 15.19). The chosen axis is divided into a number of intervals. All the data points that fall into this interval are collapsed onto the same two-dimensional graph. The two-dimensional plots for each interval are then placed side by side, and the analyst gets an idea of the three-dimensional nature of the data by imagining each two-dimensional plot is mounted on top of the other.

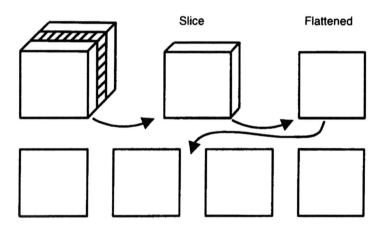

Figure 15.19   Partitioning of one axis of a graph (Chambers et al, 1983)

15.3.3 Three-dimensional graphics

The ability to concisely display the outputs of a data analysis or modelling study can be of great assistance in revealing the results of the exercise and transmitting them to other interested people. Once more than two variables are involved, the normal two-dimensional planar graphics become less effective. Three-dimensional representations are possible using various projection types, and modern computer graphics packages allow the easy application of these representations. Figure 15.20 provides an example of a three-dimensional plot of a dependent variable and two independent variables.

342 *Understanding Traffic Systems*

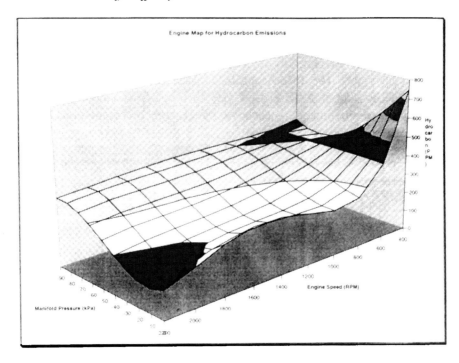

Figure 15.20   Three-dimensional display of a dependent variable and two independent variables (in this case a car engine map for hydrocarbon emissions)

15.3.4 More than three dimensions

Even more complications occur when more than three dimensions need to be viewed. The extension of the paired comparison technique is one such circumstance. Figure 15.21 shows an extension of the three variable comparison, shown in Figure 15.18, to the comparison of four variables. The extra variable is the width of the road. It can be seen that the aggregation of the number of intersections and the road widths into only seven effective points reinforces the need for more data representing variations in these variables. One further approach, of use in comparing more than three dimensions for a number of alternatives, is the *star plot* (Chambers et al, 1983). A star plot represents the values of variables along the axes of a number of polar coordinates. For instance, Figure 15.22 presents six axes that used to compare the performance of a traffic network under different speed limit regimes. This figure only indicates the axes of the plot.

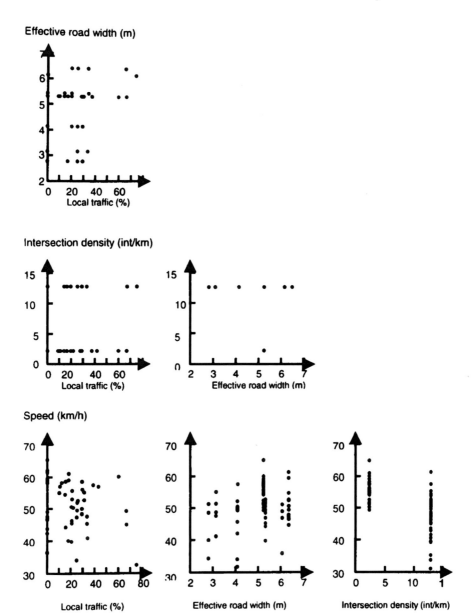

Figure 15.21 Multiple pair-wise comparisons

344 *Understanding Traffic Systems*

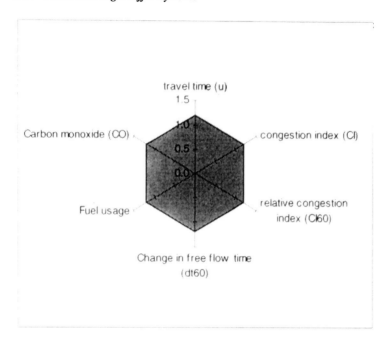

Figure 15.22 Example star plot axes

The six axes radiate at equal angles from the same central point. Along these axes are plotted the performance measures. The length $[z(i,j)]$ along the axis i for variable $x(i,j)$ (used to represent each performance measure) is given by.

$$z(i, j) = (1-c) \frac{(x(i, j) - \min_i \{x(i, j)\})}{\max_i \{x(i, j)\} - \min_i \{x(i, j)\}}$$

where c is a scale variable used to determine the size of the plot.

Comparison of the performance of three alternative speed limits is presented in Figure 15.23. The differences between the speed limits can be quickly seen from the figure. Similar comparisons can be made for different automobile performances, different transport plans, parking lot layouts, etc.

From Data to Information 345

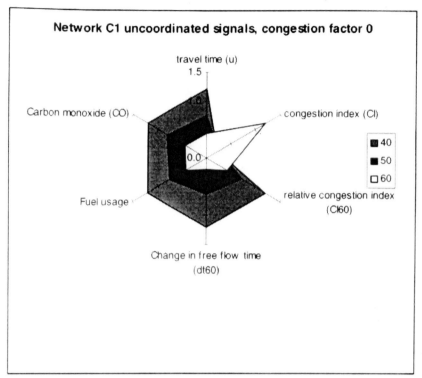

Figure 15.23 Comparison of speed limit performance using star plot

15.4 Descriptive statistics

Descriptive statistical analysis provides the basic numeric characteristics of a data set. Information on the spread, variability and central tendency of each variable assists the analyst to select the appropriate statistical techniques, and often constitutes the basic reference for a study. A number of statistical measures can be defined, to provide indications of central tendency, spread and shape of the distribution of a data set.

15.4.1 Statistics and parameters

The theory of mathematical statistics requires a distinction between population parameters and sample statistics. The measures described earlier in this chapter are all sample statistics. Their use as estimates of population parameters depends on the reliability of the sample data (see Section 15.5) and use to be made of the parameter. The normal convention is for a

346 *Understanding Traffic Systems*

symbolic distinction is made between parameters and statistics by using Greek characters for parameters and Roman characters for statistics.

15.4.2 Measures of central tendency

Several statistics may be used to represent the central tendency ('average') of a variable. The following statistics are often used: mode, median, arithmetic mean and geometric mean. Each statistic has its particular applications.

The *mode* is the value of a data variable that occurs most often. For example, consider the accident frequency histogram shown in Figure 15.24. The modal frequency in this diagram is one, i.e. the most commonly occurring number of accidents per week is one for the data in this figure.

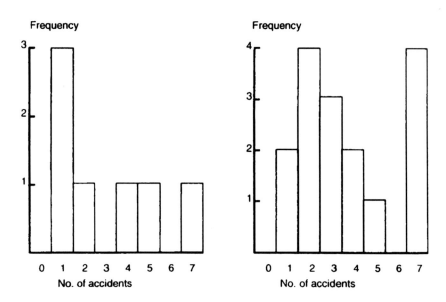

Figure 15.24 **Unimodal histogram of accident frequencies**

Figure 15.25 **Bimodal histogram of accident frequencies**

Samples of data may have more than one mode. This happens when a number of data points occur with the same most common frequency. Figure 15.25 shows such a case. When two modes are present, the distribution is said to be bimodal.; when there are three modes, it is trimodal, etc. The sensitivity of the mode to isolated groups of data makes it an unreliable estimate of central tendency. The mode can, however, be used with all levels of data (see Section 2.3).

The *median* (x_{50}) is the numerical value of the middle data point. It divides the data set into two equal parts. To determine the median, the data are ranked from lowest to highest and the points are numbered from one to n (where n is the number of data points). i.e. values $x_1, ..., x_i, ..., x_n$. The median of a sample with an odd value of n is calculated by taking the data point with ranking $i = \frac{1}{2}(n + 1)$. For a sample with an even value of n, the median is calculated by finding the value of x half way between $x_{1/2n}$ and $x_{1/2n+1}$, i.e.

$$x_{50} = \frac{1}{2}(x_{1/2n} + x_{1/2n+1})$$

For a continuous data variable, the median can be determined using the cumulative percentage to locate the middle data point. The median is equal to the 50th percentile value. Median values are stable, being little affected by outliers, and are thus commonly used as a measure of central tendency. The exploratory data analysis plots described in Section 15.3 made significant use of median values. The median cannot, however, be used for nominal level data (see Section 2.3).

The *arithmetic mean* ($\bar{x}$) is the centre of gravity of a data set. It is the most commonly used measure of central tendency of data at the interval level or above (see Section 2.3). Often referred to as the 'average', it is defined as the sum of the individual data values divided by the number of data values, i.e. as in equation (15.1):

$$\bar{x} = \frac{1}{n}\sum_{i=1}^{n} x_i \qquad (15.1)$$

An alternative method for calculating the arithmetic mean of grouped data (e.g. that in a histogram) uses the frequency of occurrence (f_i) of each data point x_i. Then the following equation may be applied:

$$\bar{x} = \frac{\sum_{i=1}^{n} f_i x_i}{\sum_{i=1}^{n} f_i}$$

A particularly useful application of the mean is the comparison of two sets of data that have a similar distribution. The process used for this test is described in Chapter 16.

The mean is affected by all of the observed data values. For instance, if there are five drivers with speeds 60, 60, 60, 60, 100 then the mean is 68 km/h. The single value of 100 km/h raised the mean speed by 8 km/h. Given only the mean value, it is reasonable to infer that all values are evenly scattered around this value. Hence, in this case, the arithmetic mean is not an adequate representation of the distribution. On the other hand, the median value of these five speeds is 60 km/h.

The arithmetic mean should not be used for nominal or ordinal level data (see Section 2.3).

Another definition of the mean that is occasionally used in traffic studies is the *geometric mean (GM)*. This is useful in determining the average growth rate over time. It is given by the expression

$$GM = \sqrt[n]{x_1 x_2 x_3 \cdots x_n} \quad)$$

15.4.3 Spread of the distribution

Measures of central tendency do not provide much information about the spread of a sample of data, and other measures may be introduced to help in this regard. The measures include the *minimum, maximum* and *range* of the data, the *variance* of the data, the *standard deviation* and the *standard error*.

A first indication of the distribution of a data variable can be found from the minimum, maximum and range. The minimum and maximum values are found by inspection. The range is the difference between the maximum value and the minimum value. When compared to a measure such as the mode, median or mean, the range can indicate the possible presence of outliers in a data set.

The variance of a data set provides a measure of its dispersion about the mean value. Mathematically, the *population variance* (σ^2) of a data set with mean value μ is defined as

$$\sigma^2 = \frac{1}{n} \sum_{i=1}^{n} (x_i - \bar{x})^2$$

while the *sample variance* s^2 of a data set is given by

$$s^2 = \frac{1}{n-1} \sum_{i=1}^{n} (x_i - \bar{x})^2 \qquad (15.2)$$

The $(n-1)$ term in equation (15.2) is related to the concept of degrees of freedom. Since the deviations about the mean of the n observations must sum to zero (see equation (15.2)), then only $(n-1)$ independent deviations can exist. The n data points and the variance thus have $(n-1)$ degrees of freedom. Equation (15.2) may be rewritten as

$$s^2 = \frac{1}{n-1} \left\{ \sum_{i=1}^{n} x_i^2 - n\bar{x}^2 \right\}$$

which is often a more convenient relation to use for computations.

Variance plays an important part in many statistical tests and procedures. Indeed, one of the chief goals of data analysis is to explain 'variance'. Usually this means determining those variables which account for (explain) variance in other variables (see Chapters 16 and 17).

The standard deviation s (sample statistic) or σ (population parameter) is another commonly used measure of dispersion about the mean. It is defined as the square root of the variance. The advantage of the standard deviation is that is has a more intuitive interpretation than the variance, as it has the same units of measurement as the original variable.

An important measure of variation in sample means is the *standard error*. This measures the potential degree of discrepancy between sample means, or a sample mean and a population mean. It is used in some tests of statistical significance and in estimating confidence levels (see Chapter 16).

Proper determination of the standard error is based on the existence of an infinite set of equally sized samples drawn from a given population. The mean of each sample would be an estimate of the true population mean, but these sample means would not be identical. The standard deviation of the sample means is the standard error. An estimate of the standard error (se) of the mean of a variable is given by:

$$se_{\bar{x}} = \frac{s}{\sqrt{n}}$$

The measures of dispersion described above provide absolute measures of the dispersion in a data set. The *coefficient of variation* provides a measure of the relative variation in a data set. It is defined as the ratio of the standard deviation to the mean value, i.e. $s/\bar{x}$ or σ/μ.

15.4.4 Shape of the distribution

The shape of the distribution of a variable about its mean value is often of importance. In particular, the degree of symmetry in the spread of the distribution may be an important consideration. If a data set is not symmetrically distributed, it is said to be *skewed*. The direction and magnitude of the skew is then important.

A comparison of the three measures of central tendency (mode, median and mean) is useful is suggesting an indication of the degree of *symmetry* of a distribution. When the three measures coincide, then the data may well be symmetric. The bell-shaped curve or normal distribution (see Figure 15.26(a)) is of this type. If the mean falls to the right of the median which in turn falls to the right of the mode, then the distribution is skewed to the right (see Figure 15.26(b)). The opposite situation indicates a skew to the left (see Figure 15.26(c)).

Skewness is a statistic that measures deviation from symmetry for interval and ratio level variables (Section 2.3). It is used to determine the degree to which an observed distribution approximates a bell-shaped (normal) distribution. The coefficient of skewness (*SK*) is defined by equation (15.3), in which s is the sample standard deviation for the data set.

$$SK = \frac{1}{ns^3} \sum_{i=1}^{n} (x_i^3 - \bar{x})^3 \tag{15.3}$$

SK takes a value of zero when the distribution is completely symmetrical. A positive value indicates a skew to the right, and a negative value indicates a skew to the left.

Kurtosis provides a measure of the relative peakedness or flatness of the distribution. The kurtosis coefficient K is defined by equation (15.4):

$$K = \frac{1}{ns^4} \sum_{i=1}^{n} (x_i - \bar{x})^4 \tag{15.4}$$

The bell-shaped (normal) curve will have a kurtosis coefficient of three (Figure 15.27(a)). A value of K greater than three indicates that the distribution is more peaked ('lepto-kurtosis') than the normal curve (Figure 15.27(b)). A value less than three indicates that the distribution is flatter ('platy-kurtosis'), see Figure 15.27(c). Some statistical packages subtract three from the K value computed using equation (15.4), so that the bell shaped curve of the normal distribution will have a kurtosis value of zero.

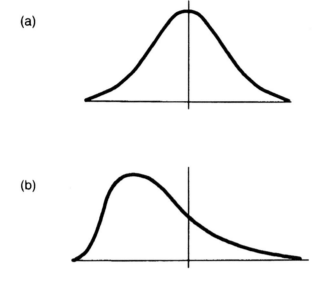

Figure 15.26 Symmetry of a distribution:
(a) bell-shaped curve (b) right skew (c) left skew

15.4.5 Proportions

The previous descriptive statistics are for measured data variables. In addition, proportions of observations in selected ranges are often of importance.

For example, in a road safety study, the proportion (p) of drivers who have a blood alcohol concentration (BAC) above 0.05 might be of interest. This proportion would be given by

$$p = (Number\ of\ drivers\ with\ BAC > 0.05) / n$$

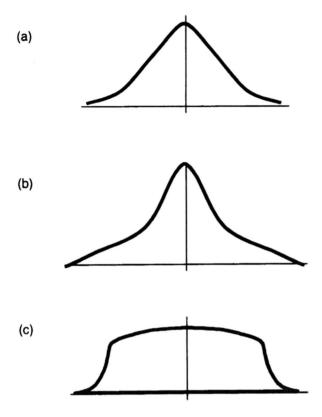

Figure 15.27 Peakedness or flatness of a distribution: (a) bell-shaped curve (b) lepto-kurtosis (c) platy-kurtosis

where n is the total number of drivers surveyed. This proportion can then be used to represent the proportion of drivers with BAC exceeding 0.05 in the total population, noting that it is only a sample statistic. The standard deviation of the population can be calculated by dividing the population into those with a BAC over 0.05 (p) and those with a BAC of 0.05 or less ($1 - p$). The standard deviation (s_p) of the proportion is then given by the binomial distribution and has the result $s_p = \sqrt{p(1-p)}$. This expression can be used to provide an estimate of the standard error in the sample proportion (se_p), i.e. as in equation (15.5)

$$se_p = \sqrt{\frac{p(1-p)}{n}} \qquad (15.5)$$

16 Statistical analysis

This chapter presents an overview of statistical analysis procedures relating to traffic analysis. In this way it takes the process of traffic data analysis to a further stage, beyond mere description of a data set, to introduce the opportunities for inference of effects, impacts and interactions between observed variables. The chapter starts by considering the nature of possible faults in the data. It then considers hypothesis testing and non- parametric statistical testing, as the tools by which statistical inference can be made.

16.1 Faulty data

Good statistical analysis is little use if the existence and nature of errors in a data set have not been considered. Variations in observations due to disturbing factors are called experimental errors. Such errors are common, and if they are not systematic are of small concern, for there are well developed methods to cope with them. Systematic errors are, however, of considerable concern. One of the most common sources of systematic error in analysis is not the statistical techniques used, but rather the data they are used on. Sometimes it may be the measuring instruments that are at fault, but more often than not, it is the experimental or sample design that provides the wrong information (see Section 7.3).

Sample designs should be set up as to avoid systematic bias in the selection procedure and to achieve maximum precision for a given outlay of resources. Unreliability in the selection of the data can arise from:
- *non-random selection*, where the sample is influenced by human choice, and each element of the population does not have an equal chance of being chosen;
- *non-representatives* of the population used to obtain the sample, and
- *difficulty in obtaining all* the *information* on all the sample population, which may result from refusal of people to answer the questions asked, or from a difficulty in locating and studying certain groups within the same population.

These factors are likely to cause systematic, non-compensating errors (bias), which can not be eliminated or reduced by an increase in sample size.

The unreliability of data can also influence models, particularly if inappropriate experimental designs are used. Some possible problems are:
- *false projection*. Prediction is often based on historical data but it is possible that events occurring in the future will be outside the range of the values used to develop the model in the first place, and
- *the influence of extraneous factors*. When experiments have to be conducted in situations where a number of external factors may influence the factor of concern. The magnitude of the effect of the extraneous factors must be recognised and determined. Experimental design in such situations may be unavoidable.

16.1.1 What to do about data errors

Many errors will be unknown to the analyst, and therefore, will be overlooked until they are perhaps found at a later date. Other errors will be known to the analyst, but there may be little that can be done to rectify them. Such errors should be recognised in reports associated with the analysis, so that others are aware of them and may be able to rectify them in later work. Those errors that can be rectified should be rectified.

A typical fault in some data sets is the presence of outliers. Outliers (see Section 15.2.1) are data points that do not appear to be consistent with the general body of the data. Outliers should not simply be ignored. A provident analyst will return to the original data set and see if there has been an error in recording it. If so, the data point can be neglected or corrected. If not, the analyst should carry out the analysis with, and without the outlier. If the outlier has an overly great influence on the result, this should be noted.

Another typical fault is missing data. Two aspects are of concern here. The first relates to specification. It is often necessary to record a code for missing data. If a blank is used, it must be possible for the computer analysis procedure to discern the 'blank' from a zero, particularly if the zero has a specific meaning. The second relates to the analysis. The treatment here depends on the circumstances. In a study of a particular variable (e.g. speed), the absence of one data point is unlikely to have an impact and may be neglected. In a study in which all the recorded values are related (e.g. a queuing study recording arrivals and departures), it may be necessary to discard an entire set of data if one value is missing. Finally, if a number of variables are collected for a sample and one variable is not defined for one member of the sample, the data associated with this member cannot be used in the determination of relationships.

16.1.2 Data editing

The possibilities of data errors mean that the analyst's first step should be to examine the new data. Mis-recordings and palpable errors may be found by applying logical checks on the recorded data items. This process is to ensure that the data taken into the analysis are legible, distinct, complete and consistent. Data records which do not meet these specifications should be discarded. Once the data have been screened, they may then be validated by checking some results against those from other sources, such as the national census or transport study data. In this way the degree to which the data represents its parent population may be assessed, at least in part. On completion of preliminary data analysis, it may be useful to check that the survey results are consistent with the assumptions made in the analysis. Ranges of data values and the relative dispersion within the range may be assessed (as discussed in Section 15.4). Analysis of parts of the data set, split by some basic variable (e.g. occupation group, time of day, urban/rural), can be made. Comparisons between these sub-populations may be useful in assessing the validity of the coded data.

Errors in data recording and transcription happen only too often. The analyst must be prepared to subject a new data set to a thorough examination before drawing any inferences from it. The methods provide the basis of and the tools for this examination. There are three major areas for attention: errors in the permissible ranges of data items; logic and consistency checks, and missing values.

Errors in recording and typing may lead to obvious errors, where the code value for a data item falls outside the range of codes used – these are *permissible range errors*. For example, a value of '1' (one) may be mis-read as a '7'(seven). If the range of codes was 0/1 (say, for 'No' and 'Yes'), then the seven is outside the allowable range. These errors are best resolved by comparing the data set with the original recording sheets.

Logical errors arise from valid responses to individual data items. This may arise when the logical relationship between two data items for the same observed unit is not correct - for example, axle loads may be recorded for three axles of a truck, when the truck is identified as having only two axles. Cross-tabulations may be used to identify such errors (see Section 15.1). When such errors are flagged, the data should be checked against the original observations. If one of the responses has been wrongly recorded, the error is then easily rectified. If the data are consistent with the original records, then an error in observation has probably occurred. An examination of the other items recorded for that unit may indicate which of the inconsistent items is likely to be correct.

Sometimes improbable events may be observed. For example, a very fast vehicle may be seen, or a foreign registration plate may be spotted. In such cases, observers might be encouraged to note a remark to this effect on the data recording sheet. This requires some care in design of the recording sheet, to give observers the opportunity of clarifying the event.

There are two types of errors involving *missing data*. Firstly, an observation may be completely missed (e.g. a vehicle is obscured, or the observer's attention is distracted by some other event). Secondly, one or more of the data items about the observed unit may be missed. Some data items may be harder to record than others, e.g. simultaneous recording of a vehicle registration plate and the number of occupants of the vehicle. When full observations are missed, corrections can only be made on the basis of weighting the observed data in terms of some known characteristic of the population. For example, the age distribution of motor vehicles may be available, and could be used to check a sample of observations. Missing individual items of data may be treated in one of three ways:

- *ignore the missing values* and report the distribution of each variable in terms of the number of complete observations taken for the variable;
- *report the proportion of missing values* for each variable, so that results are based on the total number of observations made. In this case the 'missing value' becomes another code or value for the data item. Care is needed to exclude this code/value from any further data analysis, or
- *estimate a probable value* of the missing item, based on the information available from the other data items for the observation, and the distribution of observed values for the item.

Whichever option is chosen, the analyst must be aware of the possibility that, if the missing values were systematically different from those that were successfully observed, the resulting data set will be biased. Consider, for example, the possibility that in a speed survey, the observer was unable to record the speeds of the fastest vehicles.

16.2 Hypothesis testing

The need for clear statement of hypotheses must be stressed. What we mean by an hypothesis is some assumption about a population parameter: hence hypothesis testing involves examining a sample of data drawn from the population to see if these data will support the assumption. To test an hypothesis for a parameter (such as a mean value), we compare the sample estimate and the hypothesised value. The smaller the difference between the two, the greater the likelihood that the hypothesised value is correct. Sometimes the difference will be so large that inspection will automatically reject the hypothesis; sometimes the difference will be so small that we would accept the hypothesis with no further ado. In other cases, the results are inconclusive, and no clear cut decision to reject or accept the hypothesis will emerge. The statistical inference techniques developed for hypothesis testing provide a valuable framework, for assessing the validity of a given assertion.

16.2.1 Selecting the hypothesis

Before we dash out and collect some sample data, we must explicitly define the hypothesis to be tested. Then we select a suitable test and test condition on which to accept or reject the hypothesis, gather the data, and conduct the test. Note that the statistical rejection of an hypothesis is to assert that it is false, while the acceptance of an hypothesis is to imply that there is no conclusive evidence on which to reject it. [A useful parallel, perhaps, is the Scottish verdict 'not proven', rather than 'not guilty'.] Consequently, the analyst should always define the null hypothesis (H_0) to be the hypothesis which, it is hoped, will be rejected. The null hypothesis is statistically compared to an alternative hypothesis (H_1), and rejecting H_0 leads to the acceptance of H_1. For example, when testing two means (μ and μ_0) for equivalence, the null hypothesis could be

$$H_0 : \mu = \mu_0$$

The alternative hypothesis H_1 may be defined in a number of ways, depending on the purpose of the comparison. Some possibilities are:

$$H_0 : \mu > \mu_0 \tag{16.1}$$

$$H_0 : \mu < \mu_0 \tag{16.2}$$

$$H_0 : \mu \neq \mu_0 \tag{16.3}$$

However, only one definition of H_1 can be used in any particular test. The alternative definitions of H_1 given by relations (16.1) and (16.2) are one-tailed tests, for they imply a direction to the assertion contained in H_1. The definition of H_1 in relation (16.3) is a two-tailed test; the hypothesis is that μ is not equal to μ_0, with no assertion as to the sign of the difference between μ and μ_0. Though the testing procedures are the same for one- and two-tailed tests, some care is needed to distinguish between the two in the details of the analysis (e.g. when looking up values in statistical tables). An example may help to illustrate these principles. If we wished to test the difference in mean speeds between two sites (A and B) on a residential street, and we knew that the mean speed at A was 49 km/h, and if the hypothesis was that the mean speed at B was not 49 km/h, then the null hypothesis (two tailed) would be

$$H_0 : \mu_B = 49$$

and the alternative hypothesis H_1 would be

$$H_1 : \mu_B \neq 49$$

On the other hand, if we wished to test the proposition that μ_B exceeded 49 km/h, the null hypothesis would be one tailed and we would write the alternative hypothesis as

$$H_1 : \mu_B > 49$$

In each case the aim in testing would then be to reject H_0 on statistical grounds, and thus suggest that there was some difference between the mean speeds from the two sites. If the analysis accepted the null hypothesis, we say that we have no evidence to suggest that speeds at the two sites were not the same (the double negative is intentional).

16.2.2 Principles in hypothesis testing

The use of sampled data as input to the testing procedure means that a strict deterministic interpretation of the stated hypotheses is not possible. Random fluctuations may be expected in the sample estimates of the parameters. Even though the sampled mean speed μ_B might be 47 km/h in the example in Section 16.2.1, there is some probability that the true (population) mean speed at B is greater than 49 km/h. We need to test this probability to see to what degree the sampled result has occurred purely by chance, or whether or not its occurrence is improbable if H_0 is indeed true. What we have to do is base our decision on whether or not μ_B fell within a specified critical region ($\pm \delta x$) around μ_A. In more general terms we say that if $\mu_0 - \delta x < \mu < \mu_0 + \delta x$ then we would not reject the null hypothesis ($H_0 : \mu = \mu_0$) in a two-tailed test. Similarly, if $\mu < \mu_0 + \delta x$ then we would not reject the null hypothesis H_0 in favour of the (one-tailed) alternative hypothesis ($H_1 : \mu < \mu_0$).

The definition of the critical region δx is somewhat arbitrary, and is made on the basis of providing a workable criterion for the particular test (what is a 'significant difference'?). The smaller the size of the critical region, the more likely we are to reject the null hypothesis. A guide to the optimum size of the critical region can be obtained by considering the consequences of making a decision from the results of the test. Four possible consequences are possible, as displayed in the decision table shown in Table 16.1.

Table 16.1   Possible consequences in hypothesis testing

	True state	
Decision	H_0	H_1
Accept H_0	CORRECT	Type II error
Accept H_1	Type I error	CORRECT

Thus if H_0 is indeed true and we accept H_0 as being the description, we make a correct decision; similarly if we reject H_0 when the true state is defined by H_1, then we are also correct. The more interesting possibilities occur when we make a wrong decision, and it is here that some guidance for setting the size of the critical region will be found. What are the consequences of an incorrect decision? If we reject H_0 when it is actually true, then we call this a 'Type I error'. A 'Type II error' occurs when we accept H_0 when it is not true. The broad aim in hypothesis testing is to minimise the chances of making either a Type I or a Type II error, and the particular error type we try to avoid depends on the relative consequences of making either type of error.

Normally, we express a maximum probability of making each type of error as a statement of the desire to avoid the two error types. The acceptable probability of committing a Type I error is the level of significance of the test (denoted by 'α'). The acceptable probability of a Type II error is denoted as 'β' and the power of the test is defined as $(1 - \beta)$. It is only possible to explicitly define β if the alternative hypothesis H_1 is expressed as an equality, i.e. as $(H_1 : \mu = \mu_0 + \delta x)$. Thus we usually select a value for the level of significance (α), such as 'one in twenty' ($\alpha = 0.05$) or 'one in one hundred' ($\alpha = 0.01$), and try to assess the reasonableness of the power of the test, if this is possible.

Figure 16.1 may help to clarify these relationships. This shows the distribution of sample means obtained by repeated sampling from a population. Figure 16.1(a) illustrates the case where the null hypothesis (H_0) is correct. The estimate of μ at our disposal is the sampled value (X). A given proportion (α) of the sample estimates of μ will exceed ($\mu_0 + \delta x$), and would then lead us to reject H_0, even though it is true, thus leading to a Type I error (on 100α per cent of occasions).

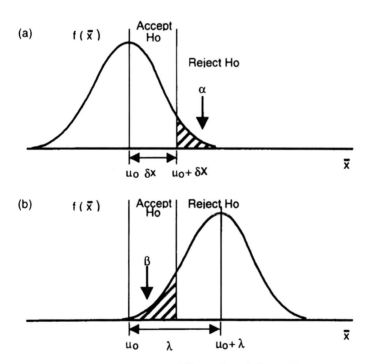

Figure 16.1 Possibilities of Type I and Type II errors

Figure 16.1(b) shows the situation where the specific alternative hypothesis H_1 is correct, in this case ($H_1 : \mu = \mu_0 + \lambda$). The sampled estimates of the mean will be distributed about ($\mu_0 + \lambda$), and if we consider the same critical region δx about μ_0, then on 100β per cent of occasions the sample estimate will be less than ($\mu_0 + \delta x$), and we will accept the null hypothesis when it is false and a Type II error will result. Figure 16.1 illustrates one further result. If we change the size of the critical region, then the values of α and β will change. Increasing δx reduces α but also increases β. This means fewer Type I errors but more Type II errors. Alternatively, decreasing δx increases α but reduces β. It is not possible to simultaneously reduce both α and β by altering δx; rather, there is always a trade-off between the frequency of the Type I and Type II errors. It is possible to reduce β while keeping the proportion α constant by increasing λ but obviously the test procedure is less useful the larger λ becomes.

The only simple way to simultaneously reduce both α and β is to increase the sample size. Increased sample size will reduce the 'standard error' (the dispersion of the sample means about the true mean), and thus reduce the area under the curve lying in the tails of each distribution. In theory, it is possible to use an alternative approach of searching for a better statistic by which to judge the difference. In basic terms the 'best' statistic is that which possesses the smallest variance when the sample size is fixed. This is the reason that the mean is commonly used as the measure of central tendency, instead of other statistics, such as the median or the mode, which also reflect central tendency in a distribution. At the same time the chosen statistic must be responsive to the quantity to be measured or assessed. This leads to the concept of 'efficacy'. The quantitative details of this process are beyond the scope of the present discussion, and the interested reader will find the information elsewhere (e.g. Walpole and Myers, 1988). The analyst can control the level of significance α in a statistical test of an hypothesis, while the value of β depends on the sample size available and the selection of the value of λ. The ability to choose an appropriate value of the level of significance permits statistical testing of hypotheses to be conducted in an orderly and useful fashion.

16.2.3 Testing the difference in means

The mean, standard deviation and standard error relate to a probability distribution for a given variable. These descriptors are the basic information used in testing the typical hypotheses about the equivalence of means drawn from different populations or samples. The null hypothesis we are concerned with when testing the difference between two means (say μ_0 and μ) may be taken as $H_0 : \mu = \mu_0$, i.e. the null hypothesis asserts that the two means are in fact the same, while the alternative hypothesis H_1 may be of any of the forms described in relations (16.1) to (16.3), depending on the particular purpose of the investigation. The actual tests to be conducted depend on the level of information available to the analyst, and draw heavily on the differences between population parameters and sample estimates of those parameters. Sample estimates are just that. They are values inferred about a population on the basis of observations of a small subset (the sample) of members of that population. Sample estimates have an inherent uncertainty about them, due to random variations in characteristics resulting from the sampling process. A population parameter such as a population standard deviation, on the other hand, is a precise value.

Much of the testing process for means depends on the concept of the 'standard error' of a mean value (see also Section 15.4.3). This standard error s_μ for a mean μ may be written as

$$s_\mu = \frac{\sigma}{\sqrt{n}} \qquad (16.4)$$

where n is the sample size, and σ is the population standard deviation. A further result of fundamental importance in hypothesis testing is the observation that the variance of the sum or difference of two independent random variables, x_1 and x_2 is equal to the sum of their variances, i.e. if random variables x_1 and x_2 have respective variances σ_1^2 and σ_2^2 then the variances of $y = x_1 + x_2$ and $z = x_1 - x_2$ are both $(\sigma_1^2 + \sigma_2^2)$. Consequently, the standard error of y or z is given by

$$s_\mu = \sqrt{\frac{\sigma_1^2}{n_1} + \frac{\sigma_2^2}{n_2}} \qquad (16.5)$$

where n_1 is the size of sample 1, and n_2 is the size of sample 2.

For *large sample* sizes (typically $n > 30$), then we can use the properties of the normal distribution to test a difference between two means, in terms of the population or sample standard deviation. If it is available, then the population standard error of the mean (s_μ) should always be used. If the population standard deviation is not known, but we have two sample estimates (e.g. from samples 1 and 2 above), then we estimate a value of the standard error from equation (16.5). Then we form the z-statistic,

$$z = \frac{\bar{x}_1 - \bar{x}_2}{s_\mu}$$

where z is a standard normal deviate, i.e. a value from a normal distribution with mean zero and unit variance. Values of z are tabulated in statistical tables (see Appendix A, Table A.1). The significance of the computed value of z may be gauged by comparing it with the values shown in that table. The five per cent significance level (two-tail test) for the value of z is 1.96, so a value of z greater than 1.96 is likely to occur by chance less than once in twenty samples. A value of z exceeding 2.58 is likely to arise by chance on less than one occasion in a hundred. So if the computed z is greater than 1.96 (2.58), we could say that there was a significant difference between the means, at the five (one) per cent level.

An example may help. Consider a set of speeds gathered on a section of rural highway. Speed observations were taken for the two directions of flow. On the basis of a sample of 158 vehicles, east-bound traffic had a mean speed of 83.3 km/h with a standard deviation of 7.8 km/h. For the west-bound flow, the mean speed of 100 vehicles was 86.7 km/h, with a standard deviation of 9.4 km/h. Were the mean speeds for the two directions of flow significantly different?

The null hypothesis H_0 (two tailed) is then $H_0 : \mu_E = \mu_W$ where μ_E and μ_W are the mean speeds for the two flow directions. The alternative hypothesis H_1 is $H_1 : \mu_E \neq \mu_W$.

Firstly, find the pooled estimate of the standard error, assuming that the two sets of speeds are samples from a common population. This is done by using equation (16.5), and yields a pooled estimate of the standard error of the mean speed of

$$\sqrt{\frac{(7.8)^2}{158} + \frac{(9.4)^2}{100}} = 1.13 \text{ km/h}$$

The observed difference in mean speeds is (86.7 - 83.3) = 3.4 km/h so that z = 3.4/1.13 = 3.00. This z-statistic indicates that the difference in the mean values is three times the standard error. Further, it exceeds the one per cent significance level ($z = 2.58$, see Table A.1), so that we can reject the null hypothesis that there is no difference in the directional mean speeds. The result is statistically significant at the one per cent level, which means that it could have arisen by chance less than one time in a hundred.

Small samples are those containing less than 30 observations. When we have such samples, we must use a different technique to that described above if we do not know the population standard deviation. In the (unusual) event that we do know the population standard deviation, then the method described for large samples would still apply. Thus if the population standard deviation is known, we use the population parameter to find the standard error of the mean, and then compute the normal deviate z and consult the tables for the normal distribution (e.g. Table A.1).

Commonly the population standard deviation will not be available. We use the small-sample approximation to the normal distribution, known as the t-distribution. Tables of the t-distribution may be found in any statistics text (e.g. Table A.2). The main reason for moving to the t-distribution is that the assumption in equation (16.6) that the sample variances are the best estimators of the population variance - no longer holds. Instead, we need to compute a pooled estimate of the population variance (σ_p^2), given by

$$\sigma_p^2 = \frac{(n_1 - 1)\sigma_1^2 + (n_2 - 1)\sigma_2^2}{n_1 + n_2 - 2} \quad (16.6)$$

from which the estimated standard error of the mean is

$$s_\mu = \sqrt{\sigma_p^2 \left(\frac{1}{n_1} + \frac{1}{n_2} \right)}$$

The t-statistic is then formed from

$$t_m = \frac{\bar{x}_1 - \bar{x}_2}{s_\mu}$$

where m is the degrees of freedom for the t-distribution, and is given by

$$m = n_1 + n_2 - 2$$

Consider the following example: 14 vehicles were observed crossing a speed control device on a local residential street. The mean speed on approach to the device was 47.9 km/h, with standard deviation 3.4 km/h. At the device, the observed mean speed was 26.7 km/h (standard deviation 5.0 km/h), and 50 m downstream from the device the mean speed was 43.2 km/h (standard deviation 56 km/h). Was the reduction in mean speeds on leaving the device significant (i.e. did the device cause drivers to maintain their lower speeds beyond the device)?

The null hypothesis is [H_0 : μ(upstream) = μ(downstream)], with the alternative hypothesis H_1 given by [H_1 : μ(downstream) < μ(upstream)], i.e. a one-tailed test. Using equation (16.6), the pooled estimate of the standard deviation of the speeds is s_p = 4.63 km/h, and the standard error of the mean is 1.75 km/h. The computed value of t, with (14 + 14 − 2) = 26 degrees of freedom, is thus

$$t_{26} = (47.9 - 43.2)/1.75 = 2.69$$

The five per cent value of t_{26} (one tailed) is 2.056 (see Table A.2), which is less than the computed value of 2.69. The one per cent value is 2.779 (Table B.2), which is greater than 2.69. Thus the null hypothesis H_0 could be rejected at the five per cent level but not at the one per cent level. The result could have occurred by chance less than once in twenty trials, but more than once in a hundred, if the null hypothesis were true.

The test described above is for independent samples. An alternative method employing the t-test is available when the data observations are paired (e.g. 'before' and 'after' observations from a particular treatment). One advantage of *paired sample* testing is that it reduces the influence of extraneous variables, such as the inherent variability between members of a population (see also Section 7.3.1).

If u_i and w_i are the observed values 'before' and 'after' the treatment, then the paired difference variable, $x_i = u_i − w_i$ is formed. The population mean of the x_i is μ_d, while the sample mean and standard deviation are d and s_d respectively. The statistic $t = (d − \mu_d)/s_d$ has a t-distribution with $(n − 1)$ degrees of freedom, where n is the number of pairs. The sample standard deviation is given by

$$s_d = \frac{1}{n}\sqrt{s_u^2 + s_w^2 - \frac{2}{n-1}\left(\sum_{i=1}^{n} u_i w_i - n\overline{u}\overline{w}\right)} \qquad (16.7)$$

and the term in parenthesis in equation (16.7) is the covariance between u and w. For the method to be useful, the covariance should be positive. This will decrease the variability of the statistic. If the covariance is negative (i.e. u and w are negatively correlated), then the procedure is ineffective. The null hypothesis (H_0) is that μ_d is equal to a specific value (typically $\mu_d = 0$).

16.2.4 Confidence limits

So far the discussion has considered the sample estimate of the population mean as a single value (i.e. a *point estimate*). An alternative approach is to consider an *interval estimate* of the parameter. This estimate is known as the confidence interval. With $100(1-\alpha)$ per cent confidence, the actual value of the mean lies in the range $\mu - \delta x < \mu_0 < \mu + \delta x$, where μ is the sample estimate of the parameter and δx is the calculated range of the estimate at the significance level α (using a two-tailed test). The sample rules as before apply for treating large and small samples. In the small sample case ($n < 30$), δx is found by selecting a significance level and then finding the corresponding value of $t_{m\alpha}$, where m is the number of degrees of freedom. The value of $t_{m\alpha}$ is found by consulting statistical tables. Then δx is given by

$$\delta x = t_{m\alpha} s_\mu \qquad (16.8)$$

so that the confidence interval on the estimate is $\mu - t_{m\alpha} s_\mu < \mu_0 < \mu + t_{m\alpha} s_\mu$. In other words, we are $100(1-\alpha)$ per cent sure that the real value of μ_0 lies within $\pm t_{m\alpha}$ standard errors of the sample estimate μ.

By way of example, return to the case of the vehicle speeds 'upstream' and 'downstream' of the 'slow-point' speed control device, as used in the section on testing differences between means for small samples. Here 14 vehicles were observed, and the upstream mean speed was 47.9 km/h. The standard error of estimate of the mean speed was estimated as 1.75 km/h, based on two samples of 14 vehicles each. (See Section 16.2.3 for the method of estimation, using equation (16.6).) The 99 per cent confidence interval can then be computed, given that $t_{26, 0.01} = 2.779$ (Table A.2). Using the result of equation (16.8), the calculated range δx is given by

$$\delta x = t_{26, 0.01} s_\mu = 2.779 \times 1.75 = 4.86 \text{ km/h}$$

Thus the 99 per cent confidence interval for the mean speed is 47.9 ± 4.86 km/h, or (43.04, 52.76).

For large samples ($n > 30$), the normal deviate $z(\alpha)$ is used instead of $t_{m\alpha}$. Had the speed data used above been extracted from a large sample, then the 99 per cent confidence interval would have been given as follows. From Table A.1, $z(0.01) = 2.58$, so that $\delta x = 2.58 \times 1.75 = 4.52$ km/h. Thus the 99 per cent confidence interval would be 47.9 ± 4.52 km/h, or (43.08, 52.12) - a slightly tighter band width than that found for the small sample case.

16.2.5 Testing the difference in dispersion

So far, our comparisons and tests have considered measures of central tendency (the mean values) only, and we have made some implied assumptions about the measures of dispersion (or variability about the mean), i.e. the standard deviation or the variance of a distribution. In the same way that we can test the difference between two means, so there are methods for comparing two estimates of a population variance. The particular test is known as the F-test. Given two estimates of variance σ_1^2, drawn from a sample of size n_1, and σ_2^2, drawn from a sample of size n_2, where $\sigma_2^2 < \sigma_1^2$ then the ratio

$$F(n_1-1, n_2-1) = \frac{\sigma_1^2}{\sigma_2^2} \qquad (16.9)$$

has an F-distribution with ($n_1 - 1$) degrees of freedom in the numerator and ($n_2 - 1$) degrees of freedom in the denominator. A table of the F-distribution is presented in Table A.3 of this book. A null hypothesis and its alternative are set for the test e.g.

$$H_0 : \sigma_1^2 = \sigma_2^2$$
$$H_1 : \sigma_1^2 > \sigma_2^2$$

which is a one-tailed test. It is necessary to compute the F-statistic as a value greater than unity, as the tables of the F-distribution are constructed for values exceeding one. The only catch is to ensure that the correct degrees of freedom for both the denominator and the numerator are specified.

If we consider the variances of the approach and departure speeds around the slow point as used in the previous sections, then we have two estimates of a population variance:
$\sigma_1^2 = 3.4 \times 3.4 = 11.56$ ($n_1 = 14$), and $\sigma_2^2 = 5.6 \times 5.6 = 31.36$ ($n_2 = 14$).
The F-statistic is then $F(13,13) = 31.36/11.56 = 2.71$. Consulting tables of the F-distribution (e.g. Table A.3) shows the following significant values of F:

$F(13,13) = 2.63$ (five per cent level)
$F(13,13) = 3.91$ (one per cent level)

Thus the computed value of F lies between the five per cent and the one per cent significance values, and we conclude that the computed value is significant at the five per cent level. The null hypothesis could then be rejected at the five per cent level. Had we chosen the one per cent level for our test, we would accept the null hypothesis.

The F-statistic, as described in equation (16.9) and required to be at least unity in value, is that used for upper tail testing. Most published tables of the F-distribution, and Table A.3 is no exception, provide only the upper tail values of F Values for the lower tail may be found from the observation that if variable y has an F-distribution $F(m,n)$, where m and n are the degrees of freedom, the variable $v = 1/y$ has an F-distribution given by $F(n,m)$ and $F(n,m)$ is equal to $1/F(m,n)$. To illustrate this result, consider the one per cent significance value of $F(13,13) = 3.91$, as quoted above. Then the 99 per cent significance value of $F(13,13)$ is $1/3.91 = 0.256$.

The existence of the F-distribution and the test of equivalence of variances opens the way to one of the most powerful of all statistical testing procedures, the 'analysis of variance' The following section deals with this technique.

16.3 Analysis of variance

The previous sections made inference about one population mean and then compared two means. This section presents a technique for comparing several means. The technique is called analysis of variance. This is a powerful technique for testing differences in experimental data collected under different combinations of factors, e.g. as discussed under experimental design (Section 7.3.2).

16.3.1 Testing for differences

To illustrate the analysis of variance technique, consider the following example. Suppose a traffic authority wishes to compare the average speeds of vehicles as they cover measured distances of 0.5 kilometres along six roads entering a town centre during peak the period. The measured speeds are shown in Table 16.2. Because of the concerns expressed by motorists on these routes, the traffic authority wishes to determine whether there is any difference between the roads.

The first question to be answered relates to the possibility of the speeds along the equal lengths of road being different. That is, are the sample means ($\overline{X}_j$)in Table 16.2 different because of differences in the underlying population mean (μ_i)? Or may the differences in the sample means be reasonably attributed to chance?

To form a basis for the comparison, consider three samples of travel speed on just one route. These are shown in Table 16.3. Comparison of Tables 16.2 and 16.3 shows that there is a smaller difference in the sample means obtained from the same route. This is as expected since, given that no changes occur in any other causal variables, the population mean speed is likely to be the same. The question to be answered now becomes: are the differences in the sample means obtained on the six different routes of the same order as those obtained on the single route, or do they indicate a difference in the underlying population mean?

Table 16.2   Speeds along test roads entering a town centre

	\multicolumn{6}{c}{Road}
	1	2	3	4	5	6
	22.81	20.91	19.39	21.43	23.88	28.20
	25.07	27.83	21.60	14.92	25.95	30.79
	23.25	18.04	19.88	20.83	21.54	28.49
	22.24	19.07	11.90	19.13	26.99	29.23
	17.51	19.49	20.95	24.12	30.02	25.28
$\bar{x}_j$	22.18	21.07	18.74	20.09	25.68	28.40
s_j	2.82	3.92	3.92	3.40	3.20	2.01
σ_j^2	7.95	15.37	15.37	11.56	10.24	4.01

Overall mean $\bar{X} = 22.69$

Source: the data in the table were supplied by Dr Peter Gipps of the CSIRO Division of Building, Construction and Engineering, Melbourne, Australia.

Table 16.3   Average speed of vehicles on the same road

	\multicolumn{3}{c}{Run number}
	A	B	C
	22.81	23.41	19.10
	25.07	25.01	22.34
	23.25	22.36	23.54
	22.24	18.02	25.21
	17.51	22.71	23.02
Sample means ($\bar{X}_j$)	22.18	22.30	22.64
Standard deviation (s_j)	2.82	2.60	2.25

To answer the above question, the usual hypothesis of 'no difference' in the population means is the null hypothesis. That is: $H_0 : \mu_1 = \mu_2 = \mu_3 = ...$ or $H_0 : \mu_j = \mu$, for all j. To test this difference, it is necessary to obtain in a numerical measure of the degree to which the sample means differ. Given that $\overline{X}$ is the average of the sample means $\overline{X}_j$ where

$$\overline{X} = \frac{1}{m} \sum_{i=1}^{m} \overline{X}_i = 22.69$$

we can calculate the variance of the sample means, that is,

$$\sigma^2 = s_{\overline{X}}^2 = \frac{1}{m-1} \sum_{j=1}^{m} (\overline{X}_j - \overline{X})^2$$

$$= [(22.18 - 22.69)^2 + (21.07 - 22.69)^2 + ...]/5$$

$$= 13.88$$

where $m = 6$ is the number of sample means. It is now possible to compare the variation between the six routes with that for the single route. However, this approach requires the obtaining of a sample of speeds along each of the six routes. This could be very time consuming. An alternative is to look at the variation 'within' each sample and compare this with the spread between each sample. The variance within each sample is shown in Table 16.2. For route 1 the result is:

$$\sigma_1^2 = \frac{1}{m-1} \sum_{j=1}^{m} (\overline{X}_{1k} - \overline{X}_1)^2$$

$$= [(22.81 - 22.18)^2 + (25.07 - 22.18)^2 + ...]/4$$

$$= 7.95$$

where n is the number of speeds recorded on each route and X_{kl} is the kth observed value in the first sample.

The average of the sample variances is often called the 'pooled variance' and is calculated as follows:

$$\sigma_2^p = s_p^2 = \frac{1}{m} \sum_{j=1}^{m} s_j^2$$

$$= (7.95 + 15.37 + \ldots]/6$$
$$= 10.75$$

Now from each of the 6 (= m) samples the sample variance has four (i.e. $n - 1$) degrees of freedom. Hence the pooled variance has $6 \times 4 = 24$ degrees of freedom (given by $m \times (n - 1)$). The key question is now the following: is the between sample variance larger than the pooled variance? To determine this, it is necessary to calculate the ratio of the between sample variance and the pooled variance, and multiply it by the sample size (n). The sample size has been introduced purely to ensure that, when H_0 is true, this ratio will have a value near one for relatively large samples. The ratio is referred to as the F-ratio and is expressed mathematically as

$$F = \frac{n \, s_{\bar{x}}^2}{s_p^2} \qquad (16.10)$$

Because of statistical fluctuations, the critical F-ratio will sometimes be above and sometimes below one. If H_0 is true, then the top line of the equation will be relatively smaller when compared with the bottom line. The calculated F-ratio will be less than the critical one. Thus it is sometimes possible to judge if the null hypothesis is true using the F-ratio. More formally, to test H_0, it is necessary to know the distribution of the F-statistic. When H_0 is true, the exact distribution is shown in Table A.3 along with the critical value of F (0.01) that cuts off one per cent of the upper tail of the distribution. To test at the one per cent level, we reject H_0 if F exceeds the critical value of 3.89. Thus, if H_0 is true, there is only a one per cent probability that the observed F value exceeds 3.89, and consequently, we would reject H_0.

To illustrate this comparison, consider the previous example. The calculated F-ratio is, from equation (16.10), $F = 5 \times 13.38 / 10.75 = 6.2$. Since this exceeds the $F(0.01)$ value, H_0 is rejected. In this case, the difference in sample means is very large relative to the chance fluctuation.

This example has illustrated a study of differences in the speeds of vehicles travelling along a road. The number of speeds, the sample size, taken on each road was the same. It is also possible to have different sample sizes for each road and use the analysis of variance approach. The general approach discussed in the previous section is usually calculated and set out in a table called the *analysis of variance (ANOVA) table*. Table 16.4 shows the general layout of the ANOVA table.

Table 16.4 The analysis of variance table

Source of variation	Degrees of freedom	Sum of squares	Mean sum of squares	F-ratio
Between samples	$m - 1$	SSB	$MSB = SSB/(m - 1)$	MSB/MSW
Within samples	$mn - m$	SSW	$MSW = SSW/(mn - m)$	
Total	$mn - 1$	TSS		

Notes:

SSB = sum of squares between samples $= n \sum_{j=1}^{m} (\overline{X}_j - \overline{X})^2$

SSW = sum of squares within samples $= \sum_{i=1}^{m} \sum_{j=1}^{m} (X_{ij} - \overline{X}_j)^2$

TSS = total sum of squares $= \sum_{i=1}^{n} \sum_{j=1}^{m} (X_{ij} - \overline{X})^2$

The ratio of the mean sum of squares between the samples to the mean sum of squares within the samples (*MSB/MSW*) is compared with the percentage points of an F-statistic with $(m - 1)$ and $m(n - 1)$ degrees of freedom.

Having rejected the null hypothesis, it is of interest to test the difference between the routes. Consider the average speeds of the six roads. Since the variance associated with each sample is the same, the standard error associated with each road can be estimated by taking the square root of the pooled variance, divided by the number of readings in each sample, i.e.

$$s_{\overline{X}_i} = \sqrt{\frac{s_p^2}{n}} = 1.47$$

The standard error is a measure of the accuracy of the estimate and can be used to compare pairs of average speeds using the t-statistic. The five per cent critical value of t ($t_{0.05}$) is given by Table A.2. As $m = 6$ factors, so $m(n - 1) = 24$ is the number of degrees of freedom, for which

$t_{0.05} = 4.37$. When this is multiplied by the standard error (1.47), the critical difference for the test is 6.41 (= 1.47 × 4.37). Hence any pair of roads where the speeds differ by more than 6.41 are significantly different at the five per cent level.

The test can be carried out by reordering the roads in descending order of average speed and constructing a table of differences. Table 16.5 presents the comparison. The significantly different means are not underlined in Table 16.5. It can be seen that road 6 is significantly different from roads 2, 3 and 4, and that road 5 is significantly different from road 3. No other differences are significantly different.

Table 16.5   Comparison of mean speeds

Road	6	5	1	2	3	4
Mean speed	28.40	25.68	22.18	21.07	20.09	18.74
Differences						
3	9.66	6.94	3.44	2.33	1.35	
Road 4	8.31	5.59	2.09	0.98		
2	7.33	4.61	1.11			
1	6.22	3.50				
5	2.72					

Table 16.6 shows the one-way analysis of variance table for the data given in Table 16.2. This analysis is termed 'one-way' because it assumes that the variation between the speeds on each of the roads is a function of the characteristics of the roads alone. It does not cater for other factors that might also influence the speeds. The study of such situations is examined in the next section. Table 16.6 shows that the F-ratio is 6.20 and that this has a probability of 0.0008 (see Table A.3) that it arose by chance (i.e. under the null hypothesis). A reasonable conclusion is thus that there is a significant difference between the speeds on the different roads.

Table 16.6 One-way analysis of variance table for the speed data

Source of variation	Degrees of freedom	Sum of squares	Mean sum of squares	F-ratio	prob
Between roads	5	333.76	66.75	6.20	0.0008
On a given road	24	258.14	10.75		
Total	29	591.90			

16.3.2 Two-way analysis of variance

The preceding example assumed that the variation between the speeds on each of the roads was a function of the road characteristics. Other factors may also have an influence on the speeds. It may, therefore, be possible to reduce the F-statistic value calculated for the speed study by introducing other factors. Some of the variance not explained by the differences between the roads may be due to systematic variations in other factors. Suppose, for example, that the travel speeds were taken on separate days (Monday to Friday). The data now take the format shown in Table 16.7.

The introduction of the extra dimension could have two effects. The first is the reduction in the variance explained by the differences in the roads, and the second is a reduction in the within group, or unexplained, variance. The first effect means that the variance explained by the roads can be considered without the effect of the days the readings were taken. The second effect means that the introduction of the days of the week aids in the explanation of the variance given by the road factor. The analysis is an extension of the one-way ANOVA and is summarised in Table 16.8.

Comparison of Tables 16.8 and 16.4 shows that the factors in the two-way analysis of variance table are considered in rows and columns. The rows represent the road factor and the columns the days of the week factor. The sample size factor has been removed since there is only one data point for each combination of the day of the week and the road factors. In numerical terms, the two-way ANOVA table can be completed as shown in Table 16.9.

Table 16.7 Speeds on roads by day of week

Road	Day					Road Mean
	Monday	Tuesday	Wednesday	Thursday	Friday	
1	22.81	25.07	23.25	22.24	17.51	22.18
2	20.91	27.83	18.04	19.07	19.49	21.07
3	19.39	21.60	19.88	11.90	20.95	18.74
4	21.43	14.92	20.83	19.13	24.12	20.09
5	23.88	25.95	21.54	26.99	30.01	25.68
6	28.20	30.79	28.49	30.01	25.28	28.40
Daily mean	22.77	24.36	22.01	22.89	22.89	22.69
s_j	3.08	5.53	3.62	4.52	4.52	

Source: These data were supplied by Dr Peter Gipps

As stated, the two-way analysis of variance enables the testing of differences between routes and between days. In either test, the extraneous information in the other test will be taken into account. To test the difference in routes, the F-ratio can be constructed as follows:

F = Variance explained by factor / unexplained variance

If the null hypothesis is true, the ratio has an F value greater than the acceptable value. The calculated F value for the roads factor is $F = 66.73/11.56 = 5.77$. This is lower than that obtained in the one-way analysis (Table 16.6), hence some of the explanation of the variance in the speeds on the roads could be attributed to the day of the week. Referring to the F tables with 5 and 20 degrees of freedom, the critical *F (0.01)* is 4.10 (Table A.3). Hence the probability value is less than 0.01 and H_0 can still be rejected at the one per cent level. The roads factor still has a significant influence on the speeds. The day of the week factor has an F-ratio of 0.59. When this is compared with the F tables with 4 and 20 degrees of freedom, the value of *F(0.01)* is 4.43 (Table A.3). Hence the probability value is less than 0.01 and H_0 cannot be rejected. Thus, in our example, the day of the week factor does not in itself have a significant influence on the speeds of vehicles on the routes.

Table 16.8 Components of two-way ANOVA table

Source of variance	Sum of squares	Degrees of freedom	Mean sum of squares	F-ratio
Between rows	SSR	$n-1$	$SSR/(n-1)$	$SSR(m-1)/SSW$
Between columns	SSC	$m-1$	$SSC/(m-1)$	$SSC(n-1)/SSW$
Within	SSW	$(n-1)(m-1)$	$SSW/[(n-1)(m-1)]$	
Total	TSS			

Notes:

$$SSR = m \sum_{i=1}^{n} (\overline{X}_{i.} - \overline{X})^2 \qquad SSC = n \sum_{j=1}^{m} (\overline{X}_{.j} - \overline{X})^2$$

$$SSW = \sum_{i=1}^{n} \sum_{j=1}^{m} (X_{ij} - \overline{X}_{i.} - \overline{X}_{.j} + \overline{X})^2 \qquad TSS = \sum_{i=1}^{n} \sum_{j=1}^{m} (X_{ij} - \overline{X})^2$$

m = number of columns
n = number of rows

$\overline{X}_{i.}$ = row average
$\overline{X}_{.j}$ = column average

Table 16.9 Two-way ANOVA table

	Sum of squares	Degrees of freedom	Mean sum of squares	F-ratio
Between roads	333.65	5	66.73	5.77
Between days	27.24	4	6.81	0.59
Within	231.04	20	11.56	
Total	591.90	29		

16.4 Non-parametric tests

Sections 16.2 and 16.3 concentrated on what are termed 'classical' statistical tests. These tests assume that the data are distributed in a fashion similar to a normal distribution. If this assumption is incorrect, statistical techniques that are free of this requirement should be considered. Such statistical techniques are called 'distribution-free' or 'non parametric'. Non-parametric tests can be found that correspond to every parametric test. These tests are particularly useful for ordinal and nominal data. They are also preferred when the corresponding classical test is invalid, or when it is reasonably valid, but a non-parametric statistic is much more efficient and provides a narrower confidence interval. This chapter will introduce four non-parametric tests. These are the sign test, the rank order correlation test, the Wilcoxon-Mann-Whitney test and the contingency table test.

16.4.1 The sign test for small samples

This test corresponds to the t-test where a sample mean is compared to a fixed value. Suppose, for instance, that the median traffic flow on arterial roads in a city was reported as 15,000 veh/day. However, a random sample of nine arterials (Table 16.10) showed that eight arterials have flows over 15,000 veh/day, while only one has a flow below that figure. Does this evidence allow the report to be rejected?

Table 16.10   Traffic flow on arterial roads

Arterial	Traffic flow (veh/day)	Arterial	Traffic flow (veh/day)	Arterial	Traffic flow (veh/day)
1	10,000	4	16,000	7	40,000
2	17,500	5	20,000	8	17,000
3	18,000	6	25,000	9	15,000

Since it is unlikely that the distribution of traffic flow would be normally distributed, a sign test will be used to answer the question. The null hypothesis may be stated formally as

H_0 : Population median = 15,000 veh/day,

that is, half the population lie above 15,000 veh/day; or, if an observation is drawn randomly, the probability that it lies above 15,000 veh/day is 0.5. This second hypothesis is similar to the determination of the fairness of a coin. That is, the probability that random observations will fall above the median is equivalent to the probability that a coin will show a head. Hence, if H_0 is true, the sample of $n = 9$ observations is similar to the tossing of a coin nine times. The total number of successes will be a binomial distribution, and H_0 may be rejected if the number of successes is too far away from the expected value to be explained by chance. More specifically, to investigate this hypothesis, it is possible to calculate the probability that the number of successes (suc) is eight or more using the binomial distribution, eight being the number of traffic flows above 15,000 veh/day in the sample. Thus, using a binomial distribution,

$$P(suc = 8) = {}^9C_8(0.5) \times (0.5) = 0.01761$$
$$P(suc = 9) = {}^9C_9(0.5) \times (0.5) = 0.00202$$
$$P(suc \geq 8) = [P(suc = 8)] + [P(suc = 9)] = 0.0196$$

This calculated probability ($P(suc \geq 8)$) is termed the probability value and is the significance level associated with the calculated statistic. Since the probability of the number of successes is below 0.05, the null hypothesis is rejected at the five per cent level. That is, it is concluded that the population median is higher than 15,000 veh/day.

16.4.2 Two matched samples

It is possible to extend the sign test to paired samples and develop a non-parametric test to investigate differences in two populations. This test is similar to the paired sample t-test. Suppose that a sample of eight people had their perception of road noise outside their homes measured on the seven-point semantic scale, shown in Figure 16.2, before and after the introduction of an advisory truck route outside their house. The scale results are shown in Table 16.11.

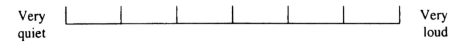

Figure 16.2 Seven-point semantic scale

Table 16.11   Perception of traffic noise

Subject No.	x (Before)	y (After)	Difference (d = y - x)
1	6	7	+1
2	5	6	+1
3	4	3	-1
4	4	6	+2
5	3	5	+2
6	5	7	+2
7	2	3	+1
8	3	1	-2

The first step in the process is to calculate the differences in the before and after readings. The differences are shown in column three of Table 16.11. A difference of zero for one respondent would result in the data point being removed and the sample size reduced by one. As in the paired sample t-test, the original data can be forgotten once the differences have been calculated. These differences form a single sample, to which the sign test can be applied. The null hypothesis is

H_0 : Population median = 0

The question to be asked is similar to that introduced in the preceding section. Are the six positive values in the sample of observations consistent with H_0? The probability is P(suc $\geq$ 6) = 0.1445. Since the probability is larger than 0.05, the null hypothesis can be accepted at the five per cent level. Hence there is no evidence of a change in people's perception of the noise level.

16.4.3 Confidence intervals from ordinal data

The sign test can also be used to provide confidence intervals for the population median. Consider the example discussed in Section 16.4.1 (i.e. the data in Table 16.10). The data, ranked in order of increasing magnitude, are given in Table 16.12. The mean flow of 15,000 veh/day is suspected of being in error in some way. It can be either too high or too low. Since it can be either too high or too low, it would be appropriate to calculate a two-sided probability for H_0 (mean flow = 15,000 veh/day). Using the binomial distribution the two-sided probability is 2 × 0.0196 = 0.039. If the classical method, described in Section 16.2, were used to test H_0, the significance level of 0.04 would just reject the null hypothesis. Turning to Table 16.12, it can be seen that any other hypothesis below the second lowest flow, 15,500 veh/day, would be rejected in the same way.

Table 16.12   Ordered traffic flows on arterial roads

Arterial no	1	9	4	2	8	3	5	6	7
Traffic flow (1000s veh/day)	10.0	15.5	16.0	17.0	17.5	18.0	20.0	25.0	40.0

Now consider the situation where the flow rate of the population is above 15,500 veh/day, say 15,700 veh/day. Since there are now seven rather than eight, traffic flows above this value, the probability of this occurring is 0.1796. This now exceeds the significance level of 0.04. If the significance level of 0.04 was the level of interest, then the flow of 15,700 veh/day could represent the median flow rate, and the null hypothesis would be an acceptable hypothesis. The use of similar logic shows that even higher flows such as 16,000 veh/day or 20,000 veh/day are also acceptable. These acceptable hypotheses are shown in Figure 16.3. Of course, at the high end of the range, a rejected hypothesis is again encountered. Finally, note that the set of hypotheses form an interval. Now it can be recalled that a confidence interval is just a set of acceptable hypotheses. Thus, in Figure 16.3 a confidence interval for the level of confidence 96 per cent has been constructed. In algebraic language, the 96 per cent confidence interval for the population median when sample size n = 9 is: 15,500 veh/day < median < 25,000 veh/day. Note that since the binomial distribution is discrete, the confidence levels are also discrete. For example, if the next highest confidence interval were considered, it would be 82 per cent. However, as the sample size increases, the binomial becomes nearly continuous and this problem tends to disappear.

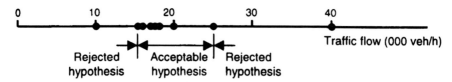

Figure 16.3   Confidence intervals for non-parametric data

16.4.4 Rank order correlation

There are many cases where dependencies may exist between two variables, say, X and Y. The relationship (or correlation) between these two variables may need to be known. The classical approach to determining such relationships is discussed in detail in Chapter 17. At this point it is useful to describe a procedure for determining the correlation between two variables of unknown distribution. The measure of correlation is termed the rank order correlation, and is based on a comparison of the rank ordering of the X and Y observations. The following example illustrates the use of rank order correlations in investigation the speeds of vehicles on two different roads. For convenience, the information is presented in tabular form (Table 16.13). The Spearman rank order correlation coefficient is defined as:

$$r_s = 1 - \frac{6}{n(n^2-1)} \sum_{i=1}^{n} d_i^2 \qquad (16.11)$$

where d_i denotes the difference between the ranks of X_i and Y_i.

Table 16.13 Sample speeds of vehicles on two roads

Speeds (km/h)		Ranking		$d = X - Y$	d^2
Road 1	Road 2	Road 1	Road 2		
X	Y	X	Y		
85	93	2	1	1	1
60	75	4	3	1	1
73	65	3	4	-1	1
40	50	5	5	0	0
90	80	1	2	-1	1
Total (Σd^2)					4

Substituting in equation (16.11) for this example gives a value for the coefficient of

$r_s = 1 - 6(4) / [5(24)] = 1 - 0.2 = 0.8$

The strength of the relationship is indicated by the size of the coefficient. A rank order correlation of zero indicates no agreement between the two variables. A rank order of one shows perfect agreement. To determine if the correlation is significantly different at the five per cent level use statistical tables such as Table A.4.1. The rank order correlation is significantly different to zero at the five per cent level if the Σd^2 value is below the lower limit or above the higher limit. In the example, the Σd^2 value is four, which is neither below nor above the critical values. Therefore, there is no significant difference between the speeds on the two roads at the five per cent level.

16.4.5 Wilcoxon-Mann-Whitney test

The Wilcoxon-Mann-Whitney test, sometimes referred to as the W-test, corresponds to the two sample t-test. It tests for differences of two independent samples. The object is to detect if two underlying populations are centred differently. For example, suppose that independent random samples of traffic flow on arterial roads were taken from two neighbouring local government areas (Table 16.14).

Table 16.14 Sample traffic flows from two local government areas

Traffic flows for local government areas (veh/day)	
North (A)	South (B)
3,000	6,000
9,000	11,000
10,000	12,000
14,000	13,000
	15,000
	20,000

The test is undertaken to see if the two underlying populations are identical. Suppose that the alternative hypothesis is that the north has a lower traffic flow than the south, so that a one-sided test is appropriate. The first step is to combine the two sets of traffic flow. The combined set of traffic flows are then ranked. Observations that have the same value should be given the same ranking. These steps are shown in Table 16.15.

Table 16.15   Combined ranking for the W-statistic

Combined ordered observations		Combined rankings	
A	B	A	B
3,000		1	
	6,000		2
9,000		3	
10,000		4	
	11,000		5
	12,000		6
	13,000		7
14,000		8	
	15,000		9
	20,000		10

The actual traffic flow levels are now discarded in favour of the ranking. This provides a test that is not influenced by skewness or other distributional peculiarities.

The W-statistic is defined as the sum of all the rankings of the smaller sample, in this case: $W\text{-}stat = 1 + 3 + 4 + 8 = 16$. The lower the W-statistic, the stronger the evidence for rejecting H_0. For this data, the W tables (Table A.4.2) show that the one-sided probability of $0.129 = 13$ per cent. This is higher than 0.05; hence at the five per cent significance level it is weak evidence that the traffic flow is lower in the north.

16.4.6 Contingency table test

One of the most commonly used non-parametric statistical tests involves the investigation of relationships between the variables presented in a two-way cross-tabulation. To discuss this test, consider the relationship between the location of accidents and the level of alcohol in the victim's blood-stream. Table 16.16 shows the results of blood alcohol concentration studies at five locations. The figures, expressed as percentages, certainly suggest differences in the distribution across locations. However, percentages may be misleading since it is difficult to tell if a percentage is taken from a small or large population. For instance, 28.6 per cent could come from 2/7 or 286/1000. The final analysis should, therefore, consider the original data (as frequencies). These are shown in Table 16.17.

Table 16.16 Alcohol level of accident victims at five locations (percentages)

Blood alcohol content	Location					Overall percentage
	A	B	C	D	E	
No reading	27.7	16.1	10.1	16.4	52.4	24.5
0.001 to 0.05	38.3	32.3	45.6	43.8	35.4	40.6
above 0.05	34.0	51.6	44.3	39.8	12.2	34.9
Total	100.0	100.0	100.0	100.0	100.0	100.0

Table 16.17 Alcohol level of accident victims at five locations (number)

Blood alcohol content	Location					Total
	A	B	C	D	E	
No reading	13	5	8	21	43	90
0.001 to 0.05	18	10	36	56	29	149
above 0.05	16	16	35	51	10	128
Total	47	31	79	128	82	367

An entry into a row i and column j (cell i,j) of Table 16.17 would be expected to come from a Poisson distribution and be independent of the entries into the other cells. The expected value of the entry E_{ij} needs to be calculated. The only evidence of what are the expected frequencies comes from the marginal totals. For example, the right hand marginal for the first row shows a total of 90 out of 367, or 24.523 per cent, for the victims with no blood alcohol concentration. Since there are a total of 47 victims at location A, the expected number of victims with no blood alcohol reading is 24.523 per cent of 47, or 11.53. A similar process can be used to calculate the expected values for the other locations and blood alcohol levels. Table 16.18 shows the results of these calculations. In general, when considering the entire table, the expected value is calculated from:

$$E_{ij} = \frac{T_i T_j}{N_T}$$

where T_i is the total frequencies in rows i, T_j is the total frequencies in column j, and N_T is the total number of observations.

Table 16.18   Expected accident number for accident locations

Blood alcohol content	Location					
	A	B	C	D	E	Total
No reading	11.53	7.60	19.37	31.39	20.11	90
0.001 to 0.05	19.08	12.59	32.07	51.97	33.29	149
above 0.05	16.39	10.81	27.55	44.64	28.60	128
Total	47	31	79	128	82	367

Since the expected values are to be independent of the data, the null hypothesis could be tested by first using the formula:

$$\chi^2 = \sum_{i=1}^{n} \sum_{j=1}^{m} \frac{(O_{ij} - E_{ij})^2}{E_{ij}}$$

where O_{ij} is the observed value in column j and row i. This formula enables comparison with a χ^2 distribution with $(n - 1) \times (m - 1)$ degrees of freedom. For the data presented in the Table 16.17 the χ^2 statistic is 56.7 and the degrees of freedom are $2 \times 4 = 8$. This value is highly significant (e.g Table 5.4 indicates that the one per cent critical value of χ^2 with eight degrees of freedom is 20.1, much below the computed value of 56.7).

It may also be of interest to see if there are any locations that show marked differences to the others: location E, for instance, may be at a signalised intersection. The frequencies for the four unsignalised intersections can, therefore, be grouped together and compared with E. This grouping provides a χ^2 value of 49.8 with two degrees of freedom and a significance level of below one per cent - thus showing a great difference between the accident distribution at signalised and unsignalised intersections. A comparison of the differences within locations A, B, C and D gives a χ^2 of 8.3 with six degrees of freedom and a significance level of 20 per cent (see Table A.5) - thus showing little difference in the accident distribution at unsignalised intersections. Most of the variation in the distribution of people with different blood alcohol concentrations can, therefore, be explained by comparing the signalised and unsignalised intersections.

17 Statistical modelling

Chapters 15 and 16 described the types of data an analyst may be considering, and the numerical procedures that provide an indication of the character of these data. Frequently, analysts may wish to go further and develop relationships between the factors contained in the data. These relationships are statistical models and provide a useful but simplified view of the process being studied. This simplicity is limited by the need to provide a reasonable representation of the process. Statistical models use sample data to develop a mathematical relationship between factors, often of the form

$$Y_i = b_0 + b_1 X_{i1} + b_2 X_{i2} + \ldots + b_m X_{im} + \varepsilon_i \qquad (17.1)$$

where the Y and X_j terms represent the variables considered in the model, the subscript i represents the ith observation, the $b_0, b_1, \ldots b_m$ terms represent the model parameters, and the ε_i is an error term. The variables can also be divided into dependent and independent variables. Dependent variables (Y) are the variables to be explained by the model. The righthand side of equation (17.1), less the error term, is the expected value $E[Y_i]$ of Y_i given a set of observed values X_{ij}, $j = 1 \ldots m$. The independent variables (X_j) are the variables providing the explanation.

The first aspect to be emphasised when describing models is that statistics is not a process for pulling a relationship out of the air. Analysts should be familiar with the data, together with their faults and strengths, before any attempt is made to develop the relationship. The techniques described in Chapters 15 and 16 should aid in gaining this insight. Some particular examples of statistical models were cited in Chapter 4 (on saturation flows at signalised junctions) and Chapter 13 (on the traffic generated by land use activities). Statistical models can be used either as descriptive or as predictive models. Descriptive models require an understanding of the process studied, i.e. there should be a theory to indicate the general form of the relationship and the model can be developed in terms of this theory.

A discussion of models also needs to indicate the types of errors that could be present in an analysis. Errors usually fall into three categories: specification, measurement and calibration errors. The specification error is related to the model's ability to represent the process being studied. It is generally thought that, the more complex the model, the less specification error there will be. Measurement error relates to the analyst's ability to accurately measure the factors' inputs into the model. The total measurement error tends to increase with model complexity. As shown in Figure 17.1, the measurement error and the specification error tend to act in different directions and an optimal compromise has to be sought. Calibration error (or parameter estimation error) will be discussed further in this chapter. The statistical procedures discussed attempt to minimise this error. Its magnitude is, however, influenced by the properties of the data, the model building technique employed, and the analyst's knowledge of the data.

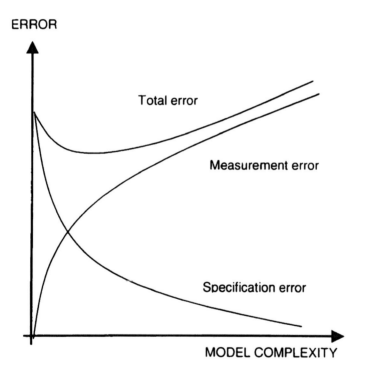

Figure 17.1 Specification error and measurement error in models

17.1 Model development process

The general process for developing a statistical model is illustrated in Figure 17.2. As in any process, none of the steps are independent, for there may be no clear differentiation from those steps before or after. Nevertheless, it is useful to discuss the process as though it is a series of discrete steps.

The initiating step (step 1) in the development is the realisation that there is a problem to be solved. A problem occurs whenever there is a mismatch between the real world and the desires and needs of the community, as described in Chapter 2 when the Systems Planning Process was considered. A clear statement of the problem is needed to aid the development of the appropriate solution. Given the problem definition it is time to derive the objectives to be met by the model (step 2). Clear definition of objectives is needed so that the analyst estimating the model can determine the level of detail required.

The objectives also lead to the definition of the criteria to be used in assessing the efficiency of the system being analysed (step 3). Traffic designers and planners most often work with ill-designed yardsticks, such as congestion, queuing, low capacity and accident hazard. The criteria on which the system is to be assessed are extremely important and deserve careful attention.

In defining criteria, three important aspects should be taken into account. These are:
- the criteria chosen should measure the effectiveness of the entire system;
- it must be possible to state the criteria in quantitative terms, and
- the criteria should be confined to those quantities that can be sampled with reasonable speed and accuracy.

Once the objectives have been determined, the analysis of the system can start. The systems analysis (step 4) highlights the essential components and interactions present in the system. One of the first aspects to be considered is the determination of the breadth of the problem. This may include the population of people to be included, the number of variables to be included in the model, the time span over which the model is valid, the range of conditions the model should be able to study, and the size of the geographic area to be considered. Furthermore, the data gathering and processing part of the systems analysis plays a more general role than one of input into the model. It often provides information on the critical elements in the process. In addition the actions and interactions of the various elements may be unknown prior to the field data collection.

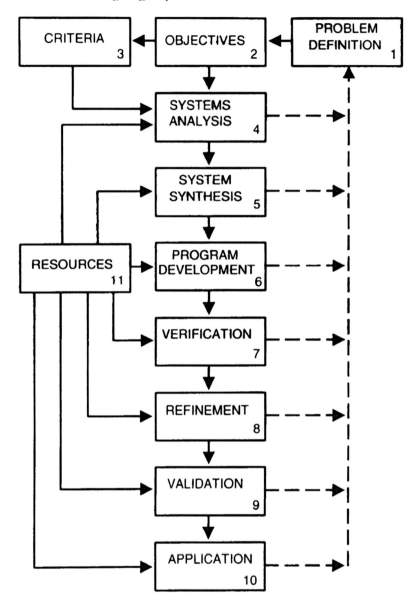

Figure 17.2 The model development process

Following the systems analysis, it is necessary to synthesise the situation of interest (step 5). This step consists of the organisation of the results into a unified and logical structure. The estimation of the model (step 6) is the main subject of this chapter.

The model estimation step produces a relationship that describes the system. Step 7 in the development is the verification stage. Though the model has been developed as carefully as possible, the modeller will still wish to experiment with the model. Verification is, therefore, concerned with determining if the model performs as it is designed to. The next step (step 8) is to determine if the model is sensitive to variations in each of the variables in the model. If the model is not sensitive to particular variables, these variables could be removed.

The ninth step is the validation of the model. This stage examines how closely the model mimics reality, and requires the predictions of the model to be compared to those of actual data. These data should not have been used in the calibration phase. A common way of validating the model is to divide the sample into two parts, calibrate the model on the first part, then use the calibrated model to predict the dependent variables in the other data set.

The tenth (and final) step is possibly the most exciting and challenging. Here the verified and validated model is used to make inferences regarding the behaviour of the system. Confusion may arise as to why this stage is present in the model development. Its main purpose is the investigation of the system's performance and the efficiency of system changes. However, each successful application of the model will enable the user to gain confidence in the model, while each unsuccessful application of the model will encourage new developments and refinements. The establishment of a set of case studies to which the model has been applied is extremely useful in both model development and in assessing its range of applicability.

The development of the model consists of proceeding through the above steps in an orderly fashion. If any of these steps are incomplete, then it is necessary to go back to an earlier stage and repeat the necessary steps. Another dimension in the model development process relates to the resources that are available to develop and validate a model. Knowledge of the resources will aid in the determination of the level of detail that can be included in the model and the verification that can be undertaken, as well as the level of validation that can be carried out. The remainder of this chapter introduces two general techniques often employed in the estimation of model parameters: regression and maximum likelihood.

17.2 Regression

Regression is a general statistical tool through which the analyst can develop a relationship between dependent and independent variables. It can be used to develop descriptive relationships or investigate causal relationships. Most regression models are predicted by using statistical packages. These packages enable the user to develop a model quickly but unless the user understands the underlying theory, serious mistakes can be made. the purpose of this chapter is to help the user avoid such mistakes. With every use of regression, there is a question of statistical inference. The problem of statistical inference consists of two sub-problems: parameter estimation and hypothesis testing. The purpose of parameter estimation is to find the most likely population parameters from the observed sample. For example, the analyst may estimate the regression coefficients for a population in order to determine some confidence intervals. On the other hand, the analyst may be interested in various hypotheses with respect to a particular population. Typical hypotheses might be that:
- there is a linear relationship between specific variables;
- specific variables have no linear effects on other variables, and
- the relationship between specific variables is non-linear.

Simple regression analysis deals with the relationship between a pair of variables. Multiple regression considers the situation where several variables may affect a given variable. Our discussion starts with simple regression, then considers multiple regression. It also considers the development and testing of relationships between variables.

17.2.1 Simple regression

The first step in developing a relationship is the definition of what constitutes a good fit. The answer is usually a fit that leaves a small total error. One error type is shown in Figure 17.3. It is defined as the vertical distance from the observed data point Y_i to the corresponding fitted data point $E(Y_i)$. Mathematically this is the difference ($Y_i - E(Y_i)$). This error is positive when the data point is above the line and negative when it is below the line. Several methods for minimising this error can be suggested. These include minimising the error term itself, minimising the absolute value of the error term $|Y_i - E(Y_i)|$, and minimising the square of the error term $(Y_i - E(Y_i))^2$. The last approach is preferred since it yields a unique solution, provides a reasonable weighting to all data points, overcomes the difference in signs of points below or above the line, is easily managed, and can be justified theoretically by reference to the Fauss-Markov Theorem and maximum likelihood criterion.

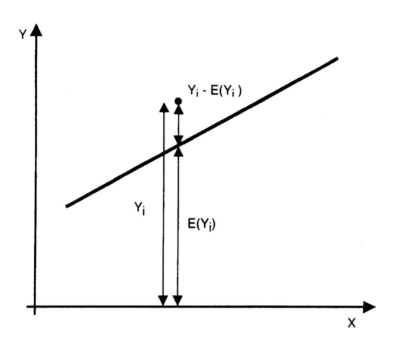

Figure 17.3   Model errors comparing observed and estimated data

To illustrate the least squares solution, consider the information provided in Table 17.1. The table presents data on the fuel consumption and vehicle mass over a standard driving cycle for a random sample of 20 vehicles. The objective is to fit a linear relationship,

$$E(Y_i) = a + bX_i \qquad (17.2)$$

to the data in Table 17.1, where X is vehicle mass and Y is fuel consumption, a is the intercept of the line with the vertical axis and b is the slope of the line. The fitting process involves three steps.

The initial step is to plot the data in the form of a scatterplot (Figure 17.4). This provides a view of the general form of the equation. This will provide support for the proposed linear model, or indicate that some other relationship might be more suitable.

394 *Understanding Traffic Systems*

Table 17.1   Vehicle mass and fuel consumption over a standard driving cycle

Data point	Fuel consumption (L)	Mass (kg)	Data point	Fuel consumption (L)	Mass (kg)
1	20.59	1245	11	15.41	1250
2	11.29	1297	12	10.17	1090
3	10.93	984	13	15.29	1203
4	14.52	1026	14	12.92	1186
5	14.77	1083	15	17.97	1253
6	6.23	1072	16	13.15	1436
7	15.99	1523	17	16.03	1400
8	17.08	1270	18	7.98	885
9	13.98	995	19	17.77	1460
10	20.93	1386	20	14.56	1157

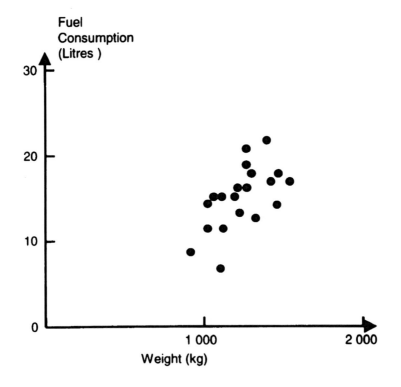

Figure 17.4   Scatterplot of vehicle mass/fuel consumption data

The second step is to determine the estimates of the parameters. To do this, it is necessary to recognise that there is an error (ε_i) associated with the fitted equation. This can be incorporated into the equation (17.2) as follows:

$$Y_i = a + bX_i + \varepsilon_i \qquad (17.3)$$

In turn this can be rewritten such that

$$\varepsilon_i = Y_i - a - bX_i$$

The sum of the squares of the errors is often referred to as the residual sum of squares (RSS) and is calculated as

$$RSS = \sum_{i=1}^{n} \varepsilon_i^2 = \sum_{i=1}^{n} (Y_i - a - bX_i)^2$$

where n is the sample size. The least squares solution states that the line of best fit is the one that minimises RSS. To determine this, take the partial derivative of RSS with respect to each parameter and equate to zero. That is,

$$\frac{\partial RSS}{\partial a} = -2 \sum_{i=1}^{n} (Y_i - a - bX_i) = 0$$

$$\frac{\partial RSS}{\partial b} = -2 \sum_{i=1}^{n} X_i(Y_i - a - bX_i) = 0$$

This provides two equations with two unknowns. Solving these two equations gives:

$$b = \frac{\sum_{i=1}^{n} X_i Y_i - n\overline{Y}\overline{X}}{\sum_{i=1}^{n} X_i^2 \ N\overline{X}^2} \quad)$$

$$a = \overline{Y} - b\overline{X}$$

For the example of the relationship between vehicle mass and fuel consumption (Table 17.1), the sample size (n) is 20, and the estimated parameter values are $b = 0.0126$ and $a = -0.8773$.

The third step in the process is to determine how good the model fit is. This can be determined by examining the amount of variance in the data set that the model explains and comparing this with the residual sum of squares (i.e. that part of the total variance which the model does not explain). The total sum of squares (*TSS*) can be calculated as follows:

$$TSS = \sum_{i=1}^{n} Y_i^2 - n\bar{Y}^2$$

The residual sum of squares (*RSS*) is that value minimised in the calculation and the model sum of squares (*MSS*) is the difference between the total sum of squares and the residual sum of squares. The calculation of the fit of the model can be summarised in an analysis of variance table. This takes the form shown in Table 17.2. To determine the mean sum of squares, it is necessary to know the degrees of freedom associated with each source of variation. The number of degrees of freedom is the number of values in a set that can be assigned arbitrarily. The model source always has one degree of freedom since there is only one parameter that can be assigned arbitrarily.

Table 17.2   The analysis of variance table

Source of variance	Degrees of freedom	Sum of Squares	Mean sum of squares	F-statistic
Model	1	TSS – RSS	(TSS - RSS)/df	MMSS/MRSS
Residual	n - 2	RSS	RSS/df	
Total	n - 1	TSS		

Notes: (1) *MMSS* = Mean Model Sum of Squares
(2) *MRSS* = Mean Residual Sum of Squares

To determine the mean sum of squares, it is necessary to know the degrees of freedom associated with each source of variation. The number of degrees of freedom is the number of values in a set that can be assigned arbitrarily. The model source has one degree of freedom since there is only one parameter that can be assigned arbitrarily. Once this value has been assigned, the other parameter can be determined. Similarly, the total source has (n - 1) degrees of freedom since, once these values have been assigned, for a given mean of Y the remaining value can be determined. The residual source has degrees of freedom equal to the total source minus the model source. A typical regression analysis output for the data set introduced in Table 17.1 is as shown in Figure 17.5.

Regression Analysis

Data: vehicle mass and fuel consumption
Number of cases: 20   Number of variables: 2

Relationship between Vehicle Weight and Fuel Consumption

Index	Name	Mean	St. Dev.
1	WEIGHT	1210.050	175.352
Dep. Var.	FUEL	14.378	3.776

Dependent Variable: FUEL

Var.	Regression Coefficient	St. Error	T(DF = 18)	Probability
WEIGHT	0.0126	0.0041	3.064	0.0067
CONSTANT	-0.8773			

St. Error of Estimate = 3.145 $r^2 = 0.343$ r = 0.585

Analysis of Variance Table

Source	Sum of Squares	D.F.	Mean Square		F Ratio	Probability
REGRESSION	92.855	1	92.855	9.385		0.0067
RESIDUAL	178.084	18	9.894			
TOTAL	270.939	19				

Figure 17.5   Regression output for the vehicle mass/fuel consumption data

The ratio of two variances has an F-distribution with degrees of freedom related to the regression and residual sum of squares. The F-statistic, with degrees of freedom 1 and 18, at the one per cent significance level is 8.28 (see Appendix A, Table A.3). This is less than the calculated value for the equation. Hence there is a significant relationship between the fuel consumption and the mass of the vehicle.

17.2.2 Multiple linear regression

The preceding section discussed simple regression where there was one independent variable and one dependent variable. Sometimes, the relationship may be a little more complicated. Consider Table 17.3. This table represents data required for the relationship between the speed of vehicles at the head of a queue, and the gap between these vehicles and the one following.

Table 17.3   Data on vehicle speeds and gaps

	Gap (m)	Mean speed (m/s)		Gap (m)	Mean speed (m/s)
1	99.5	24.84	19	134.0	29.45
2	36.4	14.20	20	83.6	83.57
3	106.0	26.26	21	107.8	24.58
4	111.5	25.25	22	139.9	29.10
5	59.8	16.89	23	94.6	24.14
6	105.3	24.77	34	153.2	32.53
7	91.5	22.50	25	87.6	23.20
8	146.4	29.16	26	53.8	18.58
9	105.0	24.83	27	117.5	26.41
10	129.2	27.61	28	89.1	22.17
11	80.6	17.72	29	96.7	25.52
12	132.7	28.99	30	58.3	17.76
13	176.9	33.02	31	68.5	20.87
14	95.5	23.93	32	111.6	25.70
15	128.9	27.39	33	16.6	26.16
16	93.9	24.32	34	102.1	23.75
17	113.6	25.31	35	141.2	29.23
18	75.8	20.04	36	92.7	22.69

Note: These data were supplied by Dr Peter Gipps, of the CSIRO Division of Building, Construction and Engineering, Melbourne, Australia

The starting point of all relationship determinations is the preparation of visual plots of the data (Figure 17.6). These plots provide guidance on the selection of the functional form of the model. It would appear from Figure 17.6 that there is a linear relationship between the speed and the gaps in the traffic. Hence the first relationship to try is the linear relationship

$$E(Y_i) = a + bX_i$$

where $E(Y_i)$ = expected gap length (m) and X_i = mean speed of vehicle (m/s). The values of a and b obtained by using least squares regression are -72.11 and 6.44 respectively. The goodness-of-fit of the relationship can be determined by using the analysis of variance approach. Table 17.4 presents a general analysis of variance table for a model with one independent variable.

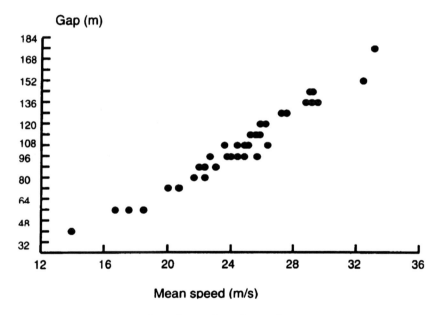

Figure 17.6 Scatterplot of speed and gap data

The F-statistic calculated from the table can be compared with that of an F-statistic with degrees of freedom 1 (model degree of freedom) and 34 (residual degrees of freedom) and significance level of 0.05. This value is approximately 4.15, which is much lower than the calculated F-statistic, and hence the model provides an acceptable fit to the data.

Table 17.4   Analysis of variance table for initial regression

Source	Degrees of freedom	Sum of squares	Mean sum of squares	F-statistic
Model	1	29 797	29 797	765.8
Residual	34	1323	38.9	
Total	35	31 120		

17.2.3 Investigation of residuals

The various significance tests associated with multiple linear regression are based on the assumptions that the sample is drawn at random, errors in the dependent and independent variables are normally distributed, and the relationship between the dependent and independent variables is linear. Hence, even though a good statistical fit is present, it may not provide the best model. The next step in the model development is to observe the variations in the deviations between the expected values and the actual data values. These variations are usually considered as the errors associated with the model. They are often termed the 'residuals' and are indicative of the error function associated with the model. These errors should be independently and identically distributed according to a normal distribution with mean zero and unknown, but fixed, variance. If this is true, it can be said that after the statistics have been calculated, no further relevant information remains in the actual data. This should, however, be checked. To check this, the residuals should be plotted against the dependent variable, the independent variables, or if the data is collected over time, the time dimension.

More specifically, the examination of the residuals provides information relevant to two basic types of question. The first is the possible lack of linearity in the data. Outliers or deviant cases are also apparent in scatterplots. The second type of question that may be answered by the scatterplot is whether the basic assumptions are met. The assumptions that can be tested are if the error components are independent, have a mean zero and the same variance throughout the range of the dependent variable.

A general inspection of the residuals could first be carried out using a dot diagram (see Chapter 15). This diagram should have a concentration of values around the zero residual value and a steady decrease in the frequency of values as one moves away from the zero residual. Considerable fluctuations will arise when the number of values are small, so any appearance of non-normality in that case is not necessarily indicative of an underlying cause. Another plot could relate that residuals to the value of the estimate of the dependent variable, a plot would take on a funnel-shaped appearance.

A useful scatterplot can be obtained by plotting the residuals against the independent variables. In the above example, the independent variable is speed. The residuals are the differences between the observed value and those predicted by the model. Figure 17.7 presents this plot and indicates that there is a systematic scatter which is roughly a crescent shape; the high values occurring at the ends and the low values in the middle. This indicates that the gaps may be a quadratic function of the speed, and a more appropriate model may be that given by equation (17.4):

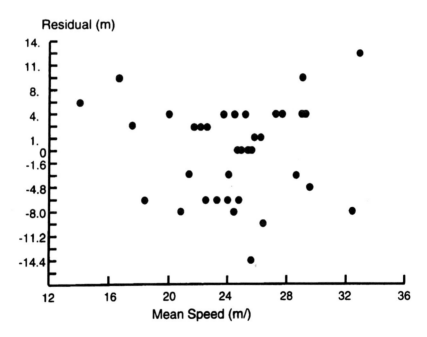

Figure 17.7 **Plot of residuals**

$$E(Y_i) = a + bX_i + cX_i^2 \qquad (17.4)$$

The parameters and analysis of variance table for this model are shown in Table 17.5. The analysis of variance table shows that this is also a good fit. The F-statistic is significant at the five percent level. The next question to be answered is: does the quadratic model provide a significant improvement over the initial model? This can be answered by constructing an analysis of variance table that looks at the difference between the two models (see Table 17.6). This table shows that the incorporation of the quadratic term in the linear model improves the F-statistic by 5.53. When this is compared with an F-statistic with 1 and 33 degrees of freedom at the five per cent significance level (4.14), we see that the quadratic model provides a significant improvement on the initial model.

Table 17.5 Parameter estimates and model fit for the quadratic model

Parameter	Estimate	Standard error
a	-18.42	23.61
b	2.54	1.97
c	0.096	0.041

Analysis of Variance table

Source	Degrees of freedom	Sum of squares	Mean sum of squares	F-statistic
Model	2	29 987	14 993.5	436.7
Residual	33	1 133	34.33	
Total	35	31 120		

17.2.4 Significance of regression coefficients

As well as the overall statistical performance of the model, it is useful to have an indication of the variation in the parameter estimates. The variances of the parameters for simple regression are:

Table 17.6 Analysis of variance table to determine model improvement

Source	Degrees of freedom	Sum of squares	Mean sum of squares	F-statistic
Initial model	1	29 797	29 797	867.9
Quadratic-Initial	1	190	190	5.53**
Quadratic model	2	29 987	14 993.5	436.7
Residual	33	1 133	34.33	
Total	35	31 120		

$$\sigma_b^2 = \frac{\sigma_Y^2}{\sum_{i=1}^{n}(X_i - \overline{X})^2} \quad \text{and} \quad \sigma_a^2 = \frac{\sigma_Y^2 \sum_{i=1}^{n} X_i^2}{n \sum_{i=1}^{n}(X_i - \overline{X})^2}$$

where $\sigma_Y^2 = \dfrac{1}{n-2} \sum_{i=1}^{n} [Y_i - E(Y_i)]^2$

Similar values can be obtained for multivariate regression. These estimates of variance can be used with the t-statistic to determine the significance of the parameters of the regression equation. The final error in assessing the regression relationship is that of the variance of the estimate. This takes the form

$$\sigma_{E(Y_k)}^2 = \frac{\sigma_Y^2}{n} + \frac{(X_k - \overline{X})^2 \sigma_Y^2}{\sum_{i=1}^{n}(X_i - \overline{X})^2}$$

This variance can also be used to plot confidence intervals around the regression line. However, it can be seen from the parameter estimates in Table 17.5 that the coefficient a is negative. This would indicate a negative headway at low speeds. Further, the standard error associated with the constant is larger than the parameter value, indicating that there is no significant difference between the parameter estimate and zero. The removal of the constant term does not significantly affect the model performance (Table 17.7). The incorporation of further variables could be investigated using the approach described above. The final model for this analysis is

$$E(Y_i) = 1.02X_i + 0.127X_i^2$$

Table 17.7   Investigation of exclusion of constant term (a)

Source	Degrees of freedom	Sum of squares	Mean sum of squares	F-statistic	
No constant	1	29 966	29 966	872.8	***
Difference	1	21	21	< 1	N.S.
With constant	2	29 987	14 993.5	436.7	***
Residual	33	1 133	34.33		
Total	35	31 120			

Notes: (1) N.S. = 'not significant'   (2) *** = 'significant at 0.1 per cent level'

17.2.5 Stepwise inclusion of variables

The previous example demonstrated one procedure for including variables into a model. In many statistical analyses it is appropriate to enter or delete independent variables one by one in some pre-established manner. These procedures are used when the analyst is looking for the best relationship. Some of the approaches used are: forward inclusion; backward elimination; stepwise solution, and combinatorial solution.

With *forward inclusion*, independent variables are entered one by one only if they meet certain statistical criteria. The order of inclusion is determined by the respective contribution of each variable to the explained variance. Usually the F-statistic is used as the basis for inclusion of a variable. The contribution to the F-statistic by every variable is calculated, and the variable that makes the greatest contribution is included in the equation. This process is repeated until the addition of one more variable will not improve the overall fit of the model by more than a given amount.

To illustrate this process, consider a forward stepwise regression on the vehicle speed data presented in Appendix B. The output of the regression analysis is presented in Figure 17.8.

_____ Regression Analysis _____
Step 1.   Variable 6: NINT   entered   Dependent Variable: SPEED

Var.	Regression Coeff	St. Error	F(1, 72)	Probability
NINT	-0.7232	0.1223	34.976	0.0001
CONSTANT	57.8361			

St. Error of Estimate = 5.228 $r^2 = 0.327$ $r = 0.522$

Analysis of Variance Table

Source	Sum of Squares	D.F.	Mean Square	F Ratio	Probability
REGRESSION	955.879	1	955.879	34.976	0.0001
RESIDUAL	1967.718	72	27.329		
TOTAL	2923.596	73			

Variables not in equation:

Name	Partial r^2	Tolerance	F to enter	Probability
FLOW	0.0184	0.8419	1.329	0.2529
OPFLW	0.0002	0.8166	0.013	0.9097
PLOC	0.2144	0.9312	19.381	0.0001
WIDTH	0.0275	0.9938	2.009	0.1607

Step 2.   Variable 5: PLOC   entered   Dependent Variable: SPEED

Var.	Regression Coeff.	St. Error	F(1, 72)	Probability	Partial r^2
LOCAL -	0.1374	0.0312	19.381	0.0001	0.2144
INTNS	-0.7232	0.1223	34.976	0.0000	0.4452
CONSTANT	60.9807				

St. Error of Estimate = 4.666 $r^2 = 0.471$ Multiple r = 0.686

Analysis of Variance Table

Source	Sum of Squares	D.F.	Mean Square	F Ratio	Probability
REGRESSION	1377.826	2	688.913	31.643	0.0000
RESIDUAL	1545.771	71	21.771		
TOTAL	2923.596	73			

Variables not in equation:

Name	Partial r^2	Tolerance	F to enter	Probability
FLOW	0.0112	0.8349	0.790	0.3771
OPFLO	0.0089	0.7984	0.627	0.4311
WIDTH	0.0202	0.9862	1.444	0.2335

Figure 17.8   Forward stepwise regression applied to the speed data in Appendix B

The first variable included is the number of intersections along the road (NINT). This has the highest contribution to the F-statistic (34.976) of any variable. The contribution to the F-statistic resulting from the inclusion of each of the remaining variables is then calculated. It can be seen that the percentage of local traffic (PLOC) is the variable that contributes the highest increase in the F-ratio. It increases the F-ratio by 19.381 compared to the next highest contribution of 2.009. It is, therefore, included next.

Once the variable PLOC is included, the contribution made by adding each of the remaining three variables individually is calculated. It can be seen that the width of the road (WIDTH) adds a further 1.444 to the F-ratio. This is, however, a relatively small increase in the size of the F-ratio and therefore the factor is not included in the model. Thus the final model has only two variables (NINT and PLOC).

In *backward elimination*, variables are eliminated one by one from a regression equation that initially contains all variables. This can be carried out in the reverse manner to that described above. It is also common to use the significance of a parameter as a basis for the deletion of variables. Those variables that have an insignificant parameter are eliminated from consideration. This is best carried out on a one by one basis. This process can again be illustrated by considering the vehicle speed data set.

The first step is to obtain a regression model with all the variables present. Figure 17.9 presents the results of a backward regression run on these data. A definite relationship is seen. The variable apparently explaining most of the variance is the number of intersections in the streets (NINT). It has the highest contribution to the total correlation and the highest t-statistic. The variable that contributes least is the flow in the direction in which the vehicle is travelling (FLOW). This is possible because of the relatively low flows being considered and the limited interaction between the vehicles. This variable could be removed from the analysis. The regression equation can be progressively refined by the removal of the flow of vehicles in the same direction as the study vehicle (FLOW), the opposing flow (OPFLO) and the width of the roadway (WIDTH). The final model contains only two variables (NINT and PLOC) and is the same as that obtained using the forward inclusion approach. Analysts often carry out the backward elimination by excluding all the variables with insignificant parameters, although this approach may result in problems if there is a large degree of correlation in the data set.

_____ Regression Analysis _____

Data: vehicle speeds on residential streets
Number of cases: 74   Number of variables: 8

Dependent variable: SPEED

Var.	Regression Coefficient	St. Error	T(DF = 68)	Prob.	Partial r^2
FLOW	-0.5837	1.3237	-0.441	0.6606	0.0029
OPFLO	-1.1478	1.5680	-0.732	0.4667	0.0078
PLOC	-0.1336	0.0320	-4.272	0.0001	0.2116
NINT	-0.9049	0.1339	-6.759	0.0001	0.4018
WIDTH	0.5774	0.5774	1.039	0.3025	0.0156
CONSTANT	59.6555				

St. Error of Estimate = 4.689 $r^2 = 0.489$ r = 0.699

Analysis of Variance Table

Source	Sum of Squares	D.F.	Mean Square	F Ratio	Probability
REGRESSION	1428.248	5	285.650	12.990	0.0001
RESIDUAL	1495.348	68	21.990		
TOTAL	2923.596	83			

Figure 17.9   Backward stepwise regression applied to the speed data in Appendix B

With a *stepwise solution*, forward inclusion is combined with deletion of variables that no longer meet the pre-established criterion at each successive step.

A *combinatorial solution* involves examination of all possible combinations. The approaches described above each take a specific route to the answer but this route may not be optimal if correlations exist in the data set. One method of obtaining an optimal solution is to therefore examine all possible combinations. The main disadvantage of the approach is the large number of combinations that must often be considered.

Statistical packages usually allow analysts to determine the procedure for entering the variables. If this is not the case, then they can carry out the task themselves. This approach can, however, be arduous.

17.2.6 Correlation

The analysis of variance of the regression equation suggests another possible measure of the goodness-of-fit. As the model explains more and more of the variance, the residual sum of squares approaches zero and the model sum of squares approaches the total sum of squares. The ratio of the model sum of squares to the total sum of squares, therefore, approaches one as the model explains more and more of the total variance. This ratio is terms the *coefficient of determination* (R^2) as defined by equation (17.5):

$$R^2 = \frac{MSS}{TSS} = \frac{\sum_{i=1}^{n}(E(Y_i) - \bar{Y})^2}{\sum_{i=1}^{n}(Y_i - \bar{Y})^2} \qquad (17.5)$$

This ratio approaches unity when the model approaches a perfect fit, and zero when the slope of the line approaches the horizontal. It also provides a measure of the percentage of variance explained by the regression. The square root of the coefficient of determination is called the *correlation coefficient* (R). The correlation coefficient is defined by equation (17.6):

$$R = \frac{\sum_{i=1}^{n}(X_i - \bar{X})(Y_i - \bar{Y})}{\sqrt{\left[\sum_{i=1}^{n}(X_i - \bar{X})^2\right]\left[\sum_{i=1}^{n}(Y_i - \bar{Y})^2\right]}} \qquad (17.6)$$

This equation allows the correlation coefficient to take a sign which indicates the slope of the relationship.

In some studies, it is possible that there will be a large and complex set of correlations between the independent variables. This is referred to as *multi-collinearity*. Multi-collinearity can cause problems in a number of ways. If at least one of the independent variables is exactly correlated with one or more variables, the coefficients for each variable may not be unique. Further, perfect collinearity can cause problems in the calculation process by introducing zero divisors and making it difficult to invert the data matrix. Another problem that may occur is that parameter estimates may fluctuate from sample to sample. Hence, when extreme multi-collinearity exists, there is not acceptable way of performing regression analysis using the given set of independent variables. Two solutions are available. The first is to remove one of the highly correlated variables. The second is to create a new variable which is made up of the correlated variables, perhaps using factor analysis (Young, Ritchie and Ogden, 1980).

A *correlation matrix* can provide useful insights into the occurrence of multi-collinearity in a data set. This matrix shows the correlation coefficients between each pair of variables in the data set. It is a symmetric matrix, with diagonal elements all equal to unity (as these elements represent the correlation of each variable with itself). The correlation matrix for the vehicle speed data (Appendix B) is given in Figure 17.10. The figure shows that the number of vehicles travelling in a particular direction (VEHS, OPVEH) and the flow (FLOW, OPFLO) are perfectly correlated. This is to be expected since one is a multiple of the other. Vehicle speed is found to have the highest correlation with the number of intersections in the street (NINT).

_____ Correlation Matrix _____

Data: vehicle speeds on residential streets
Number of cases: 74   Number of variables: 8

	SPEED	VEHS	OPVEH	FLOW	OPFLO	PLOC	NINT	WIDTH
SPEED	1.000							
VEHS	0.125	1.000						
OPVEH	0.235	0.292	1.000					
FLOW	0.125	1.000	0.292	1.000				
OPFLO	0.235	0.292	1.000	0.292	1.000			
PLOC	-0.217	0.185	-0.018	0.185	-0.018	1.000		
NINT	-0.572	-0.398	-0.428	-0.398	-0.428	-0.262	1.000	
WIDTH	0.181	-0.226	0.070	-0.226	0.070	-0.063	-0.079	1.000

Figure 17.10   Correlation matrix for the speed data

17.2.7 Generalisation of the approach

In general, multiple linear regression requires that variables are measured on interval or ratio scales and the relationship be linear and additive. It is possible that the base data used in the analysis do not fit the assumptions of linear regression. This may result from the existence of non-linearities in the data or non-additive relationships between particular variables. These problems can be overcome by a number of techniques, including transformation of variables, consideration of interactions between variables, and the use of dummy variables.

The problem of non-linear relationships between variables is often overcome through the use of *data transformations*. The independent variables can be changed by some mathematical relationship so that they are linearly related to the dependent variable. These transformations can be log transformations, polynomial transformations, square root transformations or quadratic transformations.

If some of the independent variables interact in some manner, it may be necessary to input *interactive terms* into the regression equation. One common interaction is a multiplicative one. In such a case two interacting variables X_1 and X_2 can be combined into a third variable X_3 which is a multiplicative (X_1X_2) function of X_1 and X_2. It would then be included in the model as an independent variable in place of X_1 and X_2.

Dummy variables are most commonly used when an analyst wishes to insert a nominal-scale variable into a regression equation. Since numbers assigned to categories of a nominal scale are assumed to have no order in the units of measurement, they cannot be treated as scores as they would in a conventional regression analysis. The set of dummy variables is created by treating each category as a separate variable and assigning a value to the variable, depending on the presence or absence of the category. For instance, if the gender of a respondent were to be included in a model, a binary variable 'FEMALE', which had the value of one or zero, could be introduced.

17.3 Maximum likelihood estimation

Linear regression is a technique often used for estimating model parameters. It, however, requires that assumptions be made about the distribution of errors. An alternative technique for estimating the parameters of a model that does not have this limitation is maximum likelihood estimation. To illustrate this technique, consider the withdrawal of two balls from one of five jars, where each jar contains four balls with a different ratio of white to black balls. The two balls drawn out of the jar are of different colour, one white and one black. The question is: from which jar were the balls drawn?

Obviously, they would not be drawn from the jars with only white or black balls. However, it is possible that they could have been drawn from any of the remaining three jars. It cannot be stated which of the three jars the balls came from with certainty. All that can be said is which jar is most likely. To do this, it is necessary to calculate the probability of the balls being drawn from each jar. The probability of the balls being drawn from any jar, assuming replacement, is given by

Pr{black ball and white ball} = Pr{black} Pr{(white}

where Pr{colour} is the probability of a ball of the one colour being drawn. Thus, for the jar with four black balls, the probability of selecting a white ball is zero. Hence the probability of choosing a white and black ball is zero. The same holds for the case of the jar with all white balls. For a jar with one black and three white, the probability of choosing a white and black ball is $0.25 \times 0.75 = 0.1875$. The probability of choosing a white and a black ball from the jar with three white balls and one black ball is also 0.1875. The probability of choosing one white and one black ball from the jar with two white balls and two black balls is $0.5 \times 0.5 = 0.25$.

Comparison of these probabilities shows that it is most probable that the one white ball and the one black ball are drawn from the jar with two black and two white balls.

In most analysis situations, however, the analyst is not aware of what is in each of the jars. The procedure adopted is, therefore, to collect data on the contents of each jar by carrying out a series of experiments. These experiments may involve drawing balls from each jar, recording the results and determining which jar holds the two white and two black balls.

To relate this to the estimation of parameters in a model, the drawing of balls from the jars is the data collection process, and the number of white balls and black balls in the jar are analogous to the coefficients in the model.

The use of the maximum likelihood estimation method is illustrated by three examples: determining the mean of a normal distribution; an application to a Poisson process and (3) the fitting of a normal regression line.

17.3.1 The mean of a normal distribution

To determine the mean of a normal distribution using maximum likelihood, the following procedure is adopted. The first step is to draw a number of samples from the parent population. These could be X_1, X_2 and X_3. The parent distribution (see Chapter 3) would be normally distributed with probability density function,

$$p(X_i|\mu) = \frac{1}{\sqrt{2\pi}\,\sigma} \exp\left[-\frac{1}{2\sigma^2}(X_i-\mu)^2\right] \qquad (17.7)$$

The specific values for each of the variables X_1, X_2 and X_3 drawn from this distribution would, therefore, take the form

$$p(X_1|\mu) = \frac{1}{\sqrt{2\pi}\sigma} \exp\left[-\frac{(X_1-\mu)^2}{2\sigma^2}\right]$$

$$p(X_2|\mu) = \frac{1}{\sqrt{2\pi}\sigma} \exp\left[-\frac{(X_2-\mu)^2}{2\sigma^2}\right]$$

$$p(X_3|\mu) = \frac{1}{\sqrt{2\pi}\sigma} \exp\left[-\frac{(X_3-\mu)^2}{2\sigma^2}\right]$$

If it can be assumed that the three sample points are independent, then the joint probability density is the product:

$$p(X_1, X_2, X_3|\mu) = p(X_1|\mu)\,p(X_2|\mu)\,p(X_3|\mu))$$

Since the sample values X_1, X_2 and X_3 are fixed while the mean μ is thought to vary, then the above expression is termed the likelihood function $\Theta(\mu)$. The maximum likelihood estimate (MLE) of μ is the value that maximises the function. Solutions to this problem can be obtained by calculus, or trial and error. To determine the value using calculus, the following procedure is adopted. The first step is to change the multiplicative likelihood into an additive form. This is carried out by taking the logarithm of the joint probability function. The log likelihood [$L(\mu)$] is

$$L(\mu) = \sum_{i=1}^{n} \log[p(X_i|\mu)])$$

or, in the case of the normal distribution,

$$L(\mu) = \sum_{i=1}^{n} \left[-\frac{1}{2}\left(\log(2\pi) - \log\sigma^2\right) - \frac{(X_i - \mu)^2}{2\sigma^2} \right]$$

$$= -\frac{n}{2}\log(2\pi) - \frac{n}{2}\log\sigma^2 - \sum_{i=1}^{n} \frac{(X_i - \mu)^2}{2\sigma^2}$$

where n is the sample size. Now differentiate partially with respect to μ and equate to zero:

$$\frac{\partial L(\mu)}{\partial \mu} = \sum_{i=1}^{n} \frac{(X_i - \mu)}{\sigma^2} = 0$$

Multiplying through by σ^2, the equation becomes

$$\sum_{i=1}^{n} X_i - n\mu = 0$$

Thus $\quad \mu = \frac{1}{n}\sum_{i=1}^{n} X_i$

so that the maximum likelihood estimate of the population mean for a normal distribution is the sample mean (Section 15.4.2).

17.3.2 The Poisson process

The technique described above for studying the normal distribution can also be applied to other distributions. It may, for instance, be used in conjunction with the Poisson distribution or the negative exponential distribution (see Section 3.3.1). One application that may be of interest is to estimate the value of data censored in some way. Consider, for example, the need to calculate the average number of accidents occurring at a particular site per day, for example as in Table 17.8.

Table 17.8 Distribution of accidents

Number of accidents	0	1	2	3	4	5	6	7+
Frequency	0	29	38	22	11	4	1	0

The numbers of accidents in this table could readily follow a Poisson distribution, but the number of occasions on which no accidents occurred is 'suspicious'. If the number of accidents is described by a Poisson distribution, then the probability of r accidents ($r > 0$) is

$$P^*(X_r) = \frac{P(X_r)}{1 - P(X_o)} = \frac{\mu^r \exp(-\mu)}{r! \, [1 - \exp(-\mu)]}$$

where $P(X_r) = \mu^r \exp(-\mu)/r!$ is the Poisson probability of observing r accidents, and $[1 - P(X_o)]$ is the probability of observing at least one accident from a Poisson distribution. The probability $P(X_o)$ is given by $\exp(-\mu)$, and μ is the mean number of accidents. The likelihood function for this problem is, therefore

$$\Theta = P^*(X_1)^{29} \, P^*(X_2)^{38} \, P^*(X_3)^{22} \, P^*(X_4)^{11} \, P^*(X_5)^4 \, P^*(X_6)$$

and the log likelihood (hereafter abbreviated to L) is

$$L = 29 \log [P^*(X_1)] + 38 \log [P^*(X_2)] + 22 \log [P^*(X_3)] + 11 \log [P^*(X_4)] + 4 \log [P^*(X_5)] + \log [P^*(X_6)]$$

The partial differential of the log likelihood function is given by

$$\frac{\partial L}{\partial \mu} = \sum_{r=1}^{n} \frac{\partial \log [P^*(X_r)]}{\partial \mu} \cdot 1$$

and

$$\frac{\partial \log [P^*(X_r)]}{\partial \mu} = \frac{r}{\mu} - 1 - \frac{\exp(-\mu)}{1 - \exp(-\mu)}$$

Therefore,

$$\frac{\partial L}{\partial \mu} = \frac{241}{\mu} - \frac{105}{[1 - \exp(-\mu)]}$$

Thus the estimate of μ is $\mu/[1 - \exp(-\mu)]$ 2.2952381 = 0, and $\mu = 1.98$.

17.3.3 Normal regression

Section 17.2.1 illustrated the estimation of regression coefficients using least squares. The maximum likelihood method may also be used to determine these coefficients. The derivation starts with the equation (17.3):

$$Y_i = a + bX_i + \varepsilon_i$$

where ε_i is an independent random variable representing a normally distributed error function. Use of maximum likelihood to determine the most likely estimates of parameters a and b follows the approach as outlined above. We need to find the joint probability function, i.e. $P(Y_1, Y_2, ..., Y_N | a, b)$. First, consider the probability density of the ith value of Y (i.e. Y_i). This is analogous to equation (17.7)

$$P(Y_i) = \frac{1}{\sqrt{2\pi}\sigma} \exp\left[-\frac{[Y_i - (a + bX_i)]^2}{2\sigma^2}\right]$$

This is simply the normal distribution with the mean $(a + bX_i)$ and standard deviation σ. The joint probability function then becomes:

$$P(Y_1,...,Y_n | a,b) = \prod_{i=1}^{n} \frac{1}{\sqrt{2\pi}\sigma} \exp\left[-\frac{[Y_i - (a + bX_i)]^2}{2\sigma^2}\right]$$

where $\prod_{i=1}^{n}$ is the product of all the terms in i, for $i = 1, 2, ..., n$. Using the rule of exponents this can be rewritten as:

$$P(Y_1,...,Y_n | a,b) = \exp\left[-\left(\frac{1}{\sqrt{2\pi}\sigma}\right)^n \sum_{i=1}^{n} \frac{[Y_i - (a + bX_i)]^2}{2\sigma^2}\right] = L(a,b) \quad (17.8)$$

Now the question becomes: which values of a and b will produce the largest L? The answer is found by minimising the exponent in equation (17.8). Hence the maximum likelihood estimates are obtained by choosing a and b as to minimise the term

$$\sum_{i=1}^{n} (Y_i - a - bX_i)^2$$

so that the maximum likelihood estimates of a and b are determined using the same measure as the least squares approach.

17.3.4 Other applications

Maximum likelihood estimation is commonly used in transport and traffic engineering. It has been applied to calibration of choice models such as the logit model (Hensher and Johnson, 1981) and the Elimination-by-Aspects model (Young, 1982); to find gap acceptance parameters (Troutbeck, 1993), and to estimate the parameters for bunching models of traffic flow (Taylor, Miller and Ogden, 1974). The method is seldom found in any of the standard statistical analysis software packages, and usual practice is to write a special purpose computer program to solve the maximum likelihood estimation problem for a given example.

17.4 Time series analysis

Many transport problems are a combination of deterministic and stochastic processes which may vary over time. Time series analysis is the statistical technique used to study such problems.

From a stochastic point of view traffic flow on a particular road could be regarded as consisting of four components: trend (Tr), seasonal (Se), cyclic (Cy), and random (Ra) components. The trend component may result from long term growth in traffic. Seasonal variation may result from different flows at different times of the year. Cyclic components can result from long term economic changes. The random component may result from short term variations in traffic flow. When there is an additive relationship between the components they can be combined simply as:

$$X(t) = Tr(t) + Se(t) + Cy(t) + Ra$$

as discussed in Section 2.3.

17.4.1 Time plots

The first step in analysing a time series is to plot the observations against time. This will show up important features such as trend, seasonality, discontinuities and outliers. Plotting the data is not as easy as it sounds. The choice of scales, the size of intercept, and the way the points are plotted (e.g. continuous line or separate dots) may substantially affect the appearance of the plot, and so the analyst must exercise care and judgement - see Sections 15.2 and 15.3.

17.4.2 Transformations

Plotting the data may indicate the need to transform the data variables. The three main reasons for transforming time series data are: to stabilise the variance; to make the components additive (for instance a logarithmic transformation will change a multiplicative function to an additive one), or to make them conform more closely to a normal distribution.

17.4.3 Analysing series that contain a trend

The analysis of series that contain a 'long term' change in the mean (i.e. a trend) depends on whether one wants to measure the trend or remove it (i.e. make the series a stationary time series) in order to analyse the local fluctuations. With seasonal data it is a good idea to start by calculating successive yearly averages, as these will provide a simple description of the underlying trend. Another approach is to fit curves to the data to use linear filters (i.e. fit linear relationships to the data and look at the residuals for further trends).

17.4.4 Autocorrelation

An important guide to the properties of a time series is provided by its set of sample autocorrelation coefficients, which measure the correlation between observations at different distances apart. These coefficients often provide insight into the probability model which generated the data. The correlation coefficient was introduced in Section 17.2.6: given n pairs of observations of two variables x and y, the correlation coefficient is

$$r = \frac{\sum_{i=1}^{n}(x_i - E(x))(y_i - E(y))}{\sqrt{\sum_{i=1}^{n}(x_i - (E(x))^2 \sum_{i=1}^{n}(y_i - E(y))^2}}$$

A similar idea can be applied to time series to see if successive observations are correlated. For instance, if you have n observations $x_1, \ldots, x_n$, of a discrete time series we can form $(n-1)$ pairs of observations. These pairs of observation are $(x_1, x_2), (x_2, x_3), \ldots, (x_{n-1}, x_n)$. If we take the second observation as the second variable, the correlation coefficient between x_t and x_{t+1} is given by equation (17.9):

$$r_1 = \frac{\sum_{t=1}^{n-1}(x_t - E(x_1))(x_{t+1} - E(x_2))}{\sqrt{\sum_{t=1}^{n-1}(x_t - E(x_1))^2 \sum_{t=1}^{n-1}(x_{t+1} - E(x_2))^2}} \qquad (17.9)$$

by analogy with the correlation coefficient where

$$E(x_1) = \frac{1}{n-1} \sum_{t=1}^{n-1} x_t$$

is the mean of the first $(n-1)$ observations and

$$E(x_2) = \frac{1}{n-1} \sum_{t=2}^{n} x_t$$

is the mean of the second $(n-1)$ observations.

If the autocorrelation coefficient is approximately equal to zero there is no autocorrelation between the two series. If the autocorrelation is approximately equal to one then there is a high autocorrelation. As the coefficient defined above measures correlation between successive observations, it is called an autocorrelation coefficient or serial correlation coefficient.

The autocorrelation equation as specified above is rather complicated, and so as $E(x_1)$ is approximately equal to $E(x_2)$ for large samples, the autocorrelation coefficient is usually approximated by

$$r_1 = \frac{\sum_{t=1}^{n-1}(x_t - E(x))(x_{t+1} - E(x))}{\sum_{t=1}^{n-1}(x_t - E(x))^2}$$

where

$$E(x) = \frac{1}{n} \sum_{t=1}^{n} x_t$$

is the overall mean. In a similar way it is possible to find the correlation between observations that are a distance k apart, as given by equation (17.10):

$$r_k = \frac{\sum_{t=1}^{n-k}(x_t - E(x))(x_{t+k} - E(x))}{\sum_{t=1}^{n-1}(x_t - E(x))^2} \qquad (17.10)$$

This is called the autocorrelation coefficient at a lag k. A useful method of observing the autocorrelation is to plot the autocorrelation as a function of the lag k. This is called a correlogram. In practice the autocorrelation coefficients are usually calculated by computing a series of auto-covariance coefficients (c_k). The form of this coefficient for lag k is:

$$c_k = \frac{1}{n}\sum_{t=1}^{n-k}(x_t - E(x))(x_{t+k} - E(x))$$

The autocorrelation coefficient can then be computed using

$$r_k = \frac{c_k}{c_0}$$

for $k = 1, 2, 3, ..., m$ and $m < n$. There is usually little point in calculating r_k for values greater than about $\frac{1}{4}n$.

17.4.5 Cross-correlation

The above discussion has concentrated on time series for a single variable. Sometimes it may also be necessary to determine time series relationships between two variables. The cross-correlation coefficient provides an indication of the strength of the correlation between two time series. Suppose that there are two variables x and y each containing n observations at unit time intervals over the same period. The observations are (x_1, y_1), $(x_2, y_2), ..., (x_n, y_n)$, and the method of calculating the cross-correlation and cross-correlation functions is by means of the corresponding sample functions. With n pairs of observations $((x_i, y_i), i = 1, ..., n)$, the cross-correlation function is

$$c_{xy}(k) = \frac{1}{n}\sum_{t=1}^{n-k}(x_t - E(x))(y_{t+k} - E(y))$$

where $k = 0, 1, ..., (n-1)$. The sample cross-correlation function is

$$r_{xy}(k) = \frac{c_{xy}(k)}{c_{xx}(0)\,c_{yy}(0)}$$

where $c_{xx}(0)$ and $c_{yy}(0)$ are sample variances of observations of x_t and y_t respectively.

17.4.6 Probabilistic models of time series

There are many possible probabilistic models for time series, which are collectively called *stochastic processes*. Statistics is concerned with estimating the properties of a population from a sample. In time series analysis it is often impossible to make more than one observation at a given time so that only one observation of the random variable is known at time t. Nevertheless, the one sample may be regarded as part of an infinite set of time series that may be observed. This infinite set of time series is called the ensemble. Each member of the ensemble is a possible realisation of the stochastic process.

A *stationary time series* has no trend component in the mean and no systematic change in the variance. Many of the filtering, curve fitting and modelling procedures of time series analysis (Chatfield 1984) attempt to remove the trend and periodic components to develop a stationary process.

A *random process* is one in which the random variables form a sequence of mutually independent, identically distributed variables.

A *moving average process (MR)* is one where there is systematic change in the mean of the process. The simplest example of a moving average is

$$\frac{1}{2q+1} \sum_{r=-q}^{+q} x_{t+r}$$

where q is the extreme of set of data points considered.

An *autoregressive process (AR)* is one where the present value of a variable is a function of previous values of the variable.

The combination of the above process is common. A useful class of time series combine the moving average (MR) process and the autoregressive (AR) process. Further details on the analysis of *mixed models* can be found in Chatfield (1984, pp.42-48).

17.4.7 An example of time series analysis

The data in Table 17.9 show the concepts behind time series. The table shows traffic volumes in successive four-week periods in each of four consecutive years. The data can be plotted and the seasonal trend investigated (Figure 17.11). They are more usefully observed by plotting the 'trend variation' with the plotted data. Trend variation is calculated by taking the average of the value and the six values either side of it. This is calculated in Table 17.9 and plotted in Figure 17.12. The seasonal variations are now very clear. The cycle repeats itself four times. If more cycles were available a curve could be fitted to this data and random variations could be checked for. The values on either end of the trend plot can be obtained by linear interpolation.

Table 17.9 Traffic flows over consecutive four-week periods over four years

Period	Year 1	Year 2	Year 3	Year 4
1	153	133	145	111
2	189	177	200	170
3	221	241	187	243
4	215	228	201	178
5	302	283	292	202
6	223	255	220	202
7	201	238	233	163
8	173	164	172	139
9	121	128	119	120
10	106	108	81	96
11	86	87	65	95
12	87	74	76	53
13	108	95	74	94

422 *Understanding Traffic Systems*

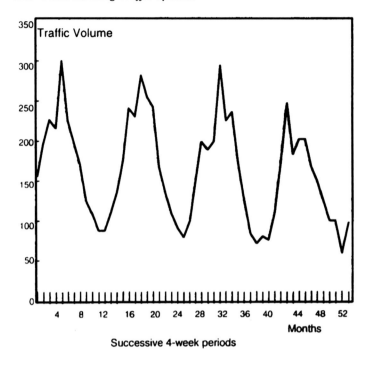

Figure 17.11 Time series plot of the traffic flows of Table 17.9

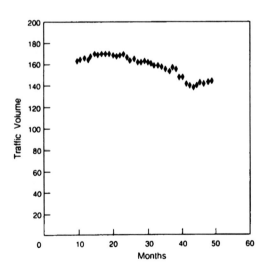

Figure 17.12 Averaged traffic flows in consecutive four-week periods over four years (see Table 17.9)

Appendix A: Statistical tables

The following tables provide critical values of sample statistics for use in statistical analysis, inference and hypothesis testing. Tables are provided for:
(1)   the normal distribution;
(2)   the t-distribution;
(3)   the F-distribution;
(4)   the χ^2 distribution;
(5)   the rank order correlation test, and
(6)   the Wilcoxon-Mann-Whitney two sample test.

Table A.1 Area under normal curve, to the right of the mean (z = 0)

z	0.00	0.01	0.02	0.03	0.04	0.05	0.06	0.07	0.08	0.09
0.0	.0000	.0040	.0080	.0120	.0160	.0199	.0239	.0279	.0319	.0359
0.1	.0398	.0438	.0478	.0517	.0557	.0596	.0636	.0675	.0714	.0753
0.2	.0793	.0832	.0871	.0910	.0948	.0987	.1026	.1064	.1103	.1141
0.3	.1179	.1217	.1255	.1293	.1331	.1368	.1406	.1443	.1480	.1517
0.4	.1554	.1591	.1628	.1664	.1700	.1736	.1772	.1808	.1844	.1879
0.5	.1915	.1950	.1985	.2019	.2054	.2088	.2123	.2157	.2190	.2224
0.6	.2257	.2291	.2324	.2357	.2389	.2422	.2454	.2486	.2517	.2549
0.7	.2580	.2611	.2642	.2673	.2704	.2734	.2764	.2794	.2823	.2852
0.8	.2881	.2910	.2939	.2967	.2995	.3023	.3051	.3078	.3106	.3133
0.9	.3159	.3186	.3212	.3238	.3264	.3289	.3315	.3340	.3365	.3389
1.0	.3413	.3438	.3461	.3485	.3508	.3531	.3554	.3577	.3599	.3621
1.1	.3643	.3665	.3686	.3708	.3729	.3749	.3770	.3790	.3810	.3830
1.2	.3849	.3869	.3888	.3907	.3925	.3944	.3962	.3980	.3997	.4015
1.3	.4032	.4049	.4066	.4082	.4099	.4115	.4131	.4147	.4162	.4177
1.4	.4192	.4207	.4222	.4236	.4251	.4265	.4279	.4292	.4306	.4319
1.5	.4332	.4345	.4357	.4370	.4382	.4394	.4406	.4418	.4429	.4441
1.6	.4452	.4463	.4474	.4484	.4495	.4505	.4515	.4525	.4535	.4545
1.7	.4554	.4564	.4573	.4582	.4591	.4599	.4608	.4616	.4625	.4633
1.8	.4641	.4649	.4656	.4664	.4671	.4678	.4686	.4693	.4699	.4706
1.9	.4713	.4719	.4726	.4732	.4738	.4744	.4750	.4756	.4761	.4767
2.0	.4772	.4778	.4783	.4788	.4793	.4798	.4803	.4808	.4812	.4817
2.1	.4821	.4826	.4830	.4834	.4838	.4842	.4846	.4850	.4854	.4857
2.2	.4861	.4864	.4867	.4871	.4875	.4878	.4881	.4884	.4887	.4890
2.3	.4893	.4896	.4898	.4901	.4904	.4906	.4909	.4911	.4913	.4916
2.4	.4918	.4920	.4922	.4925	.4927	.4929	.4931	.4932	.4934	.4936
2.5	.4938	.4940	.4941	.4943	.4945	.4946	.4948	.4949	.4951	.4952
2.6	.4953	.4955	.4956	.4957	.4958	.4960	.4961	.4962	.4963	.4964
2.7	.4965	.4966	.4967	.4968	.4969	.4970	.4971	.4972	.4973	.4974
2.8	.4974	.4975	.4976	.4977	.4977	.4978	.4979	.4979	.4980	.4981
2.9	.4981	.4982	.4982	.4983	.4984	.4984	.4985	.4985	.4986	.4986
3.0	.4987	.4987	.4987	.4988	.4988	.4989	.4989	.4989	.4990	.4990

Table A.2   Percentage points for t-distribution

				Significance			
d.f.	.20	.15	.10	.05	.025	.01	.005
1	1.376	1.936	3.078	6.314	12.706	31.821	63.657
2	1.061	1.386	1.886	2.920	4.303	6.965	9.925
3	.978	1.250	1.638	2.353	3.182	4.451	5.841
4	.941	1.190	1.533	2.132	2.776	3.747	4.604
5	.920	1.156	1.476	2.015	2.571	3.365	4.032
6	.906	1.134	1.440	1.943	2.447	3.143	3.707
7	.896	1.119	1.415	1.895	2.365	2.998	3.499
8	.889	1.108	1.397	1.860	2.306	2.896	3.355
9	.883	1.100	1.383	1.833	2.262	2.821	3.250
10	.879	1.093	1.372	1.812	2.228	2.764	3.169
11	.876	1.088	1.363	1.796	2.201	2.718	3.106
12	.873	1.083	1.356	1.782	2.179	2.681	3.055
13	.870	1.079	1.350	1.771	2.160	2.650	3.012
14	.868	1.076	1.345	1.761	2.145	2.624	2.977
15	.866	1.074	1.341	1.753	2.131	2.602	2.947
16	.865	1.071	1.337	1.746	2.120	2.583	2.921
17	.863	1.069	1.333	1.740	2.110	2.567	2.898
18	.862	1.067	1.330	1.734	2.101	2.552	2.878
19	.861	1.066	1.328	1.729	2.093	2.539	2.861
20	.860	1.064	1.325	1.725	2.086	2.528	2.845
21	.859	1.063	1.323	1.721	2.080	2.518	2.831
22	.858	1.061	1.321	1.717	2.074	2.508	2.819
23	.858	1.060	1.319	1.714	2.069	2.500	2.807
24	.857	1.059	1.318	1.711	2.064	2.492	2.797
25	.856	1.058	1.316	1.708	2.060	2.485	2.787
26	.856	1.058	1.315	1.706	2.056	2.479	2.779
27	.855	1.057	1.314	1.703	2.052	2.473	2.771
28	.855	1.056	1.313	1.701	2.048	2.467	2.763
29	.854	1.055	1.311	1.699	2.045	2.462	2.756
30	.851	1.055	1.310	1.697	2.042	2.457	2.750
40	.851	1.050	1.303	1.684	2.021	2.423	2.704
60	.848	1.046	1.296	1.671	2.000	2.390	2.660
120	.845	1.041	1.289	1.658	1.980	2.358	2.617
∞	.842	1.036	1.282	1.645	1.960	2.326	2.576

Table A.3 Percentage points for the F-distribution
(5 per cent upper line, 1 per cent lower line)

Degrees of freedom of numerator (df.1)

df.2*	1	2	3	4	5	6	7	8	9	10	100	∞
1	161	200	216	225	230	234	237	239	241	242	253	254
	4052	4999	5403	5625	5764	5859	5928	5981	6022	6056	6334	6366
2	18.5	19.0	19.2	19.3	19.3	19.3	19.4	19.4	19.4	19.4	19.5	19.5
	98.5	99.0	99.2	99.3	99.3	99.3	99.3	99.4	99.4	99.4	99.5	99.5
3	10.1	9.55	9.28	9.12	9.01	8.94	8.88	8.84	8.81	8.78	8.56	8.53
	34.1	30.8	29.5	28.7	28.2	27.9	27.7	27.5	27.3	27.2	26.2	26.1
4	7.71	6.94	6.59	6.39	6.26	6.16	6.09	6.04	6.00	5.96	5.66	5.63
	21.2	18.0	16.7	16.0	15.5	15.2	15.0	14.8	14.7	14.5	13.6	13.5
5	6.61	5.79	5.41	5.19	5.05	4.95	4.88	4.82	4.78	4.74	4.40	4.36
	16.3	13.3	12.1	11.4	11.0	10.7	10.5	10.3	10.2	10.1	9.13	9.02
6	5.99	5.14	4.76	4.53	4.39	4.28	4.21	4.15	4.10	4.06	3.71	3.67
	13.7	11.0	9.78	9.15	8.75	8.47	8.26	8.10	7.98	7.87	6.99	6.88
7	5.59	4.74	4.35	4.12	3.97	3.87	3.79	3.73	3.68	3.63	3.29	3.23
	12.3	9.55	8.45	7.85	7.46	7.19	7.00	6.84	6.71	6.62	5.75	5.65
8	5.32	4.46	4.07	3.84	3.69	3.58	3.50	3.44	3.39	3.34	3.00	2.93
	11.3	8.65	7.59	7.01	6.63	6.37	6.19	6.03	5.91	5.82	4.96	4.89
9	5.12	4.26	3.86	3.63	3.48	3.37	3.29	3.23	3.18	3.13	2.76	2.71
	10.6	8.02	6.99	6.42	6.06	5.80	5.62	5.47	5.35	5.26	4.41	4.31
10	4.96	4.10	3.71	3.48	3.33	3.22	3.14	3.07	3.02	2.97	2.59	2.40
	10.0	7.56	6.55	5.99	5.64	5.39	5.21	5.06	4.95	4.85	4.01	3.91
11	4.84	3.96	3.59	3.36	3.20	3.09	3.01	2.95	2.90	2.86	2.45	2.40
	9.65	7.20	6.22	5.67	5.32	5.07	4.88	4.74	4.63	4.54	3.70	3.60
12	4.75	3.88	3.49	3.26	3.11	3.00	2.92	2.85	2.80	2.76	2.35	2.30
	9.33	6.93	5.95	5.41	5.06	4.82	4.65	4.50	4.39	4.30	3.46	3.36
13	4.67	3.80	3.41	3.18	3.02	2.92	2.84	2.77	2.67	2.60	2.26	2.21
	9.07	6.70	5.74	5.20	4.86	4.62	4.44	4.30	4.19	4.10	3.27	3.16
14	4.60	3.74	3.34	3.11	2.96	2.85	2.77	2.70	2.65	2.60	2.19	2.13
	8.86	6.51	5.56	5.03	4.69	4.46	4.28	4.14	4.03	3.94	3.11	3.00
15	4.54	3.68	3.29	3.06	2.90	2.79	2.70	2.64	2.59	2.55	2.12	2.07
	8.68	6.36	5.42	4.89	4.56	4.32	4.14	4.00	3.89	3.80	2.97	2.87

* Degrees of freedom of denominator

Table A.3 Percentage points for the F-distribution
(5 per cent upper line, 1 per cent lower line) - *continued*

Degrees of freedom of numerator (df.1)

df.2*	1	2	3	4	5	6	7	8	9	10	100	∞
16	4.49	3.63	3.24	3.01	2.85	2.74	2.66	2.59	2.49	2.45	2.07	2.01
	8.53	6.23	5.29	4.77	4.44	4.20	4.03	3.89	3.78	3.69	2.86	2.75
17	4.45	3.59	3.20	2.96	2.81	2.70	2.62	2.55	2.50	2.45	2.02	1.96
	8.40	6.11	5.18	4.67	4.34	4.10	3.93	3.79	3.68	3.59	2.76	2.65
18	4.41	3.55	3.16	2.93	2.77	2.66	2.58	2.51	2.46	2.41	1.98	1.92
	8.28	6.01	5.09	4.58	4.25	4.01	3.85	3.71	3.60	3.51	2.68	2.57
19	4.38	3.52	3.13	2.90	2.74	2.63	2.55	2.48	2.43	2.38	1.94	1.88
	8.18	5.93	5.01	4.50	4.17	3.94	3.77	3.63	3.52	3.43	2.60	2.49
20	4.35	3.49	3.10	2.87	2.71	2.60	2.52	2.45	2.40	2.35	1.90	1.84
	8.10	5.85	4.94	4.43	4.10	3.87	3.71	3.56	3.45	3.37	2.53	2.42
21	4.32	3.47	3.07	2.84	2.68	2.57	2.49	2.42	2.37	2.32	1.87	1.81
	8.02	5.78	4.87	4.37	4.04	3.81	3.65	3.51	3.40	3.31	2.47	2.36
22	4.30	3.44	3.05	2.82	2.66	2.55	2.47	2.40	2.35	2.30	1.84	1.78
	7.94	5.72	4.82	4.31	3.99	3.76	3.59	3.45	3.35	3.26	2.42	2.31
23	4.28	3.42	3.03	2.80	2.64	2.53	2.45	2.38	2.32	2.28	1.82	1.76
	7.88	5.66	4.76	4.26	3.94	3.71	3.54	3.41	3.30	3.21	2.37	2.26
24	4.26	3.40	3.01	2.78	2.62	2.51	2.43	2.36	2.30	2.26	1.80	1.73
	7.82	5.61	4.72	4.22	3.90	3.67	3.50	3.36	3.25	3.17	2.33	2.21
25	4.24	3.38	2.99	2.76	2.60	2.49	2.41	2.34	2.28	2.24	1.77	1.71
	7.77	5.57	4.68	4.18	3.86	3.63	3.46	3.32	3.21	3.13	2.29	2.17
50	4.03	3.18	2.79	2.56	2.40	2.29	2.20	2.13	2.07	2.02	1.52	1.44
	7.17	5.06	4.20	3.72	3.41	3.18	3.02	2.88	2.78	2.70	1.82	1.68
100	3.94	3.09	2.70	2.46	2.30	2.19	2.10	2.03	1.97	1.92	1.39	1.28
	6.90	4.82	3.98	3.51	3.20	2.99	2.82	2.69	2.59	2.51	1.59	1.43
1000	3.85	3.00	2.61	2.38	2.22	2.10	2.02	1.95	1.89	1.84	1.26	1.08
	6.66	4.62	3.80	3.34	3.04	2.82	2.66	2.53	2.43	2.34	1.38	1.11
∞	3.84	2.99	2.60	2.37	2.21	2.09	2.01	1.94	1.88	1.83	1.24	1.00
	6.64	4.60	3.78	3.32	3.02	2.80	2.64	2.51	2.41	2.32	1.36	1.00

* Degrees of freedom of denominator

Table A.4.1 Rank order correlations of d^2 significant at 5 per cent level

Sample size (n)	Σd^2 Lower Limit	Upper Limit	Significance level
4	0	20	0.0417
5	2	38	0.0417
6	4	64	0.0246
7	16	96	0.0440
8	30	138	0.0469
9	48	192	0.0470
10	72	258	0.0472
11	83.6	396.4	0.050
12	117.0	455.0	0.050
13	158.0	570.0	0.050
14	207.7	702.3	0.050
15	266.7	853.3	0.050
16	335.9	1024.1	0.050
17	416.2	1215.8	0.050
18	508.4	1429.6	0.50
19	613.3	1666.7	0.50
20	732.0	1928.0	0.50

Table A.4.2 Wilcoxon-Mann-Whitney two sample test

Large sample size 2
Small sample size

1		2	
W	Pr	W	Pr
1	0.333	3	0.167
		4	0.333

Large sample size 4
Small sample size

1		2		3		4	
W	Pr	W	Pr	W	Pr	W	Pr
1	0.200	3	0.067	6	0.029	10	0.014
2	0.400	4	0.133	7	0.057	11	0.029
		5	0.267	8	0.114	12	0.057
		6	0.400	9	0.200	13	0.100
				10	0.314	14	0.171
				11	0.429	15	0.243
						16	0.343

Large sample size 6
Small sample size

1		2		3		4		5		6	
W	Pr	W	Pr	W	Pr	W	Pr	W	Pr	W	Pr
1	0.143	3	0.036	6	0.012	10	0.005	15	0.002	21	0.001
2	0.286	4	0.071	7	0.024	11	0.010	16	0.004	22	0.002
3	0.429	5	0.143	8	0.048	12	0.019	17	0.009	23	0.004
		6	0.214	9	0.083	13	0.033	18	0.015	24	0.008
		7	0.321	10	0.131	14	0.057	19	0.026	25	0.013
		8	0.429	11	0.190	15	0.086	20	0.041	26	0.021
				12	0.274	16	0.129	21	0.063	27	0.032
				13	0.357	17	0.176	22	0.089	28	0.047
				14	0.452	18	0.238	23	0.123	29	0.066
						19	0.305	24	0.165	30	0.090
						20	0.381	25	0.214	31	0.120
						21	0.457	26	0.298	32	0.155
								27	0.331	33	0.197
								28	0.398	34	0.242
								29	0.465	35	0.294

Table A.5   Percentage points for χ^2 distribution

df	\multicolumn{9}{c}{Significance}								
	0.99	0.95	0.90	0.75	0.50	0.25	0.10	0.05	0.01
1	0.032	0.023	0.016	0.102	0.455	1.323	2.71	3.84	6.63
2	0.020	0.103	0.211	0.575	1.386	2.77	4.61	5.99	9.21
3	0.115	0.352	0.584	1.213	2.37	4.11	6.25	7.81	11.34
4	0.297	0.711	1.064	1.923	3.36	5.39	7.78	9.49	13.28
5	0.554	1.145	1.610	2.67	4.35	6.63	9.24	11.07	15.09
6	0.872	1.237	1.635	2.20	3.45	5.35	7.84	12.59	16.81
7	1.239	2.17	2.83	4.25	6.35	9.04	12.02	14.07	18.48
8	1.646	2.73	3.49	5.07	7.34	10.22	13.36	15.51	20.1
9	2.09	3.33	4.17	5.90	8.34	11.39	14.68	16.92	21.7
10	2.56	3.94	4.87	6.74	9.34	12.55	15.99	18.31	23.2
11	3.05	4.57	5.58	7.58	10.34	13.70	17.28	19.68	24.7
12	3.57	5.23	6.30	8.44	11.34	14.85	18.55	21.0	26.2
13	4.11	5.89	7.04	9.30	12.34	15.98	19.81	22.4	27.7
14	4.66	6.57	7.79	10.17	13.34	17.12	21.1	23.7	29.1
15	5.23	7.26	8.55	11.04	14.34	18.25	22.3	25.0	30.6
16	5.81	7.96	9.31	11.91	15.34	19.37	23.5	26.3	32.0
17	6.41	8.67	10.09	12.79	16.34	20.5	24.8	27.6	33.4
18	7.01	9.39	10.86	13.68	17.34	21.6	26.0	28.9	34.8
19	7.63	10.12	11.65	14.56	18.34	22.7	27.2	30.1	36.2
20	8.26	10.85	12.44	15.54	19.34	23.8	31.4	34.2	37.6
21	8.90	11.59	13.24	16.34	20.3	24.9	32.7	35.5	38.9
22	9.54	12.34	14.04	17.24	21.3	26.0	30.8	33.9	40.3
23	10.20	13.09	14.85	18.14	22.3	27.1	32.0	35.2	41.6
24	10.86	13.85	15.66	19.04	23.3	28.2	33.2	36.4	43.0
25	11.52	14.61	16.47	19.94	24.3	29.3	34.4	37.7	44.3
26	12.20	15.38	17.29	20.8	25.3	30.4	35.6	38.9	45.6
27	12.88	16.15	18.11	21.7	26.3	31.5	36.7	40.1	47.0
28	13.56	16.93	18.94	22.7	27.3	32.6	37.9	41.3	48.3
29	14.26	17.71	19.77	23.6	28.3	33.7	39.1	42.6	49.6
30	14.95	18.49	20.6	24.5	29.3	34.8	40.3	43.8	50.9
40	22.2	26.5	29.1	33.7	39.3	45.6	51.8	59.3	63.7
50	29.7	34.8	37.7	42.9	49.3	56.3	63.2	67.5	76.2
60	37.5	43.2	46.5	52.3	59.3	67.0	74.4	79.1	88.4
70	45.4	51.7	55.3	61.7	69.3	77.6	85.5	90.5	100.4
80	53.5	60.4	64.3	71.1	79.3	88.1	96.6	101.9	112.3
90	61.8	69.1	73.3	80.6	89.3	98.6	107.6	113.1	124.1
100	70.1	77.9	82.4	90.1	99.3	109.1	118.5	124.3	135.8

Appendix B: Database of vehicle speeds on residential streets

The following vehicle speed data were observed at different sites on residential streets, and have been related to other traffic and physical variables pertaining to the data collection sites. For each data record in the table, the observed variables are:

(1) SPEED (km/h), the mean time speed observed at the site in a five minute period for traffic flow in one direction;
(2) VEHS, the number of vehicles travelling the one direction in the five minute period;
(3) OPVEH, the number of vehicles travelling in the opposite direction in the five minute priod;
(4) FLOW (veh/min), the traffic volume in the five minute period for the flow direction considered;
(5) OPFLO (veh/min), the traffic volume in the five minute period for the opposite flow direction;
(6) PLOC (%), the percentage of local traffic in the five minute period for the flow direction considered, where local traffic is taken as a vehicle observed to park, unpark, or turn into or out off a property along the road section;
(7) NINT (intersections/km), the density of intersections along the road section, and
(8) WIDTH (m), the effective width of the road at the observation site, assessed in terms of the available width for the passage of vehicles (and thus not including that part of the physical road width occupied by parked vehicles at the observation site).

There are 74 data records in the table.

SPEED	VEH	OPVEH	FLOW	OPFLO	PLOC	NINT	WIDTH
50.2	9	9	1.8	1.8	22.2	2.39	5.28
61.1	8	4	1.6	0.8	0.0	2.39	5.28
56.0	7	6	1.4	1.2	28.6	2.39	5.28
55.4	7	4	1.4	0.8	14.3	2.39	5.28
54.7	10	6	2.0	1.2	30.0	2.39	5.28
51.2	8	2	1.6	0.4	37.5	2.39	5.28
61.1	6	1	1.2	0.2	16.7	2.39	5.28
56.8	7	6	1.4	1.2	42.9	2.39	5.28
58.7	6	4	1.2	0.8	16.7	2.39	5.28
60.3	5	6	1.0	1.2	60.0	2.39	5.28
57.9	9	2	1.8	0.4	11.1	2.39	5.28
51.5	3	6	0.6	1.2	0.0	2.39	5.28
52.6	6	0	1.2	0.0	30.0	2.39	5.28
49.1	6	2	1.2	0.4	66.7	2.39	5.28
54.8	8	5	1.6	1.0	0.0	2.39	5.28
58.4	7	4	1.4	0.8	28.6	2.39	5.28
58.4	7	2	1.4	0.4	14.3	2.39	5.28
51.8	5	3	1.0	0.6	20.0	2.39	5.28
64.8	1	3	0.2	0.6	0.0	2.39	5.28
56.3	10	7	2.0	1.4	10.0	2.39	5.28
54.6	11	2	2.2	0.4	9.1	2.39	5.28
55.6	5	1	1.0	0.2	20.0	2.39	5.28
58.7	8	6	1.6	1.2	0.0	2.39	5.28
57.1	7	4	1.4	0.8	37.5	2.39	5.28
59.5	3	6	0.6	1.2	0.0	2.39	5.28
52.8	5	2	1.0	0.4	20.0	2.39	5.28
54.1	7	1	1.4	0.2	0.0	12.80	3.16
48.8	4	2	0.8	0.4	25.0	12.80	3.16
50.8	8	2	1.8	0.4	0.0	12.80	3.16
47.7	4	1	0.8	0.2	0.0	12.80	3.16
41.1	12	1	2.4	0.2	33.3	12.80	3.16
50.4	4	1	0.8	0.2	0.0	12.80	4.16
50.4	1	3	0.2	0.6	0.0	12.80	4.16
49.9	7	2	1.4	0.4	28.6	12.80	4.16
36.0	1	0	0.2	0.0	0.0	12.80	6.16
51.2	4	7	0.8	1.4	0.0	12.80	6.16
46.8	4	7	0.8	1.4	0.0	12.80	6.16
48.8	5	2	1.0	0.4	0.0	12.80	6.16
34.4	4	3	0.8	0.6	25.0	12.80	2.80
48.6	7	4	1.4	0.8	28.6	12.80	2.80
50.8	8	3	1.6	0.6	0.0	12.80	2.80
40.0	6	2	1.2	0.4	16.7	12.80	2.80

SPEED	VEH	OPVEH	FLOW	OPFLO	PLOC	NINT	WIDTH
43.0	5	2	1.0	0.4	20.0	12.80	4.16
52.0	4	2	0.8	0.4	25.0	12.80	4.16
31.6	4	1	0.8	0.2	75.0	12.80	6.16
57.6	2	2	0.4	0.4	0.0	12.80	6.16
44.0	8	5	1.6	0.0	0.0	12.80	6.16
49.3	3	1	0.6	0.2	0.0	12.80	6.16
42.4	4	3	0.8	0.6	0.0	12.80	6.16
50.1	6	3	1.2	0.6	0.0	12.80	6.16
52.4	4	0	0.8	0.0	25.0	12.80	6.45
44.5	3	1	0.6	0.2	66.7	12.80	6.45
47.2	4	4	0.8	0.8	0.0	12.80	6.45
47.8	6	1	1.2	0.2	33.3	12.80	6.45
54.8	4	2	0.8	0.4	0.0	12.80	6.45
59.5	4	2	0.8	0.4	0.0	12.80	6.45
50.9	4	2	0.8	0.4	0.0	12.80	6.45
46.6	5	5	1.0	1.0	20.0	12.80	6.45
52.0	6	2	1.2	0.4	0.0	12.80	6.45
52.0	6	1	1.2	0.2	0.0	12.80	6.45
49.2	5	2	1.0	0.4	25.0	12.80	6.45
61.6	3	2	0.6	0.4	0.0	12.80	6.45
54.4	2	0	0.4	0.0	0.0	12.80	5.38
49.6	2	2	0.4	0.4	0.0	12.80	5.38
44.5	7	1	1.4	0.2	0.0	12.80	5.38
39.7	5	2	1.0	0.4	20.0	12.80	5.38
51.0	5	2	1.0	0.4	0.0	12.80	5.38
48.8	1	3	0.2	0.6	0.0	12.80	5.38
47.2	6	4	1.2	0.8	0.0	12.80	5.38
53.0	8	3	1.6	0.6	0.0	12.80	5.38
52.0	6	2	1.2	0.4	0.0	12.80	5.38
52.0	3	0	0.6	0.0	0.0	12.80	5.38
45.6	3	1	0.6	0.2	33.3	12.80	5.38
52.8	2	1	0.4	0.2	0.0	12.80	5.38

References

Adams, J C and Hummer, J E (1993). Effects of u-turns on left-turn saturation flow rates. *Transportation Research Record 1398*, pp.90-100.

Adler, J and McNally, M (1994). In-laboratory experiments to investigate driver behaviour under Advanced Traveller Information Systems. *Transportation Research C 2C* (3), pp.129-148.

Akcelik, R (1981). *Traffic signals: capacity and timing analysis*. Research Report ARR123, Australian Road Research Board, Melbourne.

Akcelik, R (1988). The Highway Capacity Manual delay formula for signalized intersections. *ITE Journal 58* (3), pp.23-7.

Akcelik, R (1991). Travel time functions for transport planning purposes: Davidson's function, its time-dependent form and an alternative travel time function. *Australian Road Research 21* (3), pp.49-59.

Almond, J (1963). Sampling vehicles after a given instant of time. *Traffic Engineering and Control 6*, pp.408-9.

Apelbaum, J and Richardson, A J (1978). An airport ground traffic model. *Australian Road Research 8* (2), pp.3-12.

Asakura, Y, Hato, E, Nishibe, Y, Daito, T, Tanabe, J and Koshima, H (1999). Monitoring travel behaviour using PHS based location positioning service system. *Proc 6th World Congress on Intelligent Transport Systems*. Toronto, November. (ITS America: Washington, DC and ITS Canada: Toronto).

AUSTROADS (1988). *Guide to traffic engineering practice- Part 2: roadway capacity*. (AUSTROADS: Sydney).

Biggs, D C and Akcelik, R (1986). Car fuel consumption in urban traffic. *Proc Australian Road Research Board Conf 13* (7), pp.123-32.

Bonsall, P W (1991). The changing face of parking related data collection and analysis: the role of new technologies. *Transportation 18*, pp.83-106.

Bonsall, P W, Clarke, R, Firmin, P and Palmer, I (1994). VLADIMIR and TRAVSIM, powerful aids for route choice research. *Proc 22nd European Transport Forum*, Seminar H, pp.65-76. PTRC, London.

Bonsall, P W, Firmin, P E, Anderson, M E, Palmer, I A and Balmforth, P J (1997). Validating the results of a route choice simulator. *Transportation Research C 5C* (6), pp.371-387.

Bonsall, P W, Ghari-Saremi, F, Tight, M and Marler, N (1988). The performance of hand held data-capture devices in traffic and transport surveys. *Traffic Engineering and Control 29* (1), pp.10-19.

Bonsall, P W and McKimm, J (1993). Non-response bias in roadside mailback surveys. *Traffic Engineering and Control 34* (12).

Bonsall, P W and Montgomery, F O (1984). Bias and error in estimates of day-to-day variability in driver behaviour. *Proc 12th PTRC Summer Annual Meeting.* Seminar H, pp.191-200. PTRC, London.

Bonsall, P W and Palmer, I (1993). A cheap source of data on accident risk? *Final report to sponsors (TRL) on project on Accidents and Deceleration Data*, Institute for Transport Studies, University of Leeds.

Bonsall, P W and Parry, T (1991). Using an interactive route-choice simulator to investigate drivers' compliance with route guidance advice. *Transportation Research Record 1306*, pp.59-68.

Bonsall, P W, Pearman, A and Cobbett, M (1991). MASCOT: a decision support systems to help in the definition, development and ranking of scheme options. *Proc 19th PTRC Summer Annual Meeting.* Seminar G. PTRC, London.

Boyle, A J and Wright, C C (1984). Accident migration after remedial treatment at accident blackspots. *Traffic Engineering and Control 25* (5).

Brackstone, M, McDonald, M, Sultan, B and Mould, B (1999). Five years of the instrumented vehicle, what have we learned? *Traffic Engineering and Control 40* (11), pp.537-540.

Brindle, R E and Barnard, P O (1985). Traffic generation estimation suggestions on new research directions. *Papers of the Australasian Transport Research Forum 10* (1), pp.43-62.

Brown, A L (1980). The noise-response relationship near a freeway. *Proc Australian Road Research Board Conf 10* (5), pp.144-53.

Buchanan, C D (1963). *Traffic in Towns.* (Penguin Books: Harmondsworth).

Burrow, I (1989). A note on traffic delay formulas. *ITE Journal 59* (10), pp.29-32.

Carterette, and Friedman, (1974). *Handbook on Perception. Vol II: Psychophysical Judgement and Measurement.* (Academic Press: New York).

Chambers, J M, Cleveland, W S, Kleiner, B and Tukey, P A (1983). *Graphical Data Analysis.* (Duxbury Press: Boston).

Chatfield, C (1984). *The Analysis of Time Series: An Introduction.* (3rd ed) (Chapman and Hall: London).

Cochran, W G (1977). *Sampling Techniques* (3rd ed.). (John Wiley and Sons: New York).

CTS (1998). *Proc Int Conf on Transportation into the Next Millennium.* Singapore, September. (Centre for Transportation Studies, Nanyang Technological University: Singapore).

Cuddon, A P (1993). Saturation flows at signalised intersections. PhD Thesis, Monash University.

Davidson, K B (1978). The theoretical basis of a flow travel-time relationship for use in transportation planning. *Australian Road Research 8* (1), pp.32-5.

D'Este, G M (1997). A technique for incorporating the effect of changing patterns of travel behaviour into the traditional transport planning paradigm. *Journal of the Eastern Asia Society for Transportation Studies 2* (4), pp.1099-1111.

Dew, A M and Bonsall, P W (1991). Evaluation of automatic transcription of audio tape for registration plate surveys. In Ampt, E S, Richardson, A J and Meyburg, A N (eds). *Selected Readings in Transport Survey Methodology.* (Eucalyptus Press: Melbourne), pp.147-161.

Dia, H and Rose, G (1997). A variable threshold incident detection model. *Journal of the Eastern Asia Society for Transportation Studies 2* (4), pp.1329-1345.

Ettema, D F and Timmermans, H J P (1996). *Activity-Based Approaches to Travel Analysis.* (Elsevier-Pergamon: Oxford).

Fowkes, A S, Bristow, A L, Bonsall, P W and May, A D (1998). A shortcut method for optimising strategic transport models. *Transportation Research A 32A* (2), pp.149-157.

Gipps, P G (1981). A behavioural car-following model for computer simulation. *Transportation Research B 15B*, pp.105-11.

Gordon, C S, Galloway, W J, Kluger, B A and Nelson, D J (1971). Highway noise: a design guide for highway engineers. *NCHRP Report 117*, US Department of Transportation, Washington DC.

Grayson, G B and Hakkert, A S (1987). Accident analysis and conflict behaviour. In Rothengatter, V and de Bruin, R (eds) *Road Users and Traffic Safety.* (Van Gorcum: Assen/Maastricht.).

Greenwood, I B and Taylor, M A P (1990). Computer analysis of origin-destination traffic surveys. *Proc Australian Road Research Board Conf 15* (6), pp.189-208.

Hallam, C E (1980). Land use traffic impact prediction. *Proc Australian Road Research Board Conf 10* (5), pp.12-29.

Hauer, E (1979). Correction of license plate surveys for spurious matches. *Transportation Research A* 13A (1), pp.71-8.

Hauer, E, Pagitsas, E and Shin, B T (1981). Estimation of turning flows from automatic counts. *Transportation Research Record 795*, pp.1-8.

Hayashi, Y and Roy, J R (eds) (1996). *Transport, Land-Use and the Environment*. (Kluwer Academic Publishers: Dordrecht).

HCM (1995). Highway capacity manual. Special Report 209 update. US Transportation Research Board, Washington, DC.

Hensher, D A and Johnson, L W (1981). *Applied Discrete Choice Modelling*. (Croom Helm: New York).

Hidas, P (1998). A car following model for urban traffic simulation. *Traffic Engineering and Control 39* (5), pp.300-309.

Hothershall, D C and Salter, R J (1977). *Transport and the Environment*. (Crosby Lockwood Staples: London).

Hunt, P B, Robertson, D I, Bretherton, R D and Winton, R I (1981). SCOOT a traffic responsive method of coordinating signals. *TRRL Report LR1014*. Transport and Road Research Laboratory, Crowthorne, Berks.

Hyden, C (1987). The development of a method for traffic safety evaluation; the Swedish traffic conflicts technique. *Bulletin 70*. Department of Traffic Planning and Engineering, Lund Institute of Technology.

ITE (1997). *Trip Generation* (6th edition). (Institute of Transportation Engineers: Washington, DC).

James, H F (1991). Underreporting of road traffic accidents. *Traffic Engineering and Control 32* (12).

Jewell, W S (1967). Models for traffic assignment. *Transportation Research 1*, pp.31-46.

Johnston, R R M, Trayford, R S and van der Touw, J W (1982). Fuel consumption in urban traffic a twenty car designed experiment. *Transportation Research A 16A*, pp.173-84.

Jones, P M, Dix, M C, Clarke, M I and Heggie, I G (1983). *Understanding Travel Behaviour*. Oxford Studies in Transport. (Gower: Aldershot).

Kimber, R M, McDonald, M and Hounsell, N (1985). Passenger car units in saturation flows: concepts, definition, derivation. *Transportation Research B 19B* (1), pp.39-61.

Kimber, R M, McDonald, M and Hounsell, N B (1986). The prediction of saturation flows for road junctions controlled by traffic signals. *Research Report 67*, Transport and Road Research Laboratory, Crowthorne, Berks.

Kitamura, R and Fujii, S (1998). Two computational process models of activity-travel choice. *In* Gärling, T, Laitla, T and Westin, K (eds). *Theoretical Foundations of Travel Choice Modeling*. (Elsevier: New York).

Klungboonkrong, P and Taylor, M A P (1999). An integrated planning tool for evaluating road environmental impacts. *Computer-Aided Civil and Infrastructure Engineering 14*, pp.335-345.

Koutsopoulos, H, Lotan, T. and Yang, Q (1994). A driving simulator and its application for modelling route choice in the presence of information. *Transportation Research C 2C* (2), pp.91-108.

Layfield, R E, Summersgill, I, Hall, R D and Chatterjee, K (1996). Accidents at priority crossroads and staggered junctions. TRL Report 185. Transport Research Laboratory, Crowthorne, Berks.

Lebacque, J P and Lesort, J B (1999). Macroscopic traffic flow models: a question of order. In Ceder, A (ed). *Transportation and Traffic Theory*. (Elsevier-Pergamon: Oxford), pp.3-25.

Lighthill, M J and Whitham, G B (1955). On kinematic waves: a theory of traffic flow on long crowded roads. *Proc Royal Society, Series A 229*, pp.317-345.

Mackenzie, N B and McCallum, D G (1985). A locationally references highway management information systems, the CHIPS database in Scotland. *Traffic Engineering and Control 26* (10).

Maher, M J (1985). The analysis of partial registration-plate data. *Traffic Engineering and Control 26* (10), pp.495-7.

Maher, M J (1987). Accident migration; a statistical explanation. *Traffic Engineering and Control 28* (9).

Mahmassani, H and Herman, R (1989). Interactive experiments for the study of trip maker behaviour dynamics in congested commuting conditions. In Jones, P M (ed). *New Developments in Dynamic and Activity Approaches*. (Gower: Aldershot).

Massey, F J (1951). The Kolmogorov-Smirnov test for goodness-of-fit. *Journal of American Statistical Association 46*, pp.68-78.

May, A D, Milne, D, Smith, M, Ghali, M and Wisten, M (1996). A comparison of the performance of alternative road pricing systems. *In* Hensher, D A, King, J and Oum, T (eds). *World Transport Research 3: Transport Policy*. (Elsevier: New York), pp.335-346.

OECD (1986). Strengthening noise abatement policies. *OECD Observer 140*, pp.31-3.
Older, S J and Spicer, B R (1976). Traffic conflicts: a development in accident research. *Human Factors 18.*
Payne, H J (1979). FREFLO: a macroscopic simulation model of freeway traffic. *Transportation Research Record 772*, pp.68-75.
Perkins, S R and Harris, J L (1968). Traffic conflict characteristics: accident potential at intersections, *GMR report 718.* General Motors Research Laboratory, Warren, Michigan.
Peters, J, Mammano, F, Dennard, D and Inman, V (1993). TravTek, evaluation overview and recruitment statistics. *Proc Int Conf on Vehicle Navigation and Information Systems.* Ottawa, IEEE, Piscataway, USA.
Pitcher, I K (1990). Speed profiles of isolated vehicles on residential streets. *Proc Australian Road Research Board Conf 15* (5), pp.155-80.
Pretty, R L and Troutbeck, R J (1989). The control of highway intersections. *Papers of the Australasian Transport Research Forum 14* (1), pp.323-46.
Reilly, W R, Gardner, G C and Kell, J H (1976). A technique for measuring delay at intersections. NTIS Report PB-265 701. National Academy of Sciences, Washington DC.
Richardson, A J (1974). An improved parking duration study method. *Proc Australian Road Research Board Conf 7* (2), pp.397-413.
Richardson, A J and Graham, N R (1982). QDELAY: a signalized intersection survey method. *Proc Australian Road Research Board Conf 11* (4), pp.34-52.
Richardson, A J, Ampt, E S, and Meyburg, A N (1995). *Survey Methods for Transport Planning.* (Eucalyptus Press: Melbourne).
Robertson, D R (1969). TRANSYT: a traffic network study tool. *TRRL Report LR 253.* Road Research Laboratory, Crowthorne, Berks.
RTA-NSW (1993). *Guide to traffic generating developments.* Issue 2.0. Roads and Traffic Authority of New South Wales, Sydney.
Satterthwaite, S P (1981). A survey of research into relationships between traffic accidents and traffic volumes. *TRRL Report SR692,* Transport and Road Research Laboratory, Crowthorne, Berks.
Schulman, L L and Stout, J (1970). A parking study through the use of origin-destination data. *Highway Research Record 317.*
Seco, A (1991). Driver behaviour at uncontrolled junctions. PhD thesis. Department of Civil Engineering, University of Leeds.

Seddon, P A (1972). Another look at platoon dispersion 3: the recurrence relation. *Traffic Engineering and Control 13*, pp.442-50.

Shewey, P J H (1983). An improved algorithm for matching partial registration numbers. *Transportation Research B 17B* (5), pp.391-7.

Sims, A G and Dobinson, K W (1979). SCAT: the Sydney coordinated adaptive traffic system philosophy and benefits. *Proc Int Symp on Traffic Control Systems 2B*, pp.9-42. Institute of Transportation Studies, University of California, Berkeley.

Solomon, D (1957). Accuracy of the volume-density method of measuring travel time. *Traffic Engineering 27* (3).

Stamatiadis, C and Gartner, N H (1999). Progression optimization in large scale urban traffic networks: a heuristic decomposition approach. *In* Ceder, A (ed). *Transportation and Traffic Theory*. (Pergamon-Elsevier: Oxford), pp.645-661.

Taylor, M A P (1984). A note on using Davidson's function in equilibrium assignment. *Transportation Research B 18B* (3), pp.191-9.

Taylor, M A P (1991). Traffic planning by a 'desktop expert'. *Computers, Environment and Urban Systems 15* (3), pp.165-77.

Taylor, M A P (1992). Exploring the nature of urban traffic congestion: concepts, parameters, theories and models. *Proc Australian Road Research Board 16th Conference 16* (5), pp.83-106.

Taylor, M A P (1999). An extended family of traffic network equilibria and its implications for land use and transport policies. In Meersman, H, Van Der Voorde, E and Winkelmans, W (eds). *World Transport Research: Selected Proceedings from the 8th World Conference on Transport Research. Volume 4: Transport Policy.* (Elsevier-Pergamon: Oxford), pp.29-42.

Taylor, M A P and Anderson, M (1988). Modelling the environmental impacts of urban road traffic with MULATM-POLDIF: a PC-based system with interactive graphics. *Environment and Planning B 15*, pp.192-200.

Taylor, M A P, Miller, A J and Ogden, K W. (1974). A comparison of some bunching models for rural traffic flows. *Transportation Research 8* (1), pp.1-9.

Taylor, M A P and Young, T M (1996). Developing a set of fuel consumption and emissions models for use in traffic network modelling. *In* Lesort, J-B (ed). *Transportation and Traffic Theory*. (Elsevier-Pergamon: Oxford), pp.289-314.

Teply, S (1989). Accuracy of delay at signalized intersections. *Transportation Research Record 1225*, pp.24-32.

Thompson, R G (1989). A sampling technique for origin-destination surveys. *Australian Road Research 19* (3), pp.230-3.

Thompson-Clement, S J, Woolley, J E and Taylor, M A P (1996). Applying the transport network relational database to a turning flows study and a traffic noise model. *In* Hensher, D A, King, J and Oum, T H (eds) *World Transport Research - Volume 2 Modelling Transport Systems*. (Elsevier-Pergamon: Oxford), pp.257-270.

Tight, M, Carsten, O, Kirby, H, Southwell, M and Leake, G (1990). Urban road traffic accidents: an in depth study. *Proc 18th PTRC Summer Annual Meeting*, Seminar G. PTRC, London.

Tisato, P (1991). Suggestions for an improved Davidson travel time function. *Australian Road Research 21* (2), pp.85-100.

Tracz, M (1975). The prediction of platoon dispersion based on a rectangular distribution of journey time. *Traffic Engineering and Control 16*, pp. 490-2.

TRL (1987). *Guidelines for the Traffic Conflict Technique*. Institution of Highways and Transportation, London.

Troutbeck, R J (1986). Average delay at an unsignalised intersection with the major streams each having a dichotomised headway distribution. *Transportation Science 20* (4), pp.272-86.

Troutbeck, R J (1993). Effect of heavy vehicles at Australian traffic circles and unsignalised intersections. *Transportation Research Record 1398*, pp.54-60.

Tukey, J W (1977). *Exploratory Data Analysis*. (Addison-Wesley: Reading, Ma).

Van Aerde, M, Mackinnon, G and Hellinga, B (1991). The generation of synthetic O-D demands from real-time vehicle probe data: potential and limitations. *Proc Int Conf on Vehicle Navigation and Information Systems*, Dearborn. IEEE/SAE, pp.891-900.

Van Zuylen, J H and Willumsen, L G (1980). The most likely trip matrix estimated from traffic counts. *Transportation Research B 14B* (4), pp.281-293.

Velleman, P F and Hoaglin, D C (1981). *Applications, Basics and Computing of Exploratory Data Analysis*. (Duxbury Press: Boston).

Von Tomkewitsch, R. (1987). LISB: large scale tests of 'Navigation and Information Systems Berlin'. *Proc 16th PTRC Summer Annual Meeting*. Seminar ITT. PTRC, London.

Walpole, R E and Myers, R H (1988). *Probability and Statistics for Engineers and Scientists*. (Collier-Macmillan: London).

Wardrop, J G (1952). Some theoretical aspects of road traffic research. *Proceedings of the Institution of Civil Engineers 1* (2), pp.325-78.

Wardrop, J G and Charlesworth, G (1954). A method of estimating speed and flow of traffic from a moving vehicle. *Proceedings of the Institution of Civil Engineers 3* (2), p.158.

Wigan, M R (1976). The estimation of environmental impacts for transport policy assessment. *Environment and Planning B 8*, pp.125-47.

Wilson, A J (1967). A statistical theory of spatial distribution models. *Transportation Research 1*, pp. 253-69.

Woolley, J E, Klungboonkrong, P and Taylor, M A P (1997). Use of a decision support tool and a network noise model to gauge community noise impacts. *Papers of the Australasian Transport Research Forum 21* (2), pp.773-788.

Wootton, J and Potter, R (1981). Video-recorders, micro computers and new survey techniques. *Traffic Engineering and Control 22* (4), pp.213-215.

Young, W, Ritchie, S G and Ogden, K W (1980). Factors influencing freight facility location preference. *Transportation Research Record 747*, pp.71-8.

Young, W, Taylor, M A P and Gipps, P G (1989). *Microcomputers in Traffic Engineering*. (Research Studies Press: Taunton).

Young, W and Thompson, R G (1990). Parking duration surveys: a comparative study. *Research Report ARR 179*, Australian Road Research Board, Melbourne.

Zito, R, D'Este, G M and Taylor, M A P (1995). GPS in the time domain: how useful a tool for IVHS? *Transportation Research C* (submitted for publication).

Index

absolute capacity 106
absorption capacity 76-78
acceleration 8, 12, 45, 63, 64, 131, 180, 187, 218, 309
access 3, 11, 13, 33, 111-113, 128, 138, 172, 183, 267, 281, 291
accessibility 3, 5, 110
accident analysis 132, 309, 311, 316
accident migration 319
accident propensity 132
accident rate 132, 309, 319
accident report form 311
accident statistics 310
Adams' delay 77
Adams-Hummer method, for saturation flows 252
air pollution 211-213, 219, 222, 228
air switch 175
Akcelik congestion functions 130
all-or-nothing assignment 117
Almond's method 157, 189-192
amenity 3-5
analysis of variance (ANOVA) 150, 367, 371-376, 386-392, 405
analysis of video 175, 195, 196
annoyance 215, 217, 224, 229
annual average daily traffic (AADT) 165
annual average weekday traffic (AAWT) 165
approach delay method 208
area level analysis 12
area-wide traffic control 68, 124, 186
arithmetic mean 345-348
arm, of intersection 49-51, 71, 88, 236, 315 (see also *leg, of intersection*)
assignment model 117, 121-124, 259
asynchronous method, for saturation flow estimation 256
attitudinal surveys 218, 229
automatic counters 236, 296
automatic timing 191, 195
automatic transcription of data 199, 298
automatic vehicle identification (AVI) 206
automatic vehicle location (AVL) 206
average speed 44, 49-50, 53-54, 65, 131-132, 186, 191, 210, 336, 369
average speed model of fuel consumption 131
axle detector 178
axle weight 9

back of queue 99-101, 103, 243, 251
backward elimination 404, 406
bar-code reader 171
base saturation flow 96, 256
basic noise parameter, in CORTN model 226
bell-shaped curve 350-351
bias 151, 158-159, 162, 191, 201, 207, 263, 291, 353
Biggs-Akcelik family of models 218
bimodal distribution 345-346
binomial coefficient 55
binomial distribution 55-57, 352, 377, 379, 380
blackspot sites 310, 318
blocking 149-153

Borel-Tanner distribution 55, 57, 62
bottleneck 11, 235, 267
box plot 332-338
bunch size 57, 62
bunching of traffic 62-63, 68, 186, 198, 200, 242, 415

Calculation of Road Traffic Noise (CORTN) 226
calibration error 389
capacity
    analysis 94, 236, 250
    of a movement 90
    restraint 284
car following 41, 63-65, 260, 268
    model 65
    studies 268
car park 11, 156, 179, 265, 290
Central Limits Theorem 160
central tendency, property of statistical distribution 26, 345-347, 350, 361, 366
class interval, for histogram 327-329
closed circuit television (CCTV) 174, 176
cluster sampling 157-159
coefficient of determination 408-409
coefficient of skewness 350
coefficient of variation 26, 47, 49, 85, 186-187, 350
combined lane 78
commercial vehicle 31, 43, 95, 166, 253, 276 (see also *heavy vehicle*)
common cycle time 124
community annoyance 217
community goals 25
community reactions 229
comparison of means 150
complete stops 249
component traffic streams 53
computer assisted transcription 175
concentration, of traffic 8, 44, 340, 351, 383, 400 (see also *density, of traffic*)
confidence interval 365-366, 377, 379, 392, 403
confidence level 161-162, 349, 381
confidence limits 182, 365
conflict studies 314-316
conflicting traffic streams 309
confused characters 272
congestion delay 244
congestion function 104-108, 124, 129-130
congestion level 36, 111-112, 120
congestion pricing 125, 129, 131, 202
congestion tax 129
continuity of flow 44, 49, 51, 53, 123
continuous distribution 55, 58
continuous variable 26, 58
coordinated signal system 12, 124
coordination cruise speed 125
cordon 9, 113, 263-264, 267, 270-271, 275, 295-196, 299-302
    count 296
    line 113, 115, 264, 267
correlation coefficient 382, 408, 417-419
correlation matrix 409-410
CORTN procedure, for traffic noise 225-228
covariance 365, 418
critical gap 74-75, 256-258
critical intersection 12, 109, 124
critical movement 86, 91-93
cross-tabulation 304, 323, 356, 383
cruising 105, 132
cumulative density function (cdf) 58-61, 73
cumulative distribution 223-224, 256-257, 330
cycle failure rate 100
cycle time 88, 90, 93-95, 100, 109, 124, 243, 247, 303
cyclic variations 30

data editing 355

data errors 265, 354-355
data layers 35
data presentation 321
data structure 31
database
  interrogation 21
  management 5, 31
  package 33
  query 32-33
  system 321
datalogger 171-172, 193, 198, 198-201, 205-208
Davidson function 106-107, 130
decision making 17-19, 21, 37, 129, 155
decision making process 19
degree of saturation 76-77, 80-81, 93-94, 100, 107
delay
  coefficient 107
  equitable distribution of 128
  section 207-208, 246
demand model 22-23
dense network 13
  model 36
density, of traffic 8-9, 41-48, 50-53, 58, 67, 73, 107, 186, 199, 207, 325-326, 329-330, 334, 341, 411-412
dependent variable 397-398, 400, 405, 408-409
derived pollutant 213
descriptive statistics 33, 321, 345, 351
design hour volume (DHV) 165
design of survey form 142, 167
design speed 185
desired speed 46, 49-50, 62, 208-209, 246
destination choice 72, 110
diary survey 9, 219
difference in means from large samples 362
difference in means from small samples 363
directional link counts 238
discrete data 26
discrete distribution 55
discrete probability function 58
discrete variable 26-27
distribution of a small set of data 325
distribution of bunch sizes 57, 62
distribution of headways 59
distribution of sample means 159, 360
distribution of speeds 47, 58, 156, 185
Doppler effect 181, 192, 197
dot diagram 325-326, 400
doubly-constrained gravity model, of trip distribution 119
driver behaviour 9, 65, 185, 192, 196, 256, 296, 316
driving simulator 316-317
dynamic route guidance 207

effective green time 90, 94-95
effective stops 249
electromagnetic beam 180, 191-192
electronic tagging 265-268
electronic timing 191
elemental model, of fuel consumption 131
emissions 6-7, 12, 16, 100, 108, 112, 123-124, 128-131, 139, 152-153, 211-214, 217-221, 228, 241
enforcement 154, 166, 185-186, 192-193, 199, 287-288, 310
environment class 103-104
environmental capacity 213, 230
environmental impact 5, 7, 14-15, 108, 211-212, 215-218, 230
environmental parameter, in Davidson function 100
equilibrium assignment model of traffic flow 120-122
equilibrium in a queue 22, 46, 81, 83
equivalent energy level (Leq) 224
ergonomic design 144, 170

error
- bias 159
- data editing 355
- experimental 147, 353
- logical 355
- measurement 389
- missing data 356
- model 389-391
- permissible range 355
- sampling 159
- specification 389
- systematic 353

estimating saturation flows 95
exhaust emissions 12, 139, 218-219
existing data 21, 30, 139, 287, 289
experimental design 149-151, 218, 269, 318, 354
Exploratory Data Analysis (EDA) 33, 321
exponential distribution 58-62, 74, 80-81, 413
exposure measure 16, 132, 163, 177, 215, 311, 318-319

F-distribution 367, 399
F-statistic 367, 371-377, 374, 396-397, 399-402, 404
F-test 367
factor analysis 408-409
factorial design 220-221
field sheet 252
first General Motors model 64
floating car 9, 197, 205
flow profile 82, 158, 159, 177, 269
flow rate 44, 48, 62, 73, 77, 88, 91, 94, 98, 102, 205-206, 243, 250-251, 256, 282, 380 (see also *volume*)
follow-up headway 60, 73
forward inclusion 404-406
free flow 51, 189, 210
- speed 189, 210
- travel time 189, 241
free speed 46-51, 71, 157

frequency distribution 223, 328-330
frequency table 327-329
fuel consumption 6-8, 12, 100, 108, 112, 123-124, 129, 131, 211-212, 217-219, 241, 393-395, 399
fuel cost 112, 129
Furness method 238, 265-266, 278-279

g/c ratio 94-95
gamma distribution 46
gap acceptance 73-79, 85, 104, 172, 186, 235, 243, 256-257, 331, 415
gap size 58-59, 74, 257
Gaussian plume dispersion model 218
generalised cost of travel 112, 123-124, 128
generation, of traffic 8-9, 12, 17, 24, 110-111, 116, 184, 129, 158, 217, 219, 222, 281-282, 284, 286, 289-291, 294-295, 302-307, 387, 422 (see also *traffic generation*)
Geographic Information System (GIS) 34
geometric distribution 55, 57, 84-85
geometric mean 26, 346, 348
Global Positioning System (GPS) 187, 267, 313
Goodness-of-fit 399, 408
gradient, of road 43, 158-160, 187, 226
gravity model of trip distribution 119-120
green phase (stage) 79, 101, 207, 243, 247
green time 68, 86, 88, 90-92, 94-95, 99-100, 243, 248, 254-255
green window, for signal linking 124
greenhouse gas emissions 15, 211

Hauer-Kruithof method 116, 238-240
head of the queue 73-74, 85, 100, 243
headlight survey 265

Index 449

headway, in traffic stream 44-45, 52, 59-63, 65, 73-77, 88, 124, 185-188, 192-193, 195-197, 250-251, 254-256, 403
headway ratio method, for saturation flow 250
heavy commercial vehicle 32, 43, 95, 253-254 (see also *heavy vehicle*)
heavy goods vehicle 166 (see also *heavy vehicle*)
heavy vehicle (HGV) 6, 9, 43, 166, 215, 226
hierarchy of traffic models 36
highest hourly volume (HHV) 165, 282
histogram 33, 217, 223-224, 304, 325, 327-333, 336, 345-346
household interview 25, 291, 293-294
hydrocarbons 213, 219
hypothesis testing 335, 353, 356-359, 362, 392

impact model 22-23, 217
in-vehicle time 113
in-vehicle traffic information 269
incident detection 128
independent variable 305, 398-399, 400, 410
inductive loop detector 179, 185, 195
information theory 5, 119 (see also *mathematical theory of information*)
infrared 181
input-output 202, 247, 294-296
instantaneous model, of fuel consumption 131
instrumented vehicle 196, 206, 218, 316
inter-bunch headways 62
interactive data analysis 33, 321
internal stations 268, 275

interquartile range (IQR) 325, 332
interrupted traffic flow 12, 71, 86, 104, 108
intersecting traffic streams 11, 68, 70, 86, 108, 127
intersection degree of saturation (X) 94
intersection delay 107, 246-247
intersection flow ratio (Y) 100
intersection geometry 90, 97-98, 250
intersection green time ratio 100
interval estimate 365
interview survey 219, 291, 293-294
inventory survey 288-289

jam density 51
jitter plot 327, 329, 338
journey speed 67, 186, 202
journey time 186, 203, 244
judgement sampling 156
junction level analysis 11

kurtosis, property of statistical distribution 350-351

L10 noise level 223
lag, in gap acceptance 64, 74-75, 251, 283
lag, in time series 419
land use impact 36, 217
lane changing 42, 59, 63, 65-66
lane occupancy 9
lane saturation flow 95-96
lane type 95-97
lane width 42, 95-97
latent demand 289, 293-294
latin squares 150-152
least squares 273, 393, 395, 399, 414-415
leg, of intersection 236, 243 (see also *arm*)
lepto-kurtosis 350
Leq noise level 224-225
level of measurement 25-26

level of service, of traffic flow 50, 67
line source, of pollution 221
linear regression 229, 255, 398, 400, 409-410
link level analysis 10
local area 13-14, 268
local trips 115-116, 272, 275
log-normal distribution 46, 257
logbook 293
loop detectors 138, 179-182, 191, 195, 290
lost time, at a traffic signal 88, 90, 92, 94-95

macroscopic flow 36
magnetic imaging 180, 192, 195
mailback questionnaire 261-263
main survey 142
major traffic stream 60, 76-79, 256
marginal cost of travel 128-129, 182
mathematical theory of information 117, 241 (see also *information theory*)
matrix synthesis 116
maximum back of queue 243
maximum delay 100
maximum likelihood
    estimate 412-413
    method 414
maximum stationary queue 245, 249
mean bunch size 57, 62
mean delay 76-77, 81, 84-85, 103
mean free speed 50-51
mean gap size 257
mean journey time 203
mean queue length 81, 84
mean service time 82, 84-85
mean space speed 44, 46, 49, 53-54
mean spot speed 46, 53-54
mean time speed 46, 49
mean uniform delay 102
measures of performance 12, 99-100
median, of a variable 361

microscopic simulation 36
microwave 181, 192-194, 197
mid-block capacity 11
minimum following headway 60
minor stream 60, 73-77, 256-258
misrecording 269-270, 273, 312
mixed exponential distribution 58, 61
modal choice 72, 110, 112-113, 128
mode, of a variable 361
model development process 389-391
model sum of squares 396, 408
most probable matrix 117
move-up time 73, 257-258
movement degree of saturation 94
moving observer 197, 203, 207
multi-collinearity 409
multi-lane flows 77
multicore cable 177
multilane road 73
multiple regression 398
multiple servers 80
multiple variables 153

negative exponential distribution (see *exponential distribution*)
network analysis 72, 109-111, 113, 241
network capacity 11, 186
noise index 224
noise level 217, 223-224, 227, 229
noise pollution 15, 19, 128, 213, 215, 224, 229
nominal level data 25-26, 270, 346, 348, 377, 410
non parametric statistical testing 377
normal distribution 46-47, 58, 160-162, 219, 223, 225, 257, 350-351, 362-363, 377, 400, 411-413, 415-416
null hypothesis (H0) 356-360, 363-365, 367, 369, 371-372, 374-378, 379, 386
number of stops 100, 103, 241

O-D matrix 112-118, 120-121, 238, 259-261, 263, 266-267, 276, 340
O-D matrix estimation 116, 236
O-D survey 260, 266, 269, 276
observed link count 167, 235
observer error 263, 270, 273, 275
occupancy 8-9, 44, 52-53, 165-167, 182, 195
off-street parking 288, 290
offset 109, 124, 175, 202
on-line model 267
on-street parking 288
one-dimensional plot 325
one-dimensional table 322
one-tailed test 364, 367
one-way analysis of variance 373
operating cost of the vehicle 128
opposing traffic 11, 42, 94, 236
optimal cycle time 100
ordinal level data 25-26, 348, 377, 379
origin-destination (O-D) matrix (see *O-D matrix*)
outlier 165, 207, 321, 325, 333, 346-347, 354, 416
overall delay 244, 246-247
overflow delay 103-104
overflow queues 100, 103, 243
oversaturation 107
overtaking 11, 42, 45, 50, 62-63, 65-67, 74, 186, 196, 203-205
oxides of nitrogen 152, 212-213
ozone 213

paired comparison 149-150, 338, 342
paired sample t-test 378
paired samples 349, 365
parameter estimation 389, 392, 403
parking 7-9, 11, 13, 43, 71, 90, 94, 105, 112, 147, 182, 246, 269, 281, 284, 287-302, 305, 307, 344

accumulation 295, 297
demand 8-9, 289, 294-295, 304-305
duration 9, 289-290, 296-297, 300
lot 296, 299-301, 344
person interview 291
rate 304
search 269
supply 8-9, 291
system 284, 301
partial registration plates 264, 269, 272
partial stops 103
particulate lead 212-213, 219
passenger car units (pcu) 95-96, 163, 166, 250-251, 255-256
path trace 207
patrol survey 296, 298-301
pcu equivalent 95-96, 251, 256
pdf 58-60, 329 (see also *probability density function*)
peak demand 112
peak hour factor 165
peak hour flow 165
peak period 42, 111, 127, 236, 282, 284, 300
percentage heavy 166
percentage of time spent following 62
percentage of vehicles following 62
percentile
  50th 161, 214, 333, 346
  85th 161, 185, 325
  90th 186, 223
permanent loop installation 179
phases, of traffic signal cycle 86, 90-98, 163 (see also *stages, of traffic signal cycle*)
photochemical smog 15, 213, 219
piezo-electric detector 179, 181-182, 192
pilot survey 138, 142, 144, 155, 292
planning process 17-22, 24-25, 137, 230, 389, 421-422

platoon 42, 47, 60, 63, 67-68, 109, 124-126
    dispersion 63, 124-125
    leader 125
platy-kurtosis 350
pneumatic tube detector 175, 191
point estimate 365
point sample 207, 246
point source 219-220
Poisson distribution 55-56, 59, 384, 413-414
Poisson process 411, 413
police report form 312
Pollaczek-Khintchine formula 85
pollutant
    dispersion model 216
    emissions 100, 107, 123-124, 211
    primary 213
pooled variance 370
population parameter 159, 345, 349, 356, 361, 363, 392
population variance 348, 364, 366-367
power, of a statistical test 359
practical cycle time 93, 99
priority rule, at a junction 73
probability density function (pdf) 47, 58, 329-330, 411
probability distribution 30, 47, 83, 223, 325, 329, 361
probe vehicle 196, 206, 260, 266, 268, 316
proportion of heavy vehicles 215
public transport 5-6, 110, 113, 127, 147, 149

q-k curve 51-52, 63
QDELAY method 247-248
qualitative data 27
quality of flow 62-63
quality of service 67
quantile plot 325-326, 331
quantitative data 27

questionnaire survey 32, 116, 142, 161, 219, 229, 260-263, 268-269, 289, 291, 293-294, 318
queue
    discipline 79-80
    length 9, 79-81, 84-85, 100-102, 127, 241-244, 247, 249
    management 100, 128
queuing
    delay 76, 244
    theory 30, 79-85, 104-105

radius of turn 95
random arrivals 80-81, 84-85
random component, of a time series 30, 416
random digits 157
random sampling 156-159, 207
Random Traffic Model 59-60, 74-75, 81
randomised block data 150
rank order correlation
    coefficient 382
    test 377
red time, at traffic signal 79, 94, 101, 242-243
regional (or strategic) network 13
registered owner mailback survey 263
registration number 197-199, 293, 297-298 (see also *registration plate*)
registration plate
    matching 197-199, 201-203, 260, 262, 264, 277
    survey 116, 263
regression
    coefficient 393, 406, 408
    line 305, 403, 411
    method 250
    to the mean 318
relative frequency 329-330
reliability of travel times 128
report of survey 142
residual sum of squares 395-397, 408

revealed demand 289, 293
road crashes 25, 132, 309, 313
road pricing (see *congestion pricing*)
road user charges 124
roadside interview 9, 261-263, 268
Robertson's recurrence formula 125
route choice 12, 72, 113, 120, 267-269
route choice simulator 269
route guidance 206, 267, 269, 316
running speed 131, 187
running speed model, of fuel consumption 131
running time 187

sample data 159, 345, 357, 387
sample design 140, 155-162, 353
sample estimate 161, 357-361, 365-366
sample frame 156
sample mean 349, 360-361, 365, 369-371, 377, 413
sample method 155, 207
sample size 139, 144, 147, 155, 159-162, 260, 299, 329, 354, 361-362, 371, 375, 378, 381, 395, 412
    determination 159-161
sample statistic 345, 349, 351
sample variance 349, 371
sampling 139-140, 148, 155-162, 189, 191, 201, 205-207, 263, 268, 270, 293, 360-361
    bias 159, 191
    distribution 159-160
    error 155, 159, 201
    fraction 160
    frame 156
    method 155, 207
    unit 155-156
SATFLOW method, for saturation flows 254-255
saturation flow 86, 88, 90, 93-98, 101-102, 235, 243, 249-256, 387
SCATS 109, 182, 238
scatterplot 335, 338, 393-394, 399-401
SCOOT 109, 182, 238
screenline 114, 175, 186, 196, 263
seasonal variation, in time series 28, 416, 421
second General Motors model 64
segregated traffic streams 309
selection of variables 229, 378
semantic scale 229, 378
sensitivity analysis 17, 24, 305
sensitivity coefficient, in car following model 64
service mechanism 80
service rate 79-83
service time 80-85
shape, of a distribution 350-351
shifted exponential distribution 58-61
shopping trips 119
shut-down, error in O-D survey 270
SIDRA 88, 99
sign test 377-378
signal coordination 68, 124-126
signal linking 68
signal phasing 109, 252 (see also *signal staging*)
signal plan 186, 249
signal stages 86-87, 250 (see also *signal phasing*)
signalised intersection 60, 68, 71, 86, 207, 242-243, 386 (see also *signalised junction*)
signalised junction 79, 124, 170, 243, 246, 250, 387 (see also *signalised intersection*)
simple random sampling 157-159
simple regression 392, 398, 402
singly-constrained gravity model of trip distribution 119
skewness, property of statistical distribution 257, 350, 382
sound level meter 228

space speed 44-49, 53-54, 186, 198
spacing, in a traffic stream 23, 44-45, 52-53, 59, 63-64, 166, 179, 185-187, 250
specification error 389
speed bin 166
speed-density relation 51
speed-flow envelope 50
speed-flow relation 105 (see also *speed-volume relation*)
speed limit 55, 185-186, 189, 207, 312, 330
speed range 166
speed-time profile 186-187, 196, 218, 269
speed-volume relation 71, 105
spot speed 46-49, 53-54, 185-186, 189-190, 210 (see also *time speed*)
spreadsheet 30, 33, 321, 324, 333
spurious match 198, 264, 269-272, 276
stacked dot diagram 326
stages, of traffic signal cycle 86-88, 82, 90, 93, 110, 138, 142, 153, 251
standard deviation 33, 47, 49, 67, 159-162, 186, 220, 225, 249, 348-353, 361-366, 369, 415
standard error 159-160, 349-350, 352, 361-364, 366, 372-373, 401, 403
standard normal deviate 362
star plot 342-344
start-up, error in O-D survey 270
stationary observer 54, 197, 207
stationary process, in time series 421
statistical analysis software 30, 33, 416
statistical distribution 41, 46, 54
statistical hypotheses 159, 162
statistical model 389
statistical significance 314, 349, 363
steady-state, in a queue 106-107

stem and leaf diagram 333-334
stimulus-response law 63
Stirling's approximation 117
stochastic variable 223
stop delay 249
stopped delay 100, 244-246
stopped time 186, 188, 246-247
stopwatch method 195, 198, 202-204, 252
storage-output equation 243
strategic network model 36
stratified random sampling 157-159
supply model 22
survey design 137-138, 302
survey form 170, 249, 253, 291, 298, 300
survey instrument 139-140, 142-144
survey manager 143-144, 298
survey planning 7, 137, 140, 142
survey resources 25, 165
switch tape detector 177, 191-192, 195
symmetry, property of statistical distribution 321, 325, 350-351, 408
synchronous method, for saturation flows 255-256
synthesis of O-D matrices 116
system delay 106, 246
system travel time minimisation, in equilibrium assignment 123
systematic error 353
systematic sampling 157, 159, 207
Systems Planning Process (SPP) 19-20, 25, 230, 389, 422-423

t-distribution 363-365
t-statistic 364, 373, 403, 407
t-test 149, 365, 377-378, 382
tag surveys 260, 265
tally counter 171, 297
target population 155-162
terrestrial beacons 206-207, 267
test vehicle 207-208, 218 (see also

Index 455

instrumented vehicle and *probe vehicle*)
testing the difference in dispersion 366
testing the difference in means 361
three-dimensional data 340
three-dimensional table 276-278, 323-324
through car equivalent 97, 99
through car unit (tcu) 96
through trips 115-116, 249, 272, 303
through vehicle 73, 98, 249, 301
time-dependent data 27
time in queue 100, 246
time period (Tf), in intersection delay calculations 103, 107
time series 29-30, 182-183, 185, 195, 206, 223-224, 304, 318, 334, 416-422
time speed (see *spot speed*)
toll, charge for use of road 81, 104, 112, 124, 128-129, 182, 202
total delay 85, 101-102, 244
tracking headway 61-62
traffic assignment model 120-124
traffic calming 3, 309
traffic classifier 193 (see also *datalogger*)
traffic composition 23, 95-97, 217
traffic conflict 9,172
traffic control system 13, 36, 42, 100, 124, 179, 182, 238
traffic count 30-32, 48, 117-118, 167-168, 172, 192, 226, 260, 266-267
traffic demand 79, 91-94, 107, 244
traffic generation 281-282, 304
traffic generator 109
Traffic Impact Analysis (TIA) 37, 108-113
traffic impact study 422
traffic impacts 5, 12, 108-109, 163, 282
traffic intensity 79, 81-84, 95

traffic lane 42-44, 52-53, 60, 180, 250
traffic management 6, 12, 41, 109-110, 127, 212, 217, 222, 236, 259, 267, 422
traffic network model 100, 208
traffic noise 7, 19, 23, 214, 222, 225-226, 228-229, 378
traffic signal
   linking 68
   operation 86
traffic survey planning 137
traffic volume 10-11, 22-23, 27-29, 41, 52, 59, 63, 65-66, 71, 104, 120, 129-130, 132, 163, 213, 236, 250, 421
trafficgraph 282-284
training and supervision 145
trajectory diagram 44-46, 101, 242
TRANSYT 104, 125
travel costs 72, 112-113, 120, 124, 129-130
travel demand 3, 13, 72, 106-110, 113, 117, 120-122, 129-130, 241, 282, 294
   modelling 72, 106, 1013 282
Travel Demand Management (TDM) 4-5
travel patterns 12, 217
travel time survey 263
Tribo-electric detector 179, 192
trip attraction 111-112, 119, 282 (see also *trip generation*)
trip distribution 110-111, 116, 119-120, 238
trip end 111, 261, 282, 302, 305
   interview 261
trip generation 9, 110-112, 116, 158, 282, 287, 290-291, 295, 302-305
   rate 303
trip production 111, 119, 282, 302 (see also *trip generation*)
trip purpose 289, 291
trip rate 304
trip timing 110, 112-113

tube detector 177, 180-181, 191-192, 195-196
turning movement 113, 115, 170, 231-234, 241, 314
    matrix 116, 235-237, 241
    study 170, 236
turning vehicles 73, 78, 98-99
two-dimensional graph 335, 341
two-dimensional plot 340
two matched samples 378
two sample t-test 382
two-tailed test 357, 365
two-way analysis of variance 374-375
two-way ANOVA table 375-377
two-way classification table 323
two-way cross-tabulation 383
Type I error 359
Type II error 359

u-value 92
uniform arrivals 102
uninterrupted traffic flow 10, 41, 44-45, 50, 59, 68, 71-72, 105, 108, 193, 309
unit travel time 71-72, 128
unnecessary stopped time 246
unparking 11, 295-296
urban traffic control (UTC) 86, 100, 109, 195
user travel time minimisation, in equilibrium assignment 121-122
utilisation factor, in queuing theory 79, 86

validation, of a model 6, 390-391
valuation of travel time 128
Van Zuylen-Willumsen method 240
variance 364
    of queue length 84

vehicle classification 9, 270
vehicle colour 201
vehicle detector 203, 254, 287
vehicle guidance system 186, 316
vehicle-hours of travel (VHT) 11, 121-122
vehicle-kilometres of travel (VKT) 11, 132, 311
vehicle occupancy 166
vehicle operating cost 100, 112, 123-124
vehicle trip ends 305 (see also *trip end*)
vehicle type 8, 27, 43, 53, 56, 98-99, 132, 165, 181, 192-195, 236, 250-251, 254, 256, 267, 276, 291, 297
verification, of processes in a model 6, 17, 390
visual intrusion 211, 288

W-statistic 382-383
W-test 382
waiting time 74, 112-113 (see also *delay*)
walking time 112
Wardrop-Webster model 88-90, 102
Wardrop's first principle 121-122
Wardrop's second principle 122-124
Weigh-in-motion (WIM) 182
weighted least squares 273
wheelbase 176
Wilcoxon-Mann-Whitney test 377, 382

y-value 92-93

z-statistic 362-363

χ^2 statistic 386